J. A. Campos-Ortega · V. Hartenstein
The Embryonic Development of
Drosophila melanogaster

Springer

Berlin
Heidelberg
New York
Barcelona
Budapest
Hong Kong
London
Mailand
Paris
Tokyo

J.A. Campos-Ortega   V. Hartenstein

The Embryonic Development of Drosophila melanogaster

Second Edition

With 139 Illustrations, 37 in Colour,
Consisting of over 600 Separate Figures

 Springer

Prof. Dr. José A. Campos-Ortega
Institut für Entwicklungsbiologie
der Universität zu Köln
Gyrhofstrasse 17
D-50923 Köln
Germany

Prof. Dr. Volker Hartenstein
Department of Molecular Cell and
Developmental Biology
University of California
Los Angeles
405 Hildgard Avenue
Los Angeles, CA 90024
USA

ISBN3-540-57079-9 Springer-Verlag Berlin Heidelberg New York

Library of Congress Cataloging-in-Publication Data
Campos-Ortega, José A. (José Antonio), 1940–
The embryonic development of Drosophila melanogaster/José A. Campos-Ortega, Volker Hartenstein. –– 2nd ed.
p. cm.
Includes bibliographical references and index.
ISBN 3-540-57079-9 (alk. paper)
1. Drosophila melanogaster--Development. 2. Embryology--Insects.
I. Hartenstein, Volker, 1957- . II. Title.
QL958.C36 1997
571.8'615774--dc21

© Springer-Verlag Berlin Heidelberg 1997
Printed in Germany

Cover Design: D & P GmbH, Heidelberg
Cover lllustration: Dr. Volker Hartenstein
Typesetting: Typo-Press Service GmbH Heidelberg
SPIN 10054085   39/3137 5 4 3 2 1 0 - Printed on acid free paper –

Dedicated to Donald F. Poulson

Preface

"... but our knowledge is so weak that no philosopher will ever be able to completely explore the nature of even a fly . . ." *
Thomas Aquinas
"In Symbolum Apostolorum" 079
RSV p/96

This is a monograph on embryogenesis of the fruit fly *Drosophila melanogaster,* conceived as a reference book on the morphology of embryonic development. A monograph of this extent and content is not yet available in the literature on *Drosophila* embryology, and we believe that there is a real need for it. Thanks to the progress achieved during the last ten years in the fields of developmental and molecular genetics, work on *Drosophila* development has expanded considerably, creating an even greater need for the information that we present here. Our own interest in embryonic development of the wild type arose several years ago, when we began to study the development of mutants. In the course of these studies we repeatedly had occasion to discover gaps and inadequacies in the existing literature on the embryology of the wild type, so that we undertook to investigate many of these problems ourselves. Convinced that many of our colleagues will have encountered similar difficulties, we decided to publish the present monograph.

Although not explicitly recorded, Thomas Aquinas was probably referring to the domestic fly and not to the fruit fly. Irrespective of which fly he meant, however, we know that Thomas was right. The situation today is not very different from that in the thirteenth century insofar as nobody would seriously claim to have completely investigated the "nature" of the fly. Consequently, in our work we did not intend to cover all aspects of embryology; in particular, we have purposely avoided dealing with genetic aspects of development, restricting ourselves to morphology. We repeat: our intention was only to provide basic information about the anatomy of normal embryogenesis, illustrating our description with figures of appropriate technical quality.

In this book we have used a rather conventional distribution of topics, similar to that in other embryological monographs. A few of these topics, i.e. neurogenesis, the pattern of mitotic divisions of embryonic cells, some aspects of the morphogenetic movements and the blastoderm fate map, are based on studies by the authors and their colleagues, the results of which have already appeared in different form in the scientific literature. The third chapter deals chiefly with organogenesis, although we have also considered in some detail several aspects of the anatomical organization of the larva which, in fact, do not strictly pertain to embryogenesis; for example, we have invested some effort in investigating the pattern of distribution of sensory organs, peripheral nerves, muscles, etc. in the fully developed embryo.

Several persons and institutions have contributed to this book. Chapter 2 presents a classification of embryogenetic stages which is based on similar classifications by other authors. The authorship of this chapter is actually shared by Christiane Nüsslein-Volhard and Eric Wieschaus, with whom J. A. Campos-Ortega defined the stages almost identically several years ago. Work on this manuscript was initiated at the Institut für Biologie Ill of the University of Freiburg i. Br., FRG, of which both authors were staff members. Many friends and colleagues helped us while working in Freiburg. The most important contribution was by Sigrid Krien, our collaborator of over ten years, who patiently prepared most of the material for this work. Important, too, was the "moral" support of Rainer Hertel. Many of the ideas presented in this book originated in discussions with Ursula Dietrich, Fernando Jiménez, Gerd Jürgens, Ruth Lehmann, Christiane Nüsslein-Volhard, Gerd Technau and Eric Wieschaus. We would particularly like to thank Klaus Sander, who repeatedly contributed enlightening discussions on the embryogenesis of insects, insisted on the importance of using appropriate terminology, and critically read the entire manuscript, correcting several mistakes. Finally we want to thank Gerd Technau for permission to quote his unpublished results, and Alfonso Martinez-Arias for a discussion on tracheal pits. Financial support was provided by the Deutsche Forschungsgemeinschaft (DFG, grants SFB 46, Ca 60/6-1, Ca 60/7-1 and SFB 74).

Köln, September 1985
Jose A. Campos-Ortega
Volker Hartenstein

* The Latin quotation is: ". . . sed cognitio nostra est adeo debilis quod nullus philosophus potuit unquam perfecte investigare naturam unius muscae . . .

Preface to the 2nd edition

The first edition of this book appeared in 1985, over 10 years ago. Prior to its publication, *Drosophila* embryogenesis was the field of study of only a handful of scientists. Consequently, at that time this topic was poorly understood since we all knew very little indeed. Nevertheless, many important questions related to *Drosophila* embryonic development had already been asked by our predecessors, and we added a few more. Unfortunately, most of these questions remained unanswered because of the inadequacies of the available techniques and other adverse factors.

Since then, work on *Drosophila* embryogenesis has made considerable progress, and most of the questions posed by earlier workers have already found satisfactory solutions. Today, the *Drosophila* embryo is in fact one of the best understood developmental systems. Accordingly, the task of preparing this second edition has taken much longer, since much more had to be reviewed; paradoxically, it has also been a much easier job, as the information can now be more convincingly integrated in the larger context. The mass of information available, very much increased as compared to the situation 10 years ago, has forced us to increase the number of chapters from five to eighteen. To accommodate this new material, the previous third chapter (Organogenesis) has been expanded and now comprises ten different chapters, in which the development of each different organ system is treated separately. However, the organization of the book has remained in principle the same. Most of the data discussed in the first edition were derived from our own work; this time, many data are derived from the work by other authors.

We would like to thank Detlev Buttgereit, Christian Klämbt, Renate Renkawitz-Pohl, Thomas Menne, Evelin Sadlowski, Gerhard Technau, and Andrea Wolk, who provided slides and/or unpublished photographs for inclusion in this book. We would also like to thank Paul Hardy, Elisabeth Knust and Christian Klämbt (Köln), Torsten Bossing, Gerald Udolph and Gerhard M. Technau (Mainz), Corey Goodman (Berkeley), Amelia Younossi-

Hartenstein, Judith Lengyel and Ulrich Tepass (Los Angeles), for many discussions, communication of unpublished results and reading of the manuscript. José A. Campos-Ortega would like to thank Dr. Kazumoto Imahozi, past president of the Mitsubishi-Kasei Institute of Life Sciences, Machida (Tokyo), and Dr. Daisuke Yamamoto, of the same Institute, for the hospitality they provided in February 1992 when part of the text of this book was written. Work in our laboratories was supported by grants of the Deutsche Forschungsgemeinschaft (SFB 243), the "Fonds der Chemischen Industrie", the NSF (IBN 9221407) and the NIH (Grant NS29367).

Köln and Los Angeles,   José A. Campos-Ortega
June 1996   Volker Hartenstein

Introductory Remarks

The importance of *Drosophila melanogaster* for biological research does not need to be expressly emphasized; it has been sufficiently demonstrated and documented on a multitude of occasions during the 90 years in which the fruitfly has been used in experimental work (Allen 1975, 1978). Due to its very elaborate genetics, the fruitfly is now one of the preferred models for tackling most problems in developmental biology.

The suitability of *Drosophila* for embryogenetic studies, and, in particular, the fact that it allows genetic dissection of embryonic development, was recognized almost 60 years ago by Poulson (1937a, 1940, 1943, 1945), who studied the effects of a few chromosomal deficiencies on embryonic development. In fact, more important than his description of the effects of the lack of certain genes on development was Poulson's contribution to establishing the genetic approach to embryogenesis on a firm basis. Poulson was among the first to claim explicitly that genes are involved in directing developmental processes, on the same lines as Goldschmidt's Physiological Genetics (1927, 1938), and he called the attention of experimenters to deficiencies and other chromosomal mutations with lethal effects on the homozygotes as tools for investigating the contribution of genes to embryonic development (Poulson 1943, 1945). The genetic approach to embryonic development has recently received a tremendous boost. It suffices to compare Ted Wright's (1970) review on the genetics of embryogenesis in *Drosophila* with some of the pertinent, up-to-date papers on aspects of the same topic (e.g. the contributions to the book "Development of *Drosophila melanogaster*" edited by C.M. Bate and A. Martinez-Arias, 1993), to realize how profoundly this field has been transformed, both from the conceptual and the material point of view.

Published work on normal embryogenesis of the fruitfly was very scarce until the mid-eighties of this century. The very promising foundation of a modern embryology of dipteran insects represented by Weismann's contribution (1863; see Sander 1985a, for a critical appraisal of Weismann's studies on insect development) was unfortunately followed by only a few further extensions (e.g. Poulson 1950, Sonnenblick 1950, Schoeller 1964, Anderson 1962, 1972). The standard works on organogenesis, and some other aspects of *Drosophila* embryogenesis, for many years were those of Sonnenblick (1941, 1950) and Poulson (1937b, 1950). These authors enunciated several of the embryogenetic questions with which we are still concerned, although the technical quality of their histological material was unsatisfactory. Subsequent accounts of organogenesis in the fruit fly (e.g. Fullilove and Jacobson 1978) follow Poulson's more or less literally; our account has also been heavily influenced by his.

The small size of the *Drosophila* egg, and the technical difficulties associated with its investigation (for example, the lack for many years of reliable fixation methods for histological analyses, other than those requiring pricking, or otherwise damaging the integrity of the egg), were certainly two of the reasons for the paucity of publications on normal embryogenesis of the fruitfly for a very long time. However, most of these technical difficulties were eliminated by the elaboration of reliable techniques for egg permeabilization (Zalokar 1970, 1976; Zalokar and Erk 1976, 1977). The significance of these techniques for the embryology of the fruitfly is inestimable; they permitted considerable improvement in the fixation of specimens for histological investigation, which then became an easy undertaking, and initiated a series of studies on cell biological aspects of early embryogenesis that is still in progress. For example, the excellent preparation of the material used in Turner and Mahowald's (1976, 1977, 1979) scanning electron microscope studies on morphogenetic movements was made possible by Zalokar's permeabilization techniques.

However, the most important contribution to *Drosophila* embryology has come from the combination of phenotypic and molecular analyses of morphogenetic mutants and the corresponding genes (e.g. Nüsslein-Volhard et al. 1982) that has been exploited during the last 15 years. A key element in this process was the development of a reliable and easy technique for germ line transformation (Rubin and Spradling 1982; Spradling and Rubin 1982), which opened wholly unexpected opportunities for the analysis of the genetic bases of development. Wholemount in situ hybridization (Tautz und Pfeifle 1989), antibodies and enhancer trap lines (O'Kane and Gehring 1987) contributed to facilitate the cellular analysis of the embryogenetic mutants and their phenotypes. All these elements have made the embryo of *Drosophila* one of the best understood developmental systems.

A few terminological clarifications are necessary for following our text. The egg of *Drosophila melanogaster* is a regular ovoid, slightly flattened dorsoventrally. It has become customary in embryogenetic studies to use the percentage of egg length (% EL, 0 % at the posterior pole, Krause 1939) and to the percentage of ventrodorsal circumference (% VD, 0 % at the ventral midline) as a reference system that permits developing structures to be located; this system has been used throughout this book, together with the terms dorsal and ventral, medial and lateral, cephalic and caudal or anterior and posterior. The term *anlage* is used for the progenitor cells of a given organ in the blastoderm; the term *primordium* is reserved for the progenitor cells once the organ is recognizable, prior to completion of proliferation and cytodifferentiation. The terms *stomodeum* and *proctodeum* are used to designate the primordia of foregut and hindgut, respectively, before regional differentiations have appeared; foregut or hindgut is used once regionalization appears, i.e. when pharynx and oesophagus, Malpighian tubules and hindgut have differentiated. Due to germ band elongation, abdominal levels of the *Drosophila* embryo are located in mirror-image relative to thoracic and gnathal levels during a substantial fraction of embryogenesis; this peculiarity poses some problems in orienting structures. Unless expressly stated, orientation is indicated with respect to the polarity of the embryo. The term *germ band* refers to the overtly metameric region of the embryo, as distinct from the procephalon.

CONTENTS

Chapter 1
A Summary of *Drosophila* Embryogenesis

Embryonic development is a continuous process in which profound modifications of an initial form, the fertilized egg, are accomplished within a limited period of time. Although embryonic development is a complex continuum of interwoven events, embryologists have frequently emphasized particular events in order to organize the embryogenetic process into a series of different stages. All subdivisions of embryogenesis into stages are necessarily artificial, in the sense that they imply a beginning and an end of the embryogenetic process at each stage. In spite of the artificiality that it imposes, however, staging is very useful for describing embryonic development, since it provides a temporal framework to which embryogenetic events can be referred. In animal embryology, several stages have been defined that occur in all species, e.g. the formation of the blastula and the gastrula. In this chapter we present a summary of *Drosophila* embryogenesis; in the next chapter we present a detailed staging system based on the characteristic events of the embryogenetic process in *Drosophila melanogaster*. The purpose of the present chapter is to introduce the reader to both the major embryogenetic events and the terminology used to describe them. The egg of *Drosophila melanogaster* is a regular ovoid, slightly flattened dorsoventrally, with an average length of 500 μm and a diameter of 180 μm. Developing structures can be unambiguously located within the egg by referring to the percentage of egg length (% EL, 0% at the posterior pole) and the percentage of ventrodorsal circumference (% VD, 0% at the ventral midline).

Cleavage divisions. After fertilization, the zygotic nuclei divide thirteen times before cellularization at the blastoderm stage. The first seven zygotic divisions are synchronous (Zalokar and Erk 1976), leading to a syncytium of 128 nuclei distributed as an ellipsoid centrally in the yolk. During the course of the next three divisions most of the nuclei approach the surface of the egg in a stepwise fashion (Foe and Alberts 1983) to form the somatic buds of the *syncytial blastoderm*; about 26 nuclei remain centrally after cycle 7 to form the yolk nuclei (*vitellophages*), and another 2-3 are incorporated into the posterior pole plasm after cycle 8 to form the polar buds. The yolk nuclei and the nuclei of the polar buds will divide another three and two times, respectively (cycles 8, 9 and 10, Zalokar and Erk 1976; Foe and Alberts 1983), in synchrony with the nuclei of somatic buds. Subsequently, pole cells become distinct from the syncytial blastoderm and continue to divide, albeit no longer in synchrony with the somatic nuclei. In contrast, the yolk nuclei cease dividing and become polyploid. Three further parasynchronous mitoses of the nuclei of the somatic buds bring the number of *syncytial blastoderm* nuclei to approximately 5000. Formation of the somatic cells then occurs by extension of membrane furrows between the syncytial blastoderm nuclei. At the cellular blastoderm stage, before gastrulation occurs (refer to Fig. 1.1, 5), wide cytoplasmic bridges still connect the blastoderm cells to the yolk sac. Therefore, strictly speaking, the cells of the blastoderm are still connected to each other. Cytoplasmic bridges are pinched off during gastrulation and early germ band elongation; hence, it is not until this later stage that cellularization in *Drosophila* is actually completed, i.e. that all embryonic cells are completely individualized.

The cellular blastoderm exhibits an extremely regular architecture in which no conspicuous regional differences in cell shape or size can be seen. *Gastrulation*, i.e. the formation of germ layers, occurs by the gradual invagination of a coherent midventral cell block, which will eventually form a tubular structure inside the embryo (Fig. 1.1, 5-7). Four different primordia are included in this tube; these are the primordia of: the mesoderm, the proctodeum and most of the anterior and posterior midgut.

Gastrulation

Gastrulation occurs in three steps. In the first step the *ventral furrow* invaginates midventrally, from 6% to 86% EL. Then a group of cells at the posterior pole, of roughly oval shape, begins to move rapidly in an anterodorsal direction. This movement initiates germ band elongation; later, the whole group sinks inward as a continuation of the ventral furrow to form the *amnioproctodeal invagination*, a term coined by Sonnenblick (1950). In the second step of gastrulation, the rostral tip of the ventral furrow forms a narrow transverse groove. The cells of this invagination constitute the endodermal primordium of the anterior midgut. In the third

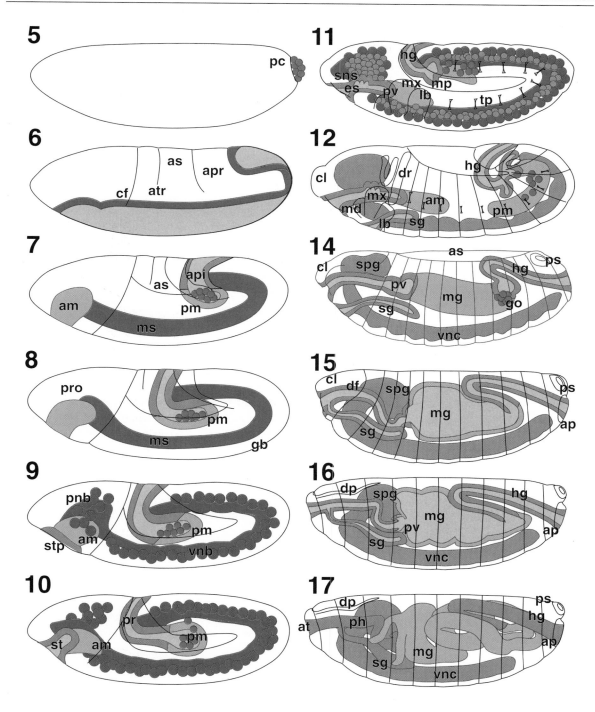

Fig. 1.1. Twelve drawings of embryos of increasing age illustrating the major events of *Drosophila* embryonic development. The number at the top left of each drawing indicates the embryonic stage (see Chapter 2). Stage 5 shows the blastoderm with the pole cells (*pc*) at the posterior pole. Stage 6 illustrates morphogenetic movements during gastrulation. Light shading indicates the larval anlagen that will invaginate during gastrulation, i.e. anlagen of the endodermal anterior midgut primordium (*am*), mesoderm (*ms*), proctodeum and posterior midgut primordium (*pm/pr*). The cephalic furrow (*cf*) separates the procephalon from the prospective metameric germ band; anterior (*atr*) and posterior (*ptr*) transverse furrows are visible. *apr* amnioproctodeal invagination; *as* amnioserosa. Mesoderm and endoderm have completed invagination, invaginated regions are shaded in lateral projection (hollow spaces -lumina- in this and following drawings are hatched), pole cells are included in the posterior midgut primordium. Stage 7 illustrates the beginning of germ band elongation. The cephalic furrow is tilted caudal-

wards; transverse furrows are approaching each other; the amnioproctodeal invagination (*api*) has deepened. During stage 8 germ band elongation proceeds further; the primordia of the posterior midgut and the proctodeum become individualized. In stage 9 germ band elongation enters its slow phase; the stomodeal plate (*stp*) becomes visible; the mesoderm transiently exhibits segmental bulges; segregation of neuroblasts (*vnb* ventral neuroblasts; *pnb* procephalic neuroblasts) begins. In stage 10 stomodeal invagination (*st*) takes place; segregation of the neuroblasts continues, first neuroblast divisions occur to give rise to ganglion mother cells. In stage 11 epidermal segmentation becomes evident; *tp* tracheal pits; *mx* maxillary bud; *lb* labial bud); *es* oesophagus; primordia of the stomatogastric nervous system (*sns*) and Malpighian tubules appear; pole cells leave the posterior midgut pocket (*mp*) and become arranged laterally. The Malpighian tubules have been omitted in this and the following drawings. *cl* clypeolabrum; *ph* pharynx; *pm* posterior midgut; *hg* hindgut; *md* mandibulary bud. The germ

band retracts during stage 12, the drawing shows the final stages of this retraction. During this process fusion of the anterior and posterior midgut primordia takes place and the definitive segmental boundaries appear. The salivary glands (*sg*) and the dorsal ridge (*dr*) appear. In stage 14 dorsal closure of the midgut and epidermis, and head involution begin. *df* dorsal fold; *hg* hindgut; *mx* maxillary bud; *spg* supraoesophageal ganglion; *ps* posterior spiracles; *vnc* ventral cord. In stage 15 head involution is well advanced. Notice growth of the hindgut. *ap* anal pad; *cl* clypeolabrum; *df* dorsal forld; *ph* pharynx; *sg* salivary glands; *t1* prothorax. In stage 16 head involution is almost complete, midgut (*mg*) constrictions appear, condensation of the ventral cord (*vnc*) begins. Stage 17 corresponds to the fully developed embryo. *at* atrium; *cl* clypeolabrum; *es* oesophagus; *hg* hindgut; *mg* midgut; *ph* pharynx; *pv* proventriculus; *sg* salivary glands; *ps* posterior spiracles. Modified from Hartenstein (1993), with permission of Cold Spring Harbor Laboratory Press

and last step of gastrulation, a complex remodelling of the amnio-proctodeal invagination occurs. This process is bound up with germ band elongation, and leads to the typical sac-like architecture of the posterior midgut primordium.

In the terminology used here, *germ band* refers to cells that will give rise to the metameric regions of the embryo. In *Drosophila* the germ band comprises the primordia of the three gnathal segments (mandible, maxilla and labium, *c1-c3*), three thoracic segments (pro-, meso- and metathorax, *t1-t3*) and at least nine abdominal segments (*a1-a9*) (ref. to Fig. 1.1, 11). Each segment anlage within the germ band has the same organization, comprising epidermal (dermomere), neural (neuromere) and mesodermal (myomere) components (Fig. 1.1, 12). The caudalmost ectoderm, which is caudal to *a9*, has been designated *a10* by Turner and Mahowald (1977). Since no mesodermal or neural components have been found to correspond to this section of ectoderm, however, it is questionable whether these ectodermal cells should be classified as a segment. The dermomeres of *a9* and the epithelial cells of the so-called *a10* contribute to the formation of the terminal region, designated the telson, which comprises the anal pads and surrounding epidermis.

Germ band elongation

Germ band elongation brings about drastic changes in the ectodermal germ band primordium by virtue of which the basic body pattern becomes evident. During germ band elongation, the width of the ectodermal primordium is reduced by approximately two-fold, while its length increases approximately two-fold; since the embryo fills the egg completely, we believe that these changes cause the germ band to fold along its own long axis while elongating, and thus to extend into dorsal levels of the embryo (Fig. 1.1, 8-11). Because the germ band extends over the dorsal surface, dorsal and ventral halves of a cross section of an embryo at the extended germ band stage are equivalent.

Germ band elongation begins immediately after ventral furrow formation. The first step in the elongation process coincides with portions of the gastrulation process, when the posterior egg cap becomes displaced dorsoanteriorly to give rise to the amnioproctodeal invagination, in which the ventral furrow ends. Initially the prospective amnioproctodeal invagination is just a groove; however, as the germ band elongates, its tip pushes the cells neighbouring the furrow forward, such that these cells come to cover the amnioproctodeal groove, forming a sort of roof. Thus, the amnioproctodeal invagination acquires the shape of a retroverted sac, which comprises the posterior midgut and hindgut primordia. Initially the germ band elongates very quickly, its tip reaching 60-62% EL in only 30 min; then elongation slows down to attain its maximal extension after another 2 h, at approximately 73% EL. During the rapid phase of germ band elongation a vertical indentation appears on either side of the proctodeal opening, which has been called the paraproctodeal fold.

During gastrulation and germ band elongation, three different transient furrows become apparent laterally in the embryo (refer to Fig. 1.1, 6-10). The cephalic furrow appears, at about the same time as the ventral furrow, as two lateral, initially almost vertical indentations. These indentations soon follow an oblique dorso-ventral course, subdividing the early embryo into two territories that roughly correspond to the procephalon and the metameric germ band. The cephalic furrow is a dynamic structure, however; various parts of the outer embryonic layer are incorporated transiently into the cephalic furrow during the period of germ band elongation. At the end of the rapid phase of germ band elongation the cephalic furrow gradually flattens and finally vanishes. The other two transient furrows, the posterior and anterior transverse furrows, appear while gastrulation is in progress and persist only during the initial, rapid phase of germ band elongation. Both furrows appear to be a consequence of the modifications that occur in the outer cell layer during elongation of the germ band, and both later flatten into the amnioserosa.

During the phase of slow elongation of the germ band a group of cells invaginates at the anteroventral tip of the egg to form the stomodeum (Fig. 1.1, 10). After invagination the stomodeum extends caudally to establish contact with the endodermal anterior midgut primordium. With further elongation, based on mitotic divisions and convergent extension movements, the foregut becomes subdivided into several domains, including the oesophagus and proventriculus. A substantial part of the growth of the foregut is due to the continued incorporation of neighbouring ectodermal regions, e.g. the hypopharynx and territory of the gnathal segments, into the stomodeum.

Neurogenesis is the development of the nervous system. The first step in neurogenesis is the segregation of the neural progenitors, the neuroblasts, from the ectodermal layer (Fig. 1.1, 9-10). Neuroblast segregation begins during the fast phase of germ band elongation and lasts for 2 h; it takes place in pulses, which give rise to different subpopulations of neuroblasts. Neuroblast segregation occurs within the limits of the neurogenic ectoderm, two well defined portions of the outer germ layer on either side of the embryo, one in the procephalic lobe and the other in the germ band. After segregation, the neuroblasts become arranged according to a characteristic, apparently constant pattern, and soon thereafter neuroblasts start dividing to produce ganglion mother cells; the progeny of these latter cells will differentiate as neurons.

Definitive segmentation in the territory of the germ band is completed by the end of germ band shortening, when deep intersegmental furrows develop at ventral and dorsal levels of the epidermis (Fig. 1.1, 12). However, in *Drosophila* embryogenesis, the process of segmentation can be viewed as a sequence of several steps, each defined by subdivision of the metameric germ band; some of these steps are transient in character, prior to the completion of intersegmental furrow formation (ref. to Chapter 17). In contrast to what is observed in the germ band, there is no clear evidence for further segmentation in the procephalic region, except for the formation of the clypeolabrum and the procephalic lobe (Fig. 1.1, 11).

The period of mitotic activity of postblastoderm cells corresponds roughly to the extended germ band stage, during which the majority of cells will have executed their mitotic program. All embryonic cells except the amnioserosa, the neuroblasts, the progenitors of the epidermal sensilla and the pole cells, divide according to a characteristic spatio-temporal pattern, on average only two or three times during the entire embryonic period (Hartenstein and Campos-Ortega 1985). The neuroblasts are thought to divide an average of nine times each (Poulson 1950, Hartenstein et al. 1987), whereas no evidence is available concerning the number of divisions of germline cells. Mitotic divisions lead to growth of the various internal organs, particularly the anterior and posterior midgut primordia and the mesoderm. Several morphogenetic events also occur during this stage of general mitotic activity: the mesodermal layer divides into somato- and splanchnopleura; the salivary glands invaginate from two special placodes, one in each labial segment; an array of segmental placodes, in *t2* to *a8*, invaginate to give rise to the tracheal pits, from which the tracheal tree will develop.

Germ band shortening

Germ band shortening follows, with far-reaching morphogenetic consequences (Fig. 1.1, 12). First of all, as far as the topology of the germ band is concerned, germ band shortening restores the normal anatomical relationships of the larva. Thus the caudal end of the hindgut comes to be located at the posterior tip of the embryo, i.e. at the definitive anal position. Second, germ band shortening is involved in important morphogenetic processes: in the fusion of the anterior and posterior midgut primordia, in the establishment of continuity among the various sections that give rise to the complete tracheal tree, and in the formation of the gonads. During germ band shortening the midgut primordia are brought together to meet at approximately 50% EL; the primordia then fuse lateral to the yolk sac and along its entire length; finally the midgut extends ventrally and dorsally to enclose the yolk sac completely (Fig. 1.1, 14-15). Similarly, germ band shortening contributes to bringing together, to form a continuous main tracheal tube, the ten tracheal pits that invaginated during the stage of extended germ band. Germ band shortening also serves to bring the germline and mesodermal cells together, to give rise to the gonads. After shortening, the embryo is open dorsally, i.e. the dorsal surface consists of the amnioserosa, which is in direct contact with the vitelline envelope. At this stage, the dorsal epidermal primordium is still located laterally.

Dorsal closure and head involution

Dorsal closure of the embryo occurs by extension of the dorsal epidermal primordium on either side, leading to fusion at the dorsal midline. At dorsal closure the dorsal epidermis extends towards the amnioserosa, which is eventually displaced into the embryo (Fig. 1.1, 15-16). At the time of germ band shortening, the complex morphogenetic movements that lead to head involution begin (Fig. 1.1, 14-17). These movements particularly affect the gnathal segments: the gnathal buds initiate atrium formation; stretching of the dorsal ridge and the dorsal fold over the procephalic lobe and the clypeolabrum terminates head involution. Atrium formation occurs due to ventro-oral movement of the labial buds to form the atrial floor, and dorso-oral displacement of the mandible and maxilla to form the lateral margins of the atrial opening. During the movement of the labial buds, the outlets of both salivary glands join and fuse at the midline into a common salivary duct, which at the end of atrium formation opens onto its floor, at the level of its boundary with the pharynx. The dorsal fold arises from the dorsal portion of each labial segment, which extend to fuse at the dorsal midline. The dorsal fold then slides over the procephalic lobe and the clypeolabrum, which becomes incorporated into the foregut cavity, forming the dorsal pouch. After dorsal closure and head involution, morphogenesis is essentially complete.

Chapter 2
Stages of *Drosophila* Embryogenesis

Accurate staging of *Drosophila* embryos is subject to various difficulties. Since females may retain developing eggs for long periods
of time before laying, elapsed time since egg laying is not a reliable measure for staging; other methods are therefore highly desi-
rable. Developmental tables based on changes in the external
morphology of the embryo are available for several species of animals. Similarly, conventions have been established to subdivide
embryogenesis of *Drosophila* into a series of stages.

Embryonic stages are defined by prominent features that are
easily distinguishable in the living *Drosophila* embryo immersed
in any of several different media (0.9% sodium chloride, paraffin
oil, fluorocarbon oil - Voltalef oil 3S or 10S). Since this method
requires only a compound microscope, and does not entail any
complicated manipulation of the embryo, it can be used by any
investigator of *Drosophila* embryonic development. Staging tables
based on observations of living embryos have already been published by Imaizumi (1958), Bownes (1975, 1982), and Wieschaus and
Nüsslein-Volhard (1986). Furthermore a cinematographic study of
Drosophila embryogenesis was carried out by Ede and Counce
(1956). In this book an attempt has been made to retain, wherever
possible, features of the classifications proposed by these earlier
authors. However, when we studied fixed and sectioned material
and compared this with living embryos, inadequacies became evident in the classifications of Imaizumi and Bownes. These inadequacies have forced us to depart in some instances from previous
classifications, by introducing new stages or modifying existing
ones, in order to harmonize the observations on living embryos
with the study of histological material (Nüsslein-Volhard, Wieschaus and Campos-Ortega, unpublished).

The following is a description of embryonic stages based mainly on observations of living embryos under Voltalef oil, fuchsin-
stained wholemounts and histological sections. Care was taken to
distinguish between the description of the different stages on the

basis of living material, where only features that can be unambiguously recognized in the embryo under Voltalef oil are emphasized, and the description of development based on histological material, in which the level of resolution is higher. The duration of the various stages indicated in the text refers to embryos developing at 25°C. We wish to stress that most of the data described here are average values obtained from observations on several different embryos. Due to difficulties in precisely defining the beginning and end of most of the relevant stages, there is some variability in the timing of stages, so that the times given here should be understood as approximations.

Stage 1

a) **Living embryos.** Corresponds to stage 1 of Bownes (1975). Stage 1 lasts for about 25 min (0-0:25 h) and begins once the egg has been laid after meiosis and fertilization, and ends when the first two cleavage divisions have been completed (Figs. 2.1, 1-1a). The living egg in stage 1 is uniformly dark in the centre and light at its periphery. This reflects the homogeneous distribution of yolk granules. Characteristically the zygote fills the egg completely without leaving any empty space.

b) **Fixed material.** In histological sections, the fertilized egg displays homogeneously distributed yolk granules and egg cytoplasm; only the periphery of the egg, occupied by a thin rim of periplasm, and a cytoplasmic island within the anterior one-third of the egg, which contains the female pronucleus, are yolk free. Meiosis and fertilization have been thoroughly studied at the cellular level by Sonnenblick (1950), who also discussed the pertinent literature. Fertilization comprises sperm transfer, sperm storage, sperm release, egg penetration, syngamy (fusion of gamete cells) and karyogamy (combination of the haploid genomes of the gametes into the diploid zygotic nucleus). A review of fertilization in insects can be found in Sander (1985b). Fertilization occurs in the uterus, where the egg is positioned so that the sperm enters at the micropilar end (Nonidez 1920). Polyspermy has been repeatedly claimed to occur in *Drosophila*, although only one sperm is the source of the male pronucleus (Huettner 1924, 1927; Sonnenblick 1950). However, this view has been challenged by experimental evidence provided by Hildreth and Lucchesi (1963). These authors detected polyspermy (actually dispermy) in 3-5% of the cases they observed.

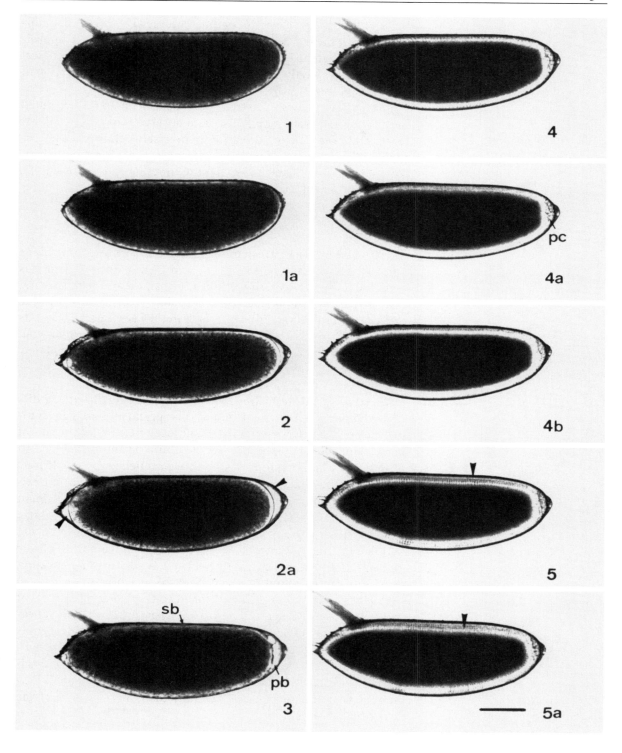

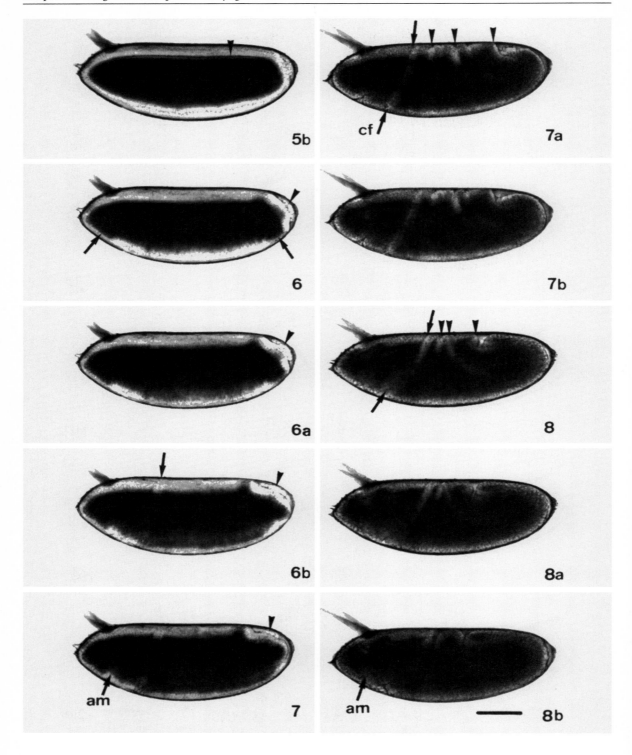

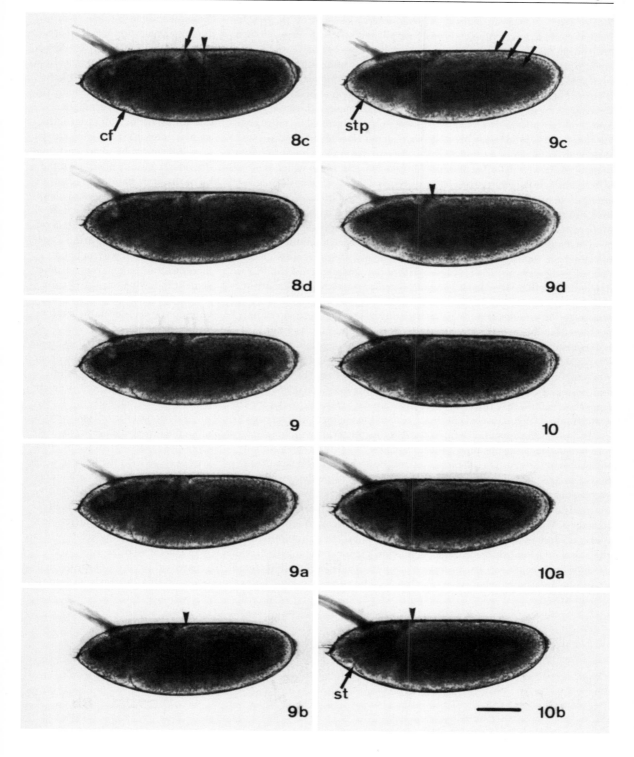

Fig. 2.1 The six plates show mid-sagittal views of developing embryos illustrating the stages of *Drosophila* embryogenesis. The stage number is indicated at the lower right hand corner of each photograph. The views in 1-1a and 5b-7 are of the same embryo, 2-5a and 7a-15b are of another embryo, and 15c-17 of a third embryo. The photographs are of embryos in Voltalef oil at about 25°C. Although the embryos develop enclosed in both the vitelline envelope and the chorion, use of Voltalef oil permits one to observe internal structures at a relatively high level of resolution. Photographs were taken at irregular time intervals, depending on the salient events in the different stages. The *bar* corresponds to 100 μm. See text for a detailed description of stages. The homogeneous structure of the egg during stage 1 is modified in stage 2, when (*arrowheads* in 2a) anterior and posterior cytoplasm retracts from the vitelline envelope and a redistribution of yolk occurs. Both events lead to the appearance of differences between the periphery and the centre of the egg. Three pole buds will appear in the posterior gap in stage 3, which undergo two divisions before pole cell formation in stage 4. View 3 shows somatic buds (*sb*) and polar buds (*pb*) after their first division (ninth cleavage), 4a shows pole cells (*pc*). The process of cellularization begins in stage 5;

arrowheads in 5, 5a and 5b point to the advancing blastoderm cell membranes during cellularization. Immediately after completion of cell formation, the movements that precede gastrulation begin. Notice the polygonal shape of the embryo, which is conditioned by morphogenetic movements during this stage. The two *arrows* in 6 point to regions of the blastoderm in which the cells are particularly thin at the beginning of gastrulation; these two *arrows* delimit the territory of the ventral furrow. Modifications of the cells at the ventral midline that accompany their invagination to form the ventral furrow can be recognized in stages 6 to 7. The *arrowhead* at the posterior pole in 6 and 7 points to the displacement of pole cells that signals the beginning of germ band elongation; they disappear into the amnioproctodeal invagination in stage 7b. The cephalic furrow (*cf*) appears in stage 6b (*arrow*), as a almost vertical groove on either side of the embryo, which progressively tilts posteriorly (*arrows* in 7a, 8 and 8c). The three *arrowheads* at dorsal levels in 7a and 8 indicate, from rostral to caudal, the anterior and posterior transverse furrows and the anterior margin of the amnioproctodeal invagination. The position of these structures changes during germ band elongation, which progresses very rapidly in its initial phase. The anterior midgut anlage

can be seen to invaginate ventrorally in stage 7 to form the corresponding primordium (*am*), which grows, while undergoing morphological modification, during subsequent stages. Germ band elongation continues, reaching its maximal extent in stage 11. Embryogenetic events occurring during the elongation of the germ band are customarily referred to the position of the tip of the germ band with respect to egg length (%EL). In several views (8c, 9b, 9c, 10b, etc.) the tip of the germ band is indicated by an *arrowhead* at dorsal levels. During stage 9 the thickness of the germ band increases and three layers become evident following the segregation of the neuroblasts (the *arrows* in 9c point to the ectoderm, the neuroblasts and the mesoderm). Also during stage 9 a transient segmentation of the mesoderm can be observed, which is less obvious in this particular embryo. In 9c the stomodeal plate (*stp*) appears, in 10b the invagination of the cells of the stomodeum (*st*) is visible. Segmentation of the epidermal anlage is apparent in stage 10 in the form of deep furrows; a few segments are indicated by the *arrows* in 10c. Definitive segmentation becomes gradually evident as germ band shortening progresses. The *arrowheads* ventrally in 12a, 12b and 12c point to the segmented territory of the germ band, which can clearly be seen to

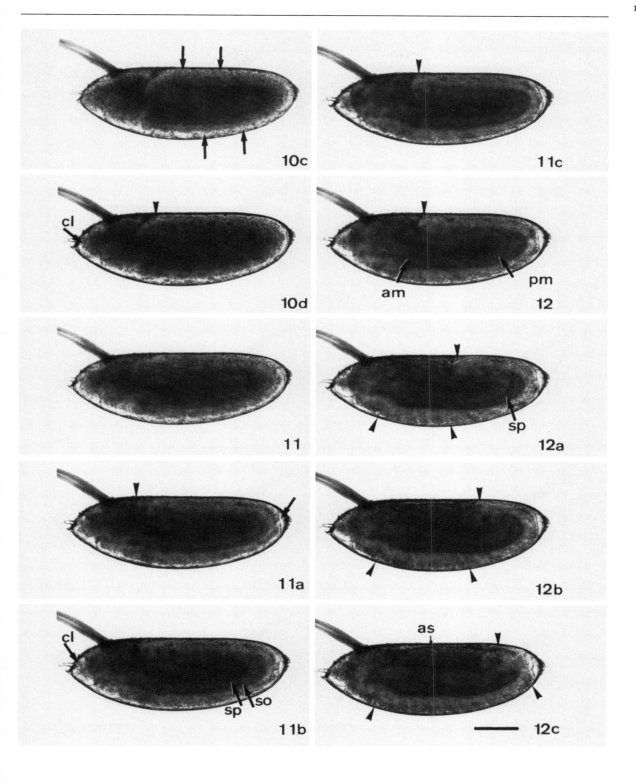

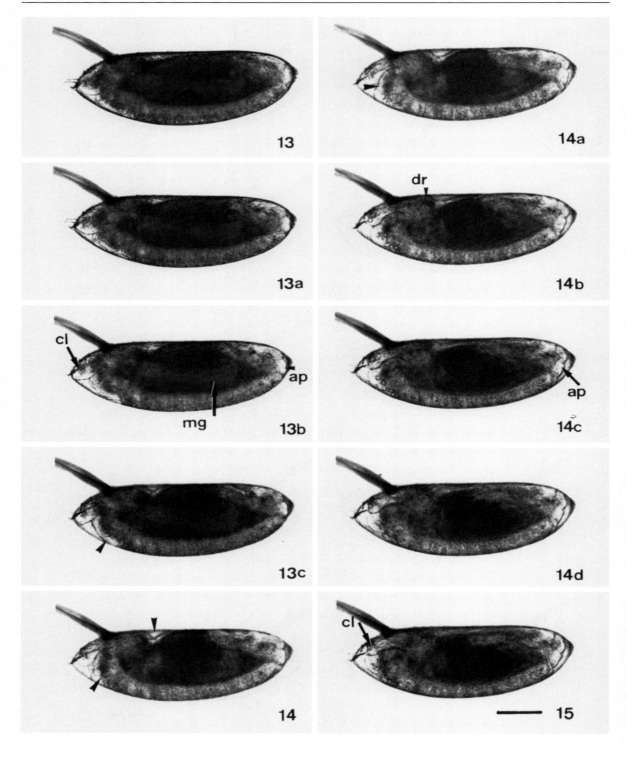

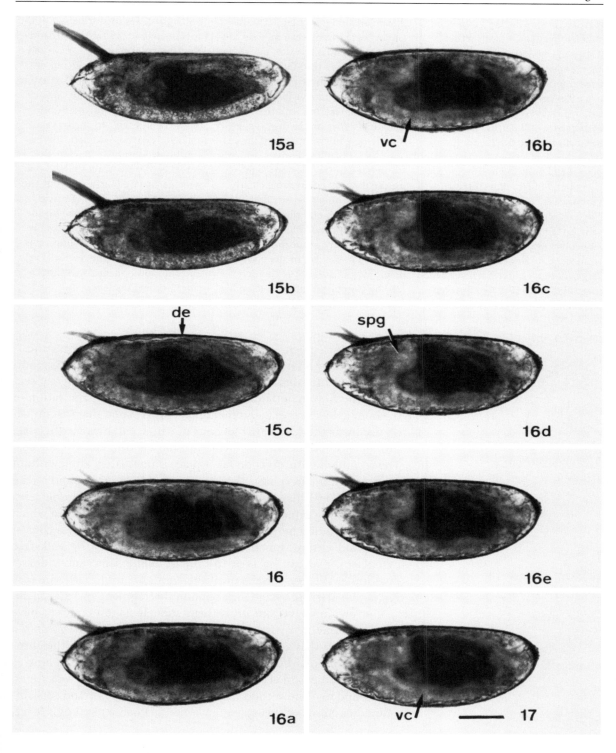

increase in size during germ band shortening. See text for further details. The beginning of germ band shortening is preceded by the appearance of a gap at the posterior egg pole (*arrow* in 11a), which will disappear when shortening is complete (see 13). The progress of germ band shortening can be observed from the displacement of the dorsal *arrowhead* pointing to the tip of the germ band in 11c, 12, 12a, 12b and 12c; when it ends at the beginning of stage 13, the embryo is covered dorsally by the amnioserosa only. The clypeolabrum (*cl*) can be recognized from stage 10 onwards (10d). During stage 11 the mesodermal layer becomes subdivided into two different layers, the somatopleura (*so*) and

the splanchnopleura (*sp*). The splanchnopleura can be seen to meet the midgut at caudal levels in panel 12a. During stage 12 the anterior and posterior midgut primordia (*am, pm*) again become visible as lighter regions contrasting with the yolk sac; this allows one to observe, in the living embryo, fusion of the primordia to form the midgut, which is completed in 12c. During stage 13 the anal pads (*ap*) become clearly visible, reaching the caudal egg pole; further growth and remodelling of the midgut primordium (*mg*) occurs. Closure of the midgut is achieved at the end of stage 14. The beginning of stage 14 is characterized by the appearance of a deep indentation dorsally (*arrowhead*),

which precedes the formation of the dorsal ridge (*dr* in 14b). Head and mouth involution can be observed by following the displacement of the clypeolabrum (see *cl* from 13b to 15) and the opening of the salivary duct (*arrowhead* ventro-anteriorly in 13c, 14 and 14a). In stage 13 the embryo is covered dorsally by the amnioserosa only. In stage 15 the dorsal epidermis reaches the dorsal midline (15 *de*). During stage 15 the midgut exhibits a characteristic heart-like shape; this shape disappears as new constrictions appear during stage 16. Head involution is completed. Ventral nerve cord (*vc*) and supraoesophageal ganglion (*spg*) become evident

The oocyte nucleus is situated dorsally at about 75% EL and is surrounded by an island of cytoplasm (an energid), which is readily visible in fuchsin-stained wholemounts and in histological sections. The first meiotic division is triggered by the penetration of the oocyte by the sperm. Sonnenblick (1950) stated that this division occurs as the sperm, the nucleus of which will become the male pronucleus, moves into the vicinity of the dividing oocyte nucleus. The second meiotic division takes place immediately after, without an intervening interphase; the innermost of the four haploid nuclei resulting from this division will become the female pronucleus. The remaining three nuclei form the polar bodies, the fate of which has been studied by Rabinowitz (1941a). The polar bodies consist of haploid chromosomal complements surrounded by yolk-free cytoplasm, and can be seen to remain at the same anterodorsal position at the egg surface until the eighth or ninth cleavage (see Fig. 2.3A); after this stage they apparently fragment and disappear.

Our own observations on fuchsin-stained stage 1 embryos have revealed some new aspects of the meiotic divisions of *Drosophila melanogaster* that are worth mentioning (Fig. 2.2). First, the chromosomes of polar bodies remain in condensed form for as long as the polar bodies are visible. Second, the first polar body was found not to undergo the second meiotic division in 199 of 200 fuchsin-stained embryos of this stage that were amenable to study. Hence, two types of polar bodies were observed in such young embryos, one with a diploid chromosomal complement, i.e. 6 chromosomes (the small fourth chromosome in nuclei of polar bodies cannot be easily distinguished in such wholemounts at the magnification used) and one with a haploid chromosomal complement. We have no explanation for the differences between our observations and Rabinowitz's (1941a) description. Third, in about one-third of the zygotes, no haploid polar body was visible, i.e. only a single polar body with a diploid chromosomal complement was found in these eggs. This latter observation is particularly striking for it raises the question of the fate of the missing, haploid polar body. Since the same situation was observed in eggs of both stages 1 and 2, premature degeneration of the haploid complement can be excluded as the reason for the absence of one polar body. Another possibility is that two female pronuclei frequently develop and that both achieve karyogamy with sperm pronuclei. This is in principle possible since polyspermy seems to occur in *Drosophila*. However, polyspermy apparently occurs at a low frequency (see above, and Hildreth and Lucchesi 1963) and therefore cannot account for the high frequency of cases in which the haploid polar body was missing. Obviously double karyogamy would have important genetic consequences, e.g. gynandromorphs and other forms of mosaicism should occur spontaneously in wild-type *Drosophila melanogaster*. Stern (1968) explicitly discusses mosaicism due to double karyogamy in *Drosophila*. Unfortunately he does not give any indication as to the frequency of the presence and fertilization of two egg pronuclei.

Stage 2

a) **Living embryos.** Corresponds to stage 2 of Bownes (1975). Stage 2 lasts for about 40 min (0:25-1:05 h). During this stage cleavage divisions 3 - 8 take place (Fig. 2.1, 2 - 2a). Characteristic of this stage is a significant retraction of the egg cytoplasm from the vitelline envelope. This leads to the appearance of two empty spaces, one at the anterior and one at the posterior pole of the egg. The posterior space will be occupied by the pole cells in stage 3 (see below) whereas the anterior space vanishes during stage 5.

b) **Fixed material.** Cleavage divisions can be conveniently studied in fuchsin-stained eggs (Fig. 2.3B, C). Zalokar and Erk (1976) have given an excellent description of cleavage in *Drosophila melanogaster* based on this technique, although the most thorough study of cleavage in *Drosophila melanogaster* has been provided by Foe and

Fig. 2.2. A to C show three stage 1 eggs (fuchsin-stained wholemounts, phase-contrast optics) to illustrate the different types of polar bodies found. As a rule the chromosomes of polar bodies remain condensed, until they disappear during the syncytial blastoderm. **A** shows the conventional type, with three haploid polar bodies. This type seems to be extremely rare, since we have found that only one of 200 embryos studied showed this distribution of polar bodies. **B** shows one haploid and one diploid polar body. This type, in which the second meiotic division has not occurred in one of the polar bodies, is the most frequent type. **C** shows the third type, in which only one diploid polar body is present. *Bar* corresponds to 10 μm

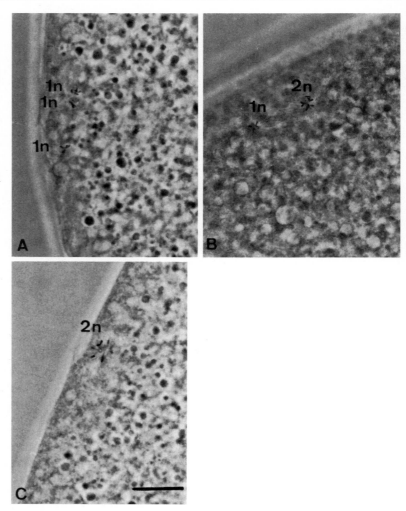

Alberts (1983) on the basis of time-lapse cinematography and the use of fixed material. The results of our own research fully confirm the descriptions by these authors. Retraction of egg cytoplasm, the characteristic feature of stage 2, begins when four zygotic nuclei are present; they are consistently located in the anterior one-third of the egg, surrounded by cytoplasmic islands and therefore clearly separated from the yolk particles. During the following three divisions, the nuclei and their surrounding cytoplasm tend to move posteriorly; by the fifth cleavage, zygotic nuclei occupy an ellipsoid field in the centre of the egg, between 20 and 80% EL. From this point on the nuclei begin to move peripherally, advancing stepwise, at a rate of about 10 μm per divisional cycle (Foe and Alberts 1983).

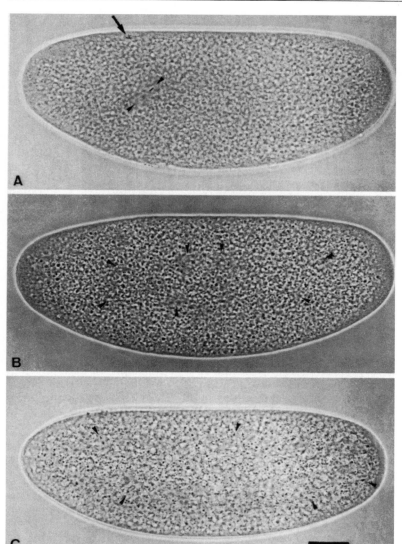

Fig. 2.3. A shows a stage 1 embryo during the second cleavage (the *arrowheads* point to the chromosomes of two nuclei, the other two are not in focus). Note the regular distribution of yolk granules and cytoplasm. A cortical rim of plasma material, particularly conspicuous at the poles, is visible. The dorsal *arrow* points to one of the nuclei of polar bodies. **B** and **C** are stage 2 embryos showing progressive peripheral displacement of zygotic nuclei, and redistribution of cytoplasm material to form the energids (*arrowheads* in **B**). The *arrowheads* in **C** point to some mitotic figures. Fuchsin-stained wholemounts (phase-contrast optics). *Bar* corresponds to 50 µm

By the end of the eighth division, the majority of nuclei are evenly arranged on an ellipsoidal surface about 35 µm beneath the membrane, whereas some remain centrally located and give rise to the vitellophages (yolk nuclei). Counts performed by Zalokar and Erk (1976) indicate that there is a doubling of the number of nuclei after each division; at the end of the eighth mitosis they found about 200 nuclei populating the periphery and about 50 presumptive yolk nuclei located in the centre of the embryo.

Stage 3

a) **Living embryos.** Corresponds to stage 3 of Bownes (1975). Stage 3 lasts for about 15 min (1:05-1:20 h). Pole bud formation and nuclear division 9 take place during this stage (Figs. 2.1, 3). The beginning of stage 3 is marked by the budding off, at the end of the eighth division, of three protuberances which jut into the space at the posterior pole. These are the polar buds (Foe and Alberts 1983). The end of stage 3 is defined by the appearance of a clear cytoplasmic rim at the periphery of the embryo. During stage 3, the polar buds divide once. During stage 4 (see below) they divide again in approximate synchrony with the somatic buds; immediately after this second division the buds pinch off, forming 12-14 pole cells. Zalokar and Erk (1976) stated that 12-18 pole cells form initially. During stage 4, most of the pole cells undergo two consecutive mitoses, giving rise to an average of 34 (Zalokar and Erk 1976) or 37 (Turner and Mahowald 1976; Underwood et al. 1980a) pole cells at the time of blastoderm formation. During stage 3, the embryo acquires a coarse, granulated appearance in a relatively broad zone at the periphery, which reflects the presence of dividing blastoderm nuclei. During this stage the empty space at the anterior pole of the embryo disappears.

b) **Fixed material.** The polar buds originate from the so-called pole plasm and are the very first buds to emerge at the surface in the *Drosophila* embryo. The ninth nuclear division occurs during stage 3 (Fig. 2.4) as the somatic nuclei approach the surface of the embryo. Somatic buds first appear when the nuclei finally reach the surface.

Stage 4

a) **Living embryos.** Corresponds to stage 4 of Bownes (1975). Stage 4 lasts approximately 50 min (1:20-2:10 h). It is the syncytial blastoderm stage in which blastoderm nuclei perform the last four cleavage divisions (Foe and Alberts 1983); stage 4 terminates at the beginning of cellularization. The living embryo has a clear rim peripherally, which increases in size during this stage (Fig. 2.1, 4 - 4b). Within the clear rim, blastoderm nuclei are readily discernible. Polar buds, prominent and clearly distinct during stage 3, have increased in number through two consecutive divisions, becoming tightly grouped by the end of stage 4. Pole cell formation takes place.

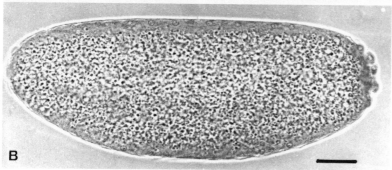

Fig. 2.4. A is a stage 3 embryo, after the emergence of the first three polar buds, showing the prophase of the ninth zygotic division. **B** is a late stage 3 embryo after completion of the ninth cleavage. Note the increase in peripheral displacement of cytoplasm. Fuchsin-stained wholemounts (phase-contrast optics). *Bar* corresponds to 50 μm

b) **Fixed material**. The last four cleavage divisions (10th, 11th, 12th and 13th) occur prior to cellularization of blastoderm nuclei. Blastoderm nuclei are located peripherally, causing the egg surface to bulge out during divisions (Turner and Mahowald 1976). After each division, the diameter of the blastoderm nuclei decreases slightly (Zalokar and Erk 1976), whereas their shape remains unchanged. The cellular basis of pole cell formation has not been studied in detail. Fragmentary observations of our own suggest that this process is very similar to that of somatic cell formation, which will be described in detail below. Pole cell mitoses also take place during stage 4; they can be readily observed in fuchsin preparations, exhibiting a slower cycle than that of the dividing syncytial blastoderm nuclei. The duration of cleavage divisions 10-13, i.e. those of the somatic buds during the syncytial blastoderm stage, increases progressively, from approximately 8 min to 20 min (Warn and Magrath 1982; Foe and Alberts 1983). The last four nuclear divisions of the preblastoderm *Drosophila* embryo have conclusively been shown to be metasynchronous by Foe and Alberts (1983). Synchrony or asynchrony of dividing syncytial blastoderm nuclei has been a matter of some debate in the past (ref.

Fig. 2.5. A to C show the last three nuclear divisions in the syncytial blastoderm stage (fuchsin-stained wholemounts, phase-contrast optics), to illustrate the high degree of local synchrony of these divisions. **A** shows the nuclei in late prophase, **B** in early prophase and **C** in anaphase. *Bar* corresponds to 10 μm

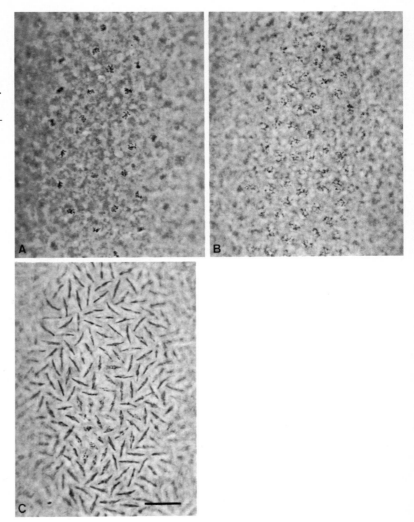

to Fig. 2.5). Asynchrony had been observed by several previous investigators (Huettner 1933; Rabinowitz 1941a; Sonnenblick 1950; Zalokar and Erk 1976), but it had been interpreted as the result of slow diffusion of fixatives, such that fixation of the different nuclei occurred at different stages of their mitotic cycle. The observations of Foe and Alberts on living material, which we fully confirm, clearly show mitoses 10-13 to progress in waves. These mitotic waves usually originate at two different sites, one near the anterior and the other near the posterior egg pole and move centripetally.

Stage 5

a) **Living material**. Corresponds to stage 5 of Bownes (1975). During this stage blastoderm cell formation takes place (Fig. 2.1, 5 - 5b), while blastoderm nuclei elongate considerably (2:10-2:50 h). This process can easily be followed in the living embryo because of the readily discernible, progressive, inward displacement of cell membranes. At the end of stage 5, pole cells begin to shift their position dorsalwards. Shifting of pole cells can be observed directly; moreover, it can be inferred from the dorsal shift of the posteriormost blastoderm cells, which is quite obvious in living material. At the same time, midventral blastoderm cells acquire an irregular, wavy appearance, darkening at their interface with the yolk. This presages their invagination to form the mesodermal and the anterior endodermal primordia.

b) **Fixed material**. Cellularization occurs by means of the introgression of membrane furrows to separate single blastoderm nuclei (Fig. 2.6B-E). This is a rapid process, and is accomplished within 30 min at 25°C. It has been studied with the electron microscope by Mahowald (1963a, b) in *Drosophila melanogaster* and by Fullilove and Jacobson (1971) in *Drosophila montana*, and a thorough description by Foe et al. (1993) is available for *Drosophila melanogaster*; we follow their descriptions closely.

After the last nuclear division has taken place, blastoderm nuclei still occupy the egg periplasm, causing the membrane to bulge outwards; they are separated from each other by plasma membrane folds, which are clearly visible with the light microscope. Cell membrane formation takes place in two phases (Foe et al. 1993). It begins when vesicles appear at the base of the plasma membrane folds. Inward progression of internuclear furrows apparently occurs through the incorporation of membrane material in the form of small vesicles, probably derived from the endoplasmic reticulum, at the base of the furrows. Specific cell junctions present in the apical portions of the intercellular membranes seem to preclude infolding of the existing plasma membrane. Thus, further infolding of egg plasma membrane does not seem to play an important role in the initial phase of cell formation. Several other organelles, including microtubules and abundant filamentous material, are visible in the immediate neighbourhood of the furrow vesicles. Mahowald (1963a) speculates that this material may participate in the synthetic processes leading to the formation of new membrane materials. However, according to Foe et al. (1993), the second phase of cell formation is triggered by microtu-

Fig. 2.6. A to **I** are sagittal sections.
A shows a stage 4 embryo, in syncytial blastoderm. After reaching the
egg periphery, the nuclei of the prospective somatic cells divide four
more times before cell formation.
Dividing somatic nuclei and pole cell
nuclei protrude to form externally
visible characteristic bulges. This
view shows clearly that the egg cytoplasm is concentrated peripherally
in the so-called cortex, whereas the
yolk grains and the vitellophages (*vg*
in **B**), surrounded by a small amount
of cytoplasm, are diffusely distributed in the centre. *pc* pole cells. **B**
shows an embryo at the end of stage
4, immediately before cell formation.
Interphase nuclei are readily visible,
having rounded up. **C** shows a stage 5
embryo. The advancing cell membranes are at about mid-nuclear level
(*arrow*), the nuclei having elongated
considerably. *Bar* corresponds to
50 μm

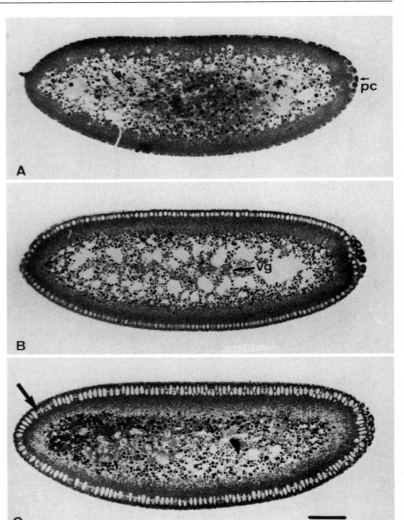

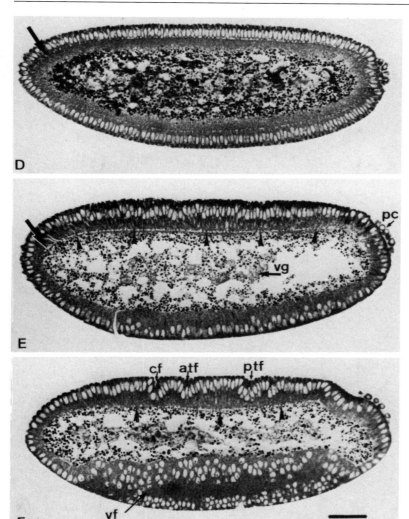

D shows a stage 5 embryo when the cell membranes have reached a level immediately below the somatic nuclei (*arrow*). E shows a stage 6 embryo. Cellularization is complete, somatic cells having incorporated most of the egg cytoplasm. The first regional morphological differences, such as irregularities in dorsal regions, become evident in the blastoderm cells. These differences foreshadow gastrulation. Notice that somatic cell nuclei exhibit a clear polarity in that nucleoli (the dark dots visible in most nuclei of this section) are always located in distal nuclear regions (compare C and D). F shows an embryo in early gastrulation. Ventral furrow (*vf*), cephalic furrow (*cf*), anterior and posterior transverse furrows (*atf, ptf*) can be distinguished. Caudally, pole cells have started to shift to dorsal levels, as a result of displacement of the underlying blastoderm cells to form the amnioproctodeal invagination.

G to **I** show the transition from stage 6 to stage 7. In **G** pole cells have attained a horizontal orientation; in **H** formation of the amnioproctodeal invagination has begun, as the posterior pole cell plate starts to sink inwards; in **I** the posterior pole cell plate has deepened further, allowing one to distinguish two different primordia, that of the posterior midgut (*pm*) and that of the proctodeum (*pr*). The anterior tip of the ventral furrow at about 85% EL broadens, acquiring the shape of a slit perpendicular to the ventral furrow, from which the anterior midgut primordium (*am*) develops. By this stage, the lumen of the ventral furrow is continuous with the anterior midgut invagination and the amnioproctodeum, as can clearly be recognized in **I**. Cephalic furrow (*cf*) and anterior and posterior transverse furrows (*atf*, *ptf*) have further deepened. *Bar* corresponds to 50 μm

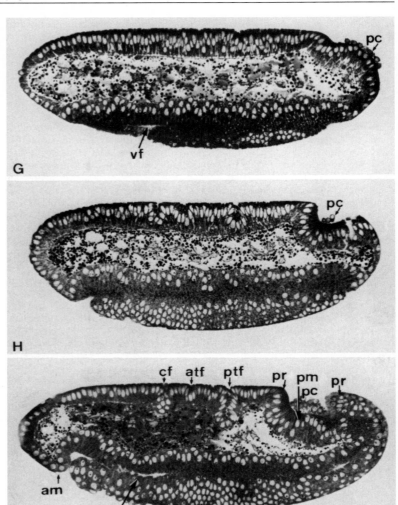

bule elongation, causing the microvilli above each nucleus to unfold, thus increasing the cell membrane surface. Fullilove and Jacobson (1971) suggest that membrane unfolding is the major source of membrane material during cell formation (see Foe et al. 1993, for discussion). Blastoderm nuclei are spherical at the onset of cellularization but elongate considerably as the process continues, increasing in length from 3-4 μm to 10-15 μm. Nuclear elongation goes hand in hand with cellularization, though it actually begins before cellularization, when the furrow vesicles are at the level of the apical pole of the nuclei, and terminates before the furrows have grown beyond the base of the nuclei to reach the yolk. The beginning of nuclear elongation is marked by the appearance of very prominent nucleoli in the apical region of the nuclei. It is very striking that nucleoli remain visible in all embryonic cells at this same apical position for a relatively long time during embryogenesis, until germ band extension is well advanced. After this, movements of embryonic cells are too intricate to permit one to relate the position of their nucleoli to that in earlier stages.

Cellularization is completed when the furrows reach the yolk. Then adjacent vesicles fuse to give rise, on the one hand, to the basal surface of blastoderm cells and, on the other hand, to a cellular envelope for the yolk - the so-called yolk sac. From this moment on, until the first instar larva hatches from the egg, the yolk sac actually consists of a large syncytium which contains the yolk granules, the yolk nuclei (vitellophages; see, for example, Fig. 2.6B) and some remnants of egg cytoplasm. Blastoderm cells around the perimeter of the entire egg at this stage are not completely isolated since they still maintain connections with the syncytial yolk cytoplasm by means of wide cytoplasmic bridges (Rickoll 1976; Turner and Mahowald 1977; Rickoll and Counce 1980; ref. to Fig. 2.6E-F). These connections are more abundant at dorsal and lateral levels than ventrally, and are lost during gastrulation (Rickoll and Counce 1980).

All blastoderm nuclei and cells have the same shape and do not show any apparent differences between particular egg regions. Both shape and size, however, will show considerable regional variations during the next stage (Fig. 2.6).

During cellularization, as the furrows progress inwards, a few pole cells can be seen to interdigitate between blastoderm cells. At the end of cellularization these pole cells have become completely internalized, lying below the posterior blastoderm cells and clearly separated from the yolk sac (Counce 1963; Underwood et al. 1980a). These pole cells will accompany the cells of the posterior pole of the blastoderm, i.e. the anlage of the posterior midgut, in their movements at germ band elongation, eventually becoming included in the midgut lumen in stage 13.

Stage 6

a) **Living embryos.** Corresponds to stage 6 of Bownes (1975). Stage 6 is a short stage, lasting approximately 10 min (2:50-3 h). It is the stage of early gastrulation (Fig. 2.1, b), in which three major morphogenetic movements are accomplished: (i) mesodermal and endodermal primordia invaginate, (ii) blastoderm cells at the posterior pole shift their position to form a dorsal plate to which the pole cells adhere, and (iii) the cephalic furrow becomes visible at a position corresponding to about 65% EL.

Stage 6 begins with the formation of the ventral furrow from which mesoderm and anterior endoderm originate. Prior to ventral furrow formation and initiation of germ band elongation, midventral cells become very thin at both poles and, consequently, the sagittal profile of the embryo exhibits a characteristically polygonal shape. The ventral furrow initiates development as a median longitudinal cleft that extends between 20 and 70% EL, along the ventral embryonic midline; over a period of approximately 10 min the ventral furrow will extend further by incorporating additional cells at its anterior and posterior tips, until it extends between 6 and 85% EL. In the living embryo observed laterally, blastoderm cells within the anterior and the posterior one-fifth of the egg, are seen to become strikingly thin, whereas ventral blastoderm cells appear slightly but clearly separated from the underlying vitelline envelope, reflecting the formation of the ventral furrow that has occurred midventrally. At the posterior pole of the embryo, a cell plate that carries the pole cells shifts dorsally. This cell plate is flanked anteriorly by a prominent piling up of blastoderm cells, which becomes more and more striking as germ band elongation proceeds. At the same time, other irregularities become evident within the dorsal half of the embryo; these anticipate the formation of the dorsal folds in stage 7. The cephalic furrow first becomes visible as a lateroventral slit at about 65% EL, later extending obliquely towards both the dorsal and the ventral egg surfaces.

b) **Fixed material.** At gastrulation, the first detectable changes concern the shape of blastoderm cells, which becomes strikingly different in dorsal and in ventral territories (Fig. 2.6E-H). Dorsally, blastoderm cells elongate further, becoming very thin and tightly packed; ventrally they show the same cylindrical shape as in the previous stage. Besides Sonnenblick's (1950) description with the light microscope, changes within the ventral blastoderm cells that lead to the invagination of the mesodermal primordium have been described by Rickoll (1976) with the transmission electron microscope, and by Turner and Mahowald (1977) based on observation of cross-fractured embryos with the scanning electron microscope; more recently Leptin and Grunewald (1990) and Kam et al. (1991) have used a variety of techniques to study these changes. The ventral furrow forms as a result of cell shape changes which affect an area about 12 cells in width centred on the ventral midline (Leptin 1994). The apical (external) portions of the six midventral cell rows constrict while their basal portions enlarge; the nuclei are thereby forced into the basal portion of the cells. Then these cells shorten. The lateral cells are stretched as the medial cells are drawn into the invagination. Invagination of the posterior midgut anlage occurs a few minutes after that of the mesodermal cells, but the cells' behavior appears to be the same. These modifications of the ventral embryonic cells are accompanied by changes in the location of myosin, which becomes concentrated at the apical side of the cells. This behavior is striking since myosin was previously restricted to the basal region of the cells (Young et al. 1991; Leptin 1994).

The anteroposterior extent of the ventral furrow measured by Turner and Mahowald (1977) on scanning electron micrographs comprises initially about 45% of the total embryonic length, from 65% EL anteriorly to 20% EL posteriorly. During subsequent development, the ventral furrow extends further anteriorly and posteriorly, eventually reaching beyond the level of the cephalic furrow to encompass 6-86% EL. Poulson (1950) estimated on the basis of cell counts in transverse sections that about one-sixth of the blastoderm circumference invaginates to form the mesoderm primordium; Turner and Mahowald (1977) have published values between 15 and 18% of the total of the blastoderm cells, i.e. a total of 1000 cells approximately. Hartenstein and Campos-Ortega (1985) counted a total of 1250 cells invaginating at gastrulation, of which 450 will form the amnioproctodeal invagination, 730 the mesoderm primordium and the remaining 70 the anterior endodermal midgut primordium (see Chapter 18).

The cephalic furrow forms at the same time as the ventral furrow (Fig. 2.6F). It extends transversely from the dorsal midline, at about 60% EL, to the ventral midline at about 75% EL, therefore following an oblique course. The furrow originates from the shortening of a row of lateral cells, forming a fold in the embryo surface; this movement then expands across the ventral and dorsal surfaces (Turner and Mahowald 1977). The width of the folded region initially comprises about 6-8 cells, and is quite regular along its whole dorsoventral extent.

During stage 6 the pole cells can be seen to shift their position from the posterior egg pole to the dorsal surface (Fig. 2.6F-G). Pole cells apparently do not move actively; they seem rather to be transported by the movements of a discoid plate of about 150 underlying blastoderm cells which had earlier formed the posterior egg pole.

Stage 7

a) **Living embryos.** Corresponds to stage 7 of Bownes (1975). Stage 7 is also a very short stage, lasting for about 10 min (3:00-3:10 h). During this stage, gastrulation is completed (Figs. 2.1, 7-7b): the endodermal primordia of the anterior and posterior midgut, and the primordium of the hindgut, invaginate; the dorsal folds appear. Stage 7 begins when the cell plate that carries the pole cells has attained a horizontal orientation on the dorsal egg surface, and ends when the anterior wall of the amnioproctodeal invagination starts to move cephalad. At the same time, the anterior midgut rudiment completes invagination anteroventrally to form a transverse groove at the tip of the ventral furrow, at 85% EL. Three folds are then clearly visible, which diverge in a dorsoventral direction: the cephalic furrow, the anterior transverse fold and the posterior transverse fold. The latter are superficial, short-lived structures which change their position, advancing anteriorly as germ band elongation progresses, whilst the cephalic furrow is a deep invagination that will persist until stage 9 (see below). Most of the cells forming the transverse furrows comprise the primordium of the amnioserosa, the extraembryonic membrane. The posterior midgut develops by invagination of the cell plate carrying the pole cells. These accompany the posterior midgut primordium in its invagination and disappear from the embryo surface; from this time on, pole cells are no longer visible without fixation and sectioning. The cells immediately caudal to the invaginating posterior midgut will form a neck at which the midgut opens superficially. These cells, together with the bunched cells anterior to the posterior midgut primordium, will give rise to the proctodeum (hindgut).

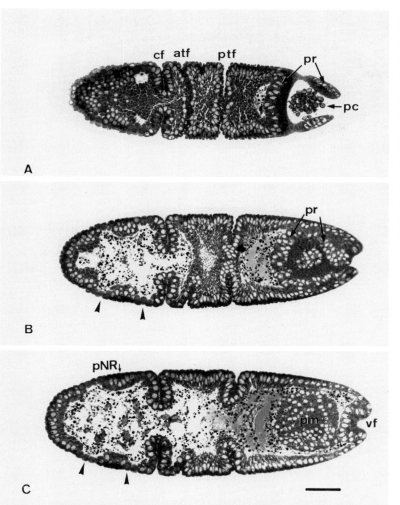

A

B

C

Fig. 2.7. A to F show horizontal sections of a late stage 7 embryo. **A** shows the dorsalmost, **F** the ventralmost level. Mitotic activity of embryonic cells has begun (*arrowheads*), localized in the ectoderm of the presumptive procephalic territory. The primordium of the procephalic neurogenic ectoderm (*pNR*) becomes distinguishable for the first time, flanked by a territory of mitotic activity. The horizontal level of sectioning permits one clearly to distinguish, first, the various morphogenetic furrows, cephalic (*cf*), anterior (*atf*) and posterior (*ptf*), transverse and ventral (*vf*); second, the organization of the amnioproctodeal invagination, in which the proctodeal primordium (*pr*) consists of an anterior wall and two latero-posterior walls, above the floor of the invagination from which the posterior midgut will develop (*pm*); and, third, the continuity of the ventral furrow (*vf*) with the anterior and posterior midgut primordia. *Bar* 50 µm

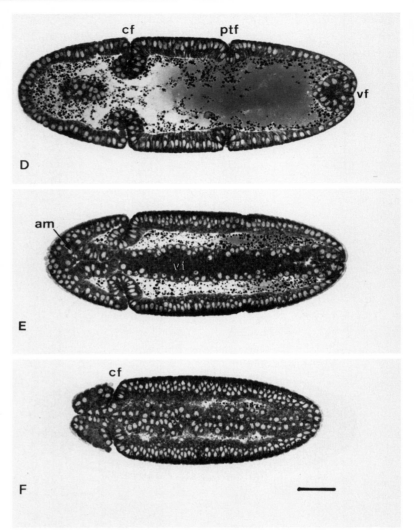

b) **Fixed material**. During stage 7 (Figs. 2.6, 2.7), gastrulation is completed by the invagination of the endoderm, i.e. the primordia of anterior and posterior midgut. The dorsal folds become visible. Cells that remain on the outside after invagination of the ventral furrow, including endodermal primordia and the prospective amnioserosa, constitute the ectoderm. All three germ layers can therefore be distinguished in the stage 7 embryo. The ectoderm accounts for approximately 3700 cells, of which 1000 are located in front of, and 2700 behind the cephalic furrow (see Chapter 18).

The anterior midgut primordium invaginates ventrally at about 85% EL and is continuous with the ventral furrow, whose anterior end it forms (Figs. 2.6I, 2.7). Thus, the ventral furrow is clearly T-shaped in the region anterior to the cephalic furrow, whereby the bar of the T gives rise to the anterior midgut primordium and the stem to mesodermal cells of the head. Whereas the ventral furrow soon closes, through the fusion of its lips, the anterior midgut primordium remains open, maintaining a connection to the outside for a relatively long time, until the end of stage 8.

At the posterior egg pole, in stage 6, a discoid plate of approximately 150 cells had shifted dorsally, carrying the pole cells with it. The cell plate tilts inward anteriorly and eventually forms a pocket that contains the pole cells. The cells immediately posterior to this deepening depression sink inwards along the midsagittal plane, to form a deep groove that is continuous with the ventral furrow. The entire structure, pocket containing pole cells and the neck of the pocket (Fig. 2.7), was called the amnioproctodeal invagination by Sonnenblick (1950). Turner and Mahowald (1977) noticed that the infolded cells posterior to the posterior midgut rudiment are clearly different from the cells of the ventral furrow, despite their being topologically continuous with it, and recognized that they will give rise to the hindgut and the proctodeal opening. We shall come back to this point later.

The dorsal folds, or anterior and posterior transverse folds, appear dorsally at the onset of gastrulation (Fig. 2.6F) and follow an oblique course, diverging from dorsal to ventral. The two transverse folds can be seen in sagittal sections to appear simultaneously; since the posterior fold is deeper than the anterior, it appears more prominent. The cephalic furrow has deepened further during stage 7, forming an oblique ring that extends continuously from middorsal to midventral levels, between 54% (dorsally) and 74% (ventrally) EL. The first mitotic divisions of embryonic cells can be distinguished at the end of stage 7. They mainly affect two clusters of superficial ectodermal cells located anterior to the cephalic furrow, in a dorsolateral position. These two cell clusters are separated from each other by a group of non-dividing cylindrical cells. The mitotically quiescent cell cluster will later give rise to the neuroblasts of the procephalic lobe. It will be discussed below.

Stage 8

a) **Living embryos.** Corresponds to part of stage 8 of Bownes (1975). Stage 8 lasts for approximately 30 min (3:10-3:40 h). The beginning of stage 8 is marked by the formation of the amnioproctodeal invagination and the rapid phase of germ band elongation (Figs. 2.1, 8 - 8d); it ends with mesodermal segmentation. Mesoderm and endoderm formation has taken place during the previous two stages by means of the invagination of the ventral furrow and of the amnioproctodeum, whereas the cells that remain at the outer surface of the embryo form the ectoderm and the amnioserosa. Elongation of the germ band consists in the gradual extension of this structure cephalad, which occurs very rapidly during its initial stages and quite slowly later. By the end of stage 8 elongation has progressed to bring the proctodeal opening to about 60% EL (Fig. 2.1, 8d), whereas during the following 2 h the tip advances a distance equivalent to only 15% EL, that is, until the tip of the germ band reaches about 75% EL. Therefore, within about 35 min, the germ band reaches more than two-thirds of its final extent, while the remaining one-third requires an additional 2 h.

The germ band consists of an inner, mesodermal, and an outer, ectodermal layer; in living embryos these two layers are almost indistinguishable from each other during the initial phase of germ band elongation. Moreover, no clear interface is visible between mesoderm and yolk sac. Dorsal folds and cephalic furrow remain visible during the first half of stage 8; however, the dorsal folds progressively vanish, and only the ventral, more pronounced part of the cephalic furrow persists for some time longer.

b) **Fixed material.** The beginning of germ band extension coincides with profound modifications of the mesodermal primordium. The ventral furrow closes off shortly after invagination, except for its anterior and posterior tips, which correspond to the anterior and posterior midgut primordia and proctodeum. The cylindrical cells of the ventral furrow are organized in a regular epithelium during the whole of stage 7 (Fig. 2.7). However, this architecture is lost at the onset of stage 8, when the epithelium disaggregates and

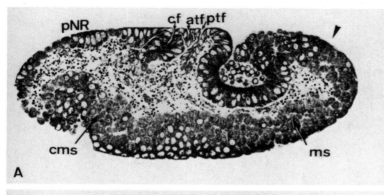

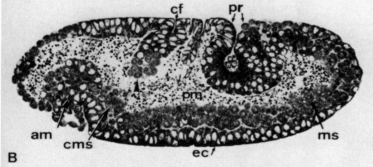

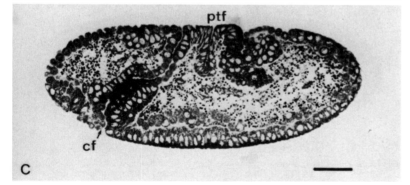

Fig. 2.8. A to **C** show parasagittal sections of an early stage 8 embryo, during the fast period of germ band elongation. The overwhelming majority of cells in the procephalic ectoderm and in the mesoderm (*ms*), as well as cells in the middle of the cephalic furrow and at caudomedial ectodermal (mesectodermal) levels (*arrowheads*), are dividing for the first time. The cells within the ventral half of the cephalic furrow do not divide at this stage and differ in form from the remaining cells of the furrow. These cells constitute the neurogenic primordium of the gnathal segments (see **C**) The remodelling of the amnioproctodeal invagination is completed when it adopts its typical shape of a sac-like fold. Anterior and posterior transverse furrows (*atf*, *ptf*) are now considerably closer together. *Bar* 50 μm

Fig. 2.9. A to **F** show horizontal sections of an early stage 8 embryo, at about the same age as the embryo in Fig. 2.8. The horizontal level of sectioning makes it possible to clearly distinguish the lateral proctodeal fold *(lpf)* on either side, flanking the proctodeal opening *(pr)*. Cell shape modifications in the territory of the posterior transverse furrow *(ptf)*, preceding the appearance of the amnioserosa, can clearly be seen in **B** and **C**. The two *arrowheads* in **B** and **D** delimit the central part of the procephalic neurogenic ectoderm *(pNR)*. Abbreviations. *am* anterior midgut primordium; *atf* anterior transverse furrow; *cf* cephalic furrow; *cms* cephalic mesoderm; *ec* ectoderm; *gne* gnathal neurogenic region; *ms.* mesoderm; *pc* pole cells; *pl* procephalic lobe; *pm* posterior midgut primordium; *pr* proctodeum. *Bar 50 μm*

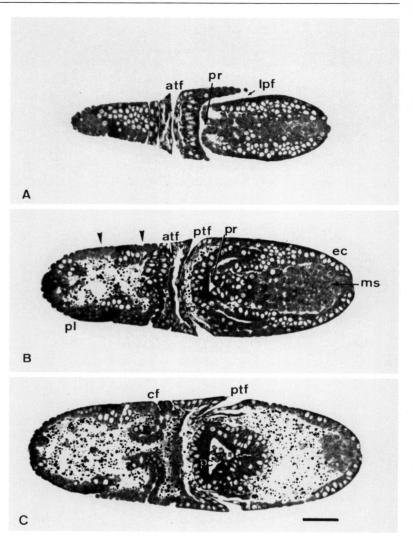

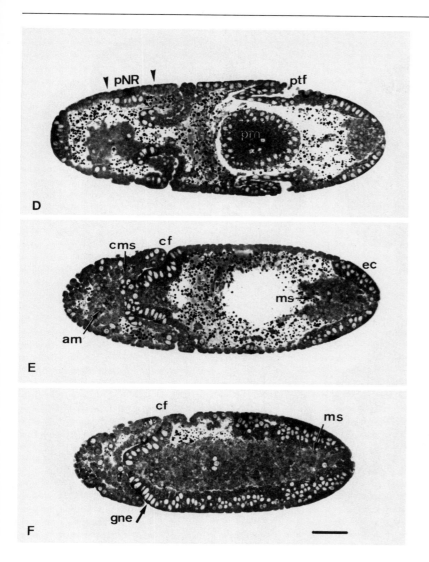

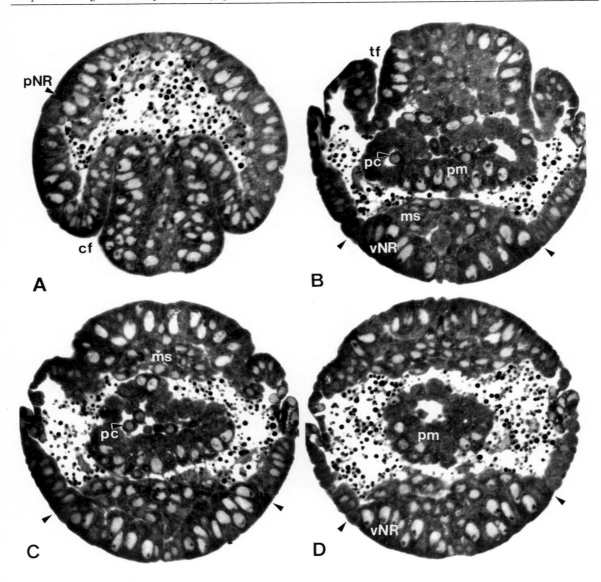

Fig. 2.10. A to D are transverse sections of a stage 8 embryo. Notice the size difference between the cells of the neuroectoderm *(pNR* procephalic neuroectoderm; *vNR* ventral neuroectoderm, between *arrowheads)* and the lateral epidermal anlagen (further lateral). The cells of the mesoderm *(ms)* have completed their first postblastodermal mitosis. Cells within the posterior midgut rudiment *(pm)* are in mitosis. *pc* pole cells

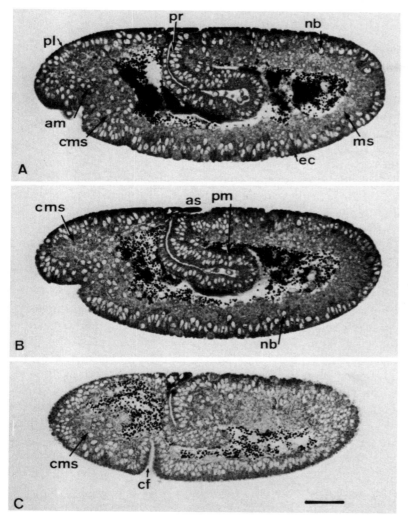

Fig. 2.11. A to **C** show parasagittal sections of a late stage 8 embryo. The fast period of germ band elongation is complete. The cells of the mesoderm (*ms*) are undergoing their second mitotic division; cell changes leading to the segregation of the neuroblasts (*nb*) have begun. The cells of the anterior midgut primordium (*am*) have become disorganized and intermingled with the cells of the cephalic mesoderm (*cms*) from which, however, they can still be distinguished because the midgut cells are mitotically quiescent by this stage. Abbreviations. *as* amnioserosa; *cf* cephalic furrow; *ec* ectoderm; *ms* mesoderm; *pl* procephalic lobe; *pm* posterior midgut primordium; *pr* proctodeum. *Bar* 50 μm

all cells start dividing within a very short time (Figs. 2.8, 2.9). During the initial period of germ band elongation, all mesodermal cells perform two consecutive, parasynchronous mitoses (Figs. 2.8, 2.9, 2.11) separated by an interphase which the first cells enter when the opening of the proctodeum has reached 55% EL. The median row of mesectodermal cells also divides at this time, while the ectodermal cells immediately adjacent to them remain mitotically quiescent.

At the time of the first mesodermal mitosis, the posterior midgut primordium, i.e. the pocket containing the pole cells, sinks further inwards (Fig. 2.8). Its posterior border consists of the walls of the groove mentioned in the previous stage, which will contribute to define the proctodeum during subsequent development. Most of the cells in the walls of the posterior midgut rudiment and proctodeum can be seen to divide at this stage (Figs. 2.9A, 2.10); it appears that these dividing cells later form the primordium of the Malpighian tubules. By now, the anterior midgut primordium consists of a thin slit, visible ventrally in the ectoderm, which is gradually separated from the closing ventral furrow (Fig. 2.8A). The cells of the anterior midgut primordium divide at the same time as the mesodermal primordium, so that the invaginating cells attain deeper levels inside the embryo. Shortly afterwards the cells disaggregate and lose their sac-like organization, and by the end of stage 8 the lumen of the slit can no longer be distinguished (Fig. 2.11). Cells of the ventral furrow become trapped at medial levels between the anterior midgut rudiment and the cephalic furrow (Figs. 2.9E, 2.11); these cells will contribute to the head mesoderm.

During the initial, rapid phase of germ band elongation the transverse folds can be seen to approach each other at the dorsal surface, obviously being displaced by the extending dorsal tip of the germ band (Fig. 2.8A). These dorsal movements of the transverse folds can only be a consequence of compression by the advancing germ band, since the cellular composition of the folds is continuously changing during rostral displacement. As displacement proceeds new cells seem to slide into the folds and even more cells seem to leave them, for the number of cells that separate dorsally the anterior border of the posterior midgut rudiment from the cephalic furrow on the midsagittal plane decreases markedly during stage 8. Simultaneously the shape of the cells in the transverse folds changes, from tall cylindrical to extremely flat (Figs. 2.8, 2.10). These changes in cell shape precede the appearance of the amnioserosa, the so-called extraembryonic membrane, which consists of 200 cells derived from a middorsal strip of the blastoderm. At the end of stage 8 the dorsal folds can barely be distinguished and cell shape modifications leading to the differentiation of the amnioserosa become clearly visible in this region (Fig. 2.11). The amnioserosa remains at the same position along the embryo flanks during the entire period of elongated germ band. Mid-dorsally, at the point where the tip of the germ band contacts the procephalon, the amnioserosa is buckled into many folds and sinks into the proctodeum, contiguous to the proctodeal epithelium; after germ band shortening the extraembryonic membrane will expand to form the dorsal covering of the embryo. It is noteworthy that the cells of the amnioserosa do not proliferate at all during the entire duration of embryogenesis.

The cephalic furrow at this stage is considerably deeper than previously (Figs. 2.8, 2.10), indeed deep enough to allow contact, at the midline inside the embryo, of the furrow on one side with that on the other side. As in other parts of the germ band a large number of cells can be seen to divide within each wall of the furrow; this intense mitotic activity accompanying germ band elongation probably contributes to further growth in depth of the cephalic furrow in stage 8. The cephalic furrow at this stage is a ring-like structure, the anterior tips having fused midventrally.

Signs of increased mitotic activity are also visible within the ectodermal germ layer, chiefly anterior to the cephalic furrow, in the territory of the procephalic lobe. The large majority of cells in the procephalic lobe can be seen to divide during stage 8 (Figs. 2.8, 2.9B). However, the dorsolateral cluster of non-dividing ectodermal cells alluded to in stage 7 can still be discerned, completely surrounded by mitotic cells. These non-dividing cells are strikingly larger than all remaining cells of the procephalic lobe and can easily be followed, though they remain mitotically quiescent into stage 9, when it will become evident that some of them develop into cephalic neuroblasts. Therefore part of the neurogenic region of the procephalic lobe consists of cells that have not divided at all after blastoderm formation. In the trunk ectoderm a lateral and a medial region can be clearly distinguished. Mitotic divisions are visible in the lateral ectoderm in a strip of cells that follows the course of the germ band quite closely and invades the invaginated posterior midgut rudiment at its posterior border. The ectoderm medial to this strip consists of large cuboidal cells, which extend throughout the expanding germ layer, all the way from the cephalic furrow to the invaginating posterior midgut rudiment. These cells are mitotically quiescent and will give rise to the neuroblasts of the trunk during stage 9. Thus, in the metameric germ band also, neuroblasts will segregate from the ectoderm without having divided since blastoderm formation. Later in stage 8, immediately before the segregation of the neuroblasts, characteristic clusters of dividing cells appear within the prospective neurogenic ectoderm of abdominal segments, which show a clearly metameric pattern (see Chapter 14). The abdominal clusters of mitotic cells represent the first visible manifestation of metamery in *Drosophila* embryogenesis.

Fig. 2.12. A to **C** show parasagittal sections of an early stage 9 embryo. Neuroblasts (*nb*) of the first subpopulation are in process of segregation. Cells of the mesoderm (*ms*) have entered their second interphase, and are rearranged into a coherent epithelium. During this process of rearrangement, the mesodermal cells show characteristic periodical bulges that transiently mark the position of the segments. Periodic bulges of the mesoderm are particularly evident in **A**. Abbreviations. *am* anterior midgut primordium; *as* amnioserosa; *cf* cephalic furrow; *cms* cephalic mesoderm; *ec* ectoderm; *pl* procephalic lobe. *Bar* 50 μm

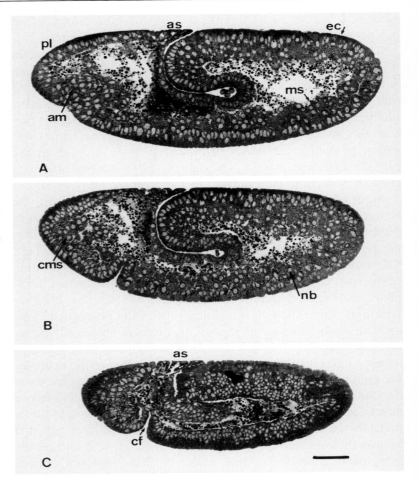

Stage 9

a) **Living embryos**. Corresponds to part of stage 8 of Bownes (1975). Stage 9 lasts for approximately 40 min (3:40-4:20 h) and in the living embryo is initiated by the transient appearance of segmentation in the mesodermal layer (Figs. 2.1, 9-9d); the end of stage 9 is defined by the onset of formation of the stomodeal invagination. At the end of stage 8 the interface between yolk sac and mesoderm becomes obvious in the living embryo; mesodermal segmentation is visible within the territory of the germ band as a series of prominent bulges protruding into the yolk sac. In stage 9 the first two populations of neuroblasts segregate from the ectoderm; therefore, by the end of stage 9 the germ band exhibits a

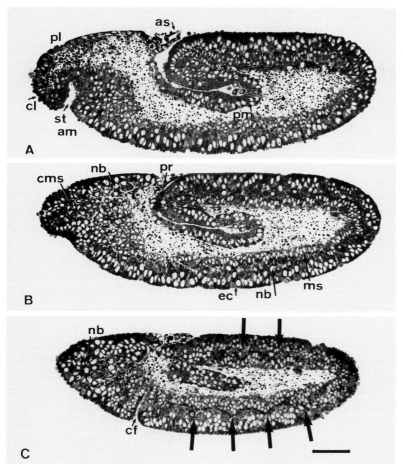

Fig. 2.13. A to C show three parasagittal sections of an early stage 10 embryo. Cells invaginate to form the stomodeum (*st*). Periodic bulges (large *arrows* in **C**) are still distinguishable in the mesodermal layer (*ms*). The clypeolabrum (*cl*) can be distinguished within the procephalic lobe (*pl*). Abbreviations. *as* amnioserosa; *am* anterior midgut primordium; *cf* cephalic furrow; *ec* ectoderm; *nb* neuroblasts; *pm* posterior midgut primordium; *pr* proctodeum. *Bar* 50 μm

clearly three-layered architecture. The proctodeal opening reaches 70% EL. The dorsal region of the procephalic lobe becomes very thin and wavy in the midsagittal plane. The invaginated anterior midgut primordium increases in size. Simultaneously, the anterior pole of the embryo starts to withdraw from the vitelline envelope ventrally, eventually to give rise to the stomodeal cell plate, from which the stomodeum will invaginate in stage 10.

b) **Fixed material**. At the beginning of stage 9, while germ band elongation proceeds, mitotic divisions in the mesoderm cease (Fig. 2.12); the mesodermal cells then reorganize, changing from the loose arrangement they showed in stage 8, during the first mitotic division of the primordium, into a regular monolayer of cuboidal epithelial cells along the whole germ layer (Fig. 2.13). During the process of reorganization, mesodermal cells exhibit clear signs of segmentation in the form of bulges that protrude slightly into the yolk sac (Fig. 2.12A). These mesodermal protuberances are obvious only for a short time and are scarcely distinguishable later. However, morphologically manifest metamery will reappear soon afterwards at the end of stage 10. Other changes affecting the mesoderm that occur during stage 9 are related to its extension into the territory of the procephalic lobe. The procephalic mesoderm derives in part from a subpopulation of cells that remained trapped between the anterior midgut primordium and the cephalic furrow; on either side of the embryo, these cells form a somewhat irregular monolayer, located between the ectoderm and the yolk sac, which moves in dorso-anterior direction. In addition, those parts of the procephalon that flank the stomodeal plate anteriorly and laterally also give rise to ingressing head mesoderm cells. During their dorsal displacement, cephalic mesoderm cells cannot easily be distinguished from the neighbouring cells of the anterior midgut primordium.

The ectoderm of the germ band in stage 9 is organized in two clearly defined regions. One lies ventromedially, immediately beneath the mesoderm, and consists of large cells; the neuroblasts and the ventrolateral epidermis derive from the cells of this region (Fig. 2.12A-B). The other ectodermal region is located laterally, directly contiguous to the amnioserosa, and consists of a palisade of narrow, tightly packed cells, derived from mitoses which took place in stage 8; the dorsal epidermis and the tracheal placodes will develop from this lateral region (Fig. 2.12C).

Segregation of neural cell progenitors, neuroblasts, from the ectodermal germ layer is one of the most important events in stage 9. Although the majority of neuroblasts segregate during stage 9, the process of neuroblast segregation actually lasts for 2 h, extending therefore into stages 10 and 11. The neurogenic ectoderm of *Drosophila* consists of two different parts. One is laterodorsal, restricted to the procephalic lobe, and will give rise to the supraoesophageal ganglion, and the other extends ventromedially along the whole of the germ band and will give rise to the suboesophageal ganglion and the thoracic and abdominal neuromeres. The two parts are clearly discontinuous, being separated from each other by a narrow band of non-neurogenic cells. Both divisions of the neurogenic ectoderm consist of large, strikingly basophilic cells that derive directly from blastoderm cells, without intervening mitoses.

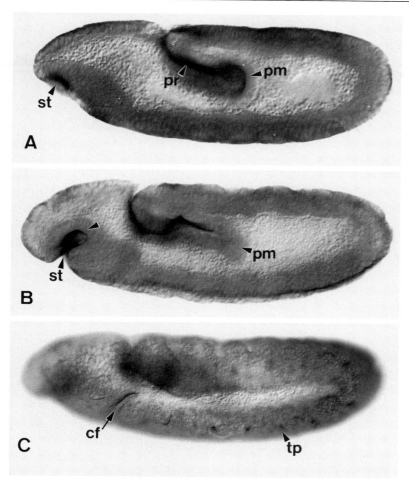

Fig. 2.14. A shows a midsagittal view of a late stage 9 embryo stained with an anti-Crumbs antibody. The anlage of the stomodeum, a horizontal cell plate (*st*), begins to differentiate. Both proctodeum (*pr*) and posterior midgut (*pm*) are organized as epithelia and, consequently, express the Crumbs protein. **B** and **C** show midsagittal and lateral planes of focus, respectively, of a stage 10 embryo stained with an anti-Crumbs antibody. The *arrowhead* points to the fundus of the stomodeum, in contact with the anterior midgut primordium. The stomodeum (*st*) has invaginated. The larval tracheal tree originates from ten ectodermal placodes, which will invaginate to form the tracheal pits (*tp*) in segments *t2 - a8*. The cells of the posterior midgut primordium (*pm*) have lost their epithelial organization and are devoid of the Crumbs protein. *cf* cephalic furrow

Observations on living embryos show the cephalic region ventrally to lift away from the vitelline envelope during stage 9. A conspicuous plate of large cylindrical cells becomes evident at this position, which will invaginate during the next stage to give rise to the stomodeum (Figs. 2.1 - 9c, 2.13). The proctodeum is already apparent as a groove formed by two cellular walls; mitotic activity invades the proctodeum, yet its cells remain arranged in a regular epithelium (Fig. 2.12, 2.13). The cells of the posterior midgut primordium divide immediately after the proctodeal cells.

The cells lying between the transverse anterior and posterior folds have given rise to the amnioserosa throughout the territory of the germ band. At the proctodeal opening the amnioserosa is continuous with the cells of the posterior midgut primordium.

Fig. 2.15. A to F are horizontal sections of a late stage 10 embryo. The arrangement of neuroblasts (*nb*) in rows immediately beneath the ectoderm (*ec*) can be seen in **A**. The two sides are symmetrically organized, being separated by the mesectodermal cells (*me*). **B** shows periodic bulges in the lateral epidermis (*arrows*) of abdominal segments, between the tracheal placodes. Proctodeum (*pr*) and posterior midgut (*pm*) are well established. Evaginations of the proctodeal wall to form the primordium of the Malpighian tubules (*mt*) can be seen in **D**. *am* anterior midgut; *as* amnioserosa; *cf* cephalic furrow; *vc* procephalic lobe; *st* stomodeum. *Bar* 50 μm

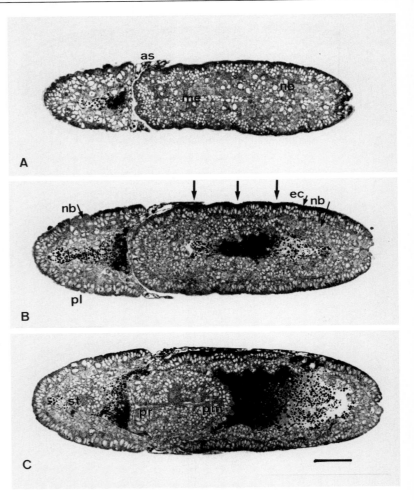

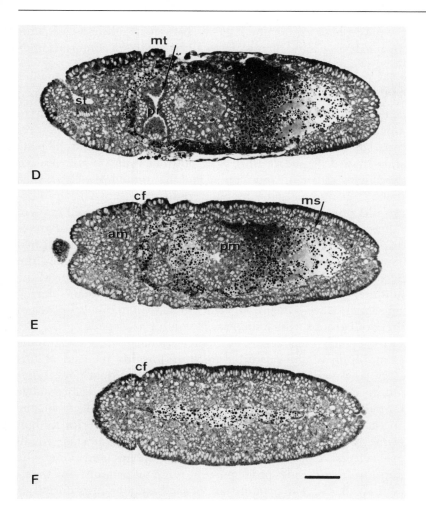

Stage 10

a) **Living embryos**. Corresponds to part of stage 8 and to stage 9 of Bownes (1975). Stage 10 lasts for about 1 h (4:20-5:20 h). The onset of this stage is characterized by the appearance of the stomodeum, which invaginates ventrally at the anterior pole, slightly anterior to the position of the earlier invagination, that of the anterior midgut primordium (Figs. 2.1, 10-10d). The latter grows further caudad and flattens out laterally. The posterior midgut primordium, however, is barely distinguishable in living embryos of this stage. The germ band continues to stretch, reaching its maximum extent towards the end of stage 10 at about 75% EL. The interior of the egg is occupied by the yolk sac, which extends into the procephalic region and is easily distinguishable as a dark, uniform mass. Periodic furrows become visible in the epidermis midway through stage 10 (Fig. 2.1, 10c).

b) **Fixed material**. The stomodeum develops from the stomodeal cell plate located ventrally near the anterior pole of the embryo, anterior to the midgut primordium. All cells of the stomodeal plate start to divide at approximately the same time and the plate sinks inward (Figs. 2.13, 2.14). After only one round of mitoses, cell division is transiently interrupted and the stomodeal cells arrange themselves into a regular monolayered epithelium. This epithelium tilts posteriorly and establishes contact with the cell mass of the anterior midgut primordium, which consequently becomes displaced caudalwards (Figs. 2.15, 2.16). After having established contact with the stomodeum, the anterior midgut primordium becomes separated into two cell groups, one on either side of the midline, which will remain attached to the stomodeum during subsequent development. All cells can be seen to divide in each of these cell groups, which increase in volume as a result.

During most of stage 10, no changes can be detected within the primordium of the posterior midgut. Towards the end of this stage, however, the cells at the bottom of the midgut pocket divide, disaggregate and, in a manner similar to that described above for the anterior midgut primordium, two groups of cells form, one on each side of the midline. Both parts of the posterior midgut primordium thereafter grow in register. During late stage 10, all pole cells leave the cavity of the posterior midgut, and come to lie dorsally on both sides of the posterior midgut primordium, outside the yolk sac. The primordium of the hindgut shows dorsally a conspicuous transverse fold (Fig. 2.15D), from which two blind tubes will form in stage 11. These are the primordia of the Malpighian tubules, whose further development will be considered later. The neuroblasts are the largest cells in the stage 10 embryo (Fig. 2.16). Within the territory of the metameric germ band, in the presumptive ventral cord, segregated neuroblasts have started dividing according to a characteristic asymmetrical pattern that does not exhibit any apparent variations within the array of neuromeres (Figs. 2.13, 2.16). Neuroblasts produce smaller daughter cells, so-called ganglion mother cells, which remain distal to the dividing neuroblasts. This arrangement is due to the polarity of divisions, which is the same in all germ band neuroblasts, with the mitotic spindle oriented perpendicular to the embryo surface. Symmetrical divisions are visible among the population of gangli-

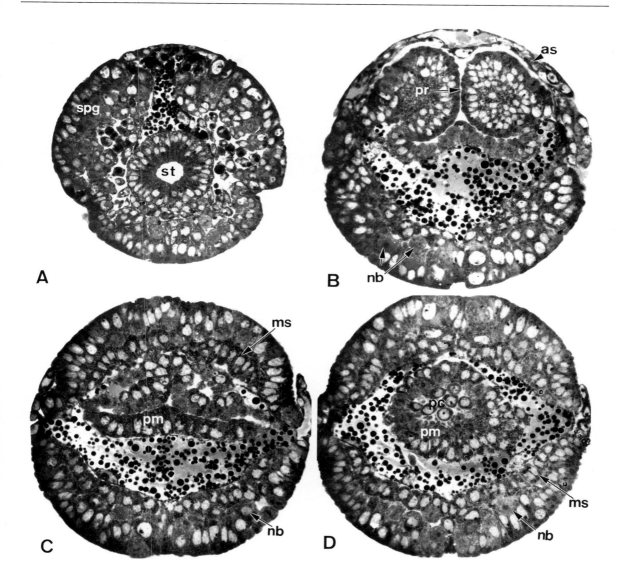

Fig. 2.16. A to D are transverse sections of a stage 10 embryo. The cells of the mesoderm (*ms*) have completed the second postblastoderm mitosis and are arranged as a regular epithelium. SI and SII neuroblasts (*nb*) have completed segregation. *as* amnioserosa; *pm* posterior midgut rudiment. *pc* pole cells; *pr* proctodeum; *spg* supraoesophageal ganglion; *st* stomodeum

on mother cells, so that the number of developing neural cells increases. Within the neurogenic region of the procephalic lobe, divisions affect both internalized neuroblasts and those still located superficially. The latter divisions are all symmetrical, at least during the initial phases of neurogenesis; later on, asymmetrical divisions are also seen in the developing procephalic central nervous system.

Segmentation became apparent transiently in the mesoderm of the germ band during stage 9. In stage 10 mesoderm segmentation reappears, now definitively (Fig. 2.13C). Within the ectodermal derivatives signs of segmentation can also be observed in stage 10. Ten tracheal placodes appear first, which are recognizable as slight concavities in the laterodorsal region of the prospective dorsal epidermis, described above (Fig. 2.14). At the end of stage 10 the cells of the placodes start to divide and invaginate, apparently all at once - without following a gradient. Immediately after the formation of the tracheal placodes, deep furrows appear in the ectoderm slightly medial and anterior to the placodes, which will define the parasegmental boundaries. The cells in these folds do not divide.

Stage 11

a) **Living embryos.** Corresponds to part of stage 10 of Bownes (1975). Stage 11 lasts for about 2 h (5:20-7:20 h). It begins with the appearance in the epidermis of the parasegmental furrows; at the end of stage 11 germ band retraction begins. Within thorax and abdomen, parasegmental boundaries appear as relatively deep folds restricted to ventralmost epidermal levels. Within the prospective head, the gnathal protuberances - mandible, maxilla and labium - become visible immediately ventral to the cephalic furrow, which becomes shallower and finally disappears; the hypopharynx appears midventral to the gnathal buds; in the procephalic lobe, the clypeolabrum becomes distinct. Germ band extension has reached its maximum extent at about 75% EL. Separation of two different layers in the mesoderm, homologous to the somato- and splanchnopleura (Fig. 2.1, 11b) of vertebrates, occurs midway through stage 11. The anterior midgut anlage continues to grow, reaching the level of the amnioserosa. During the second half of stage 11, the posterior midgut primordium again becomes visible in the living embryo; by this stage, the tip of the posterior midgut has reached the level of the posterior embryonic pole, where it will bend ventralwards to continue anteriorly. Towards the end of stage 11 the posterior pole of the embryo is withdrawn from the vitelline envelope and a conspicuous cleft appears which signals the beginning of germ band shortening.

b) **Fixed material.** The most obvious morphogenetic change that occurs at the transition between stage 10 and stage 11 is the appearance of the parasegmental furrows, clearly visible in the living embryo as described above. Parasegments in stage 11 are periodic structures, in which both ectodermal and mesodermal derivatives are included (Fig. 2.15). Fifteen metameres can be distinguished in the germ band, whereas only the procephalic lobe and the clypeolabrum can be morphologically defined in the procephalon.

Turner and Mahowald (1977) found only ten tracheal pits, and the present observations confirm their description (Figs. 2.17C-D, 2.18B, 2.20B). The anteriormost tracheal pit opens into the boundary between the prothoracic and mesothoracic segments and the posteriormost opens into the eighth abdominal segment. These anterior openings will form the anterior spiracles, the posterior spiracles develop from a separate placode in *a8*. The remaining pits will grow and their extensions eventually fuse to give rise to the tracheal tree. However, behind the tenth tracheal pit, a groove transiently appears in the ectoderm lateral to the proctodeal opening, demarcating what Turner and Mahowald (1977) have called *a10*. This groove does not invaginate to the same extent as the tracheal pits, nor are mitoses visible in it, and it soon disappears.

The segments of the gnathocephalon, corresponding in cephalocaudal order to those of the mandibular, maxillary and labial appendages, develop from the ectoderm immediately ventral and anterior to the vanishing cephalic furrow (Figs. 2.17E, 2.18E, 2.20C, 2.21E). Midway through stage 11 the salivary gland placode appears at the medial border of each labial appendage. The salivary gland placodes (Figs. 2.17D, 2.19B) soon invaginate and form a tube on either side that extends caudalwards (Figs. 2.18F, 2.19B, 2.20C). No mitotic divisions are visible during embryogenesis in the developing salivary glands. During subsequent development the gnathal segments will change their relative positions in a way that will be discussed later.

Cell death is a conspicuous phenomenon in *Drosophila* embryogenesis that becomes evident in the second half of stage 11, extending throughout most of stage 12 (see Figs. 2.19C, 2.21). Pycnosis and consequent shrinkage of dead cells lead to an increase in the local density of staining in these areas, which therefore become unambiguously distinguishable. The majority of cell

Fig. 2.17. A to F are sagittal sections of a late stage 11 embryo. Two of the three invaginations (*arrowheads*) from which the subdivisions of the stomatogastric nervous system derive are visible in the dorsal wall of the foregut (*fg*). Anterior and posterior midgut (*am, pm*) each consist of cell masses growing caudalwards. The

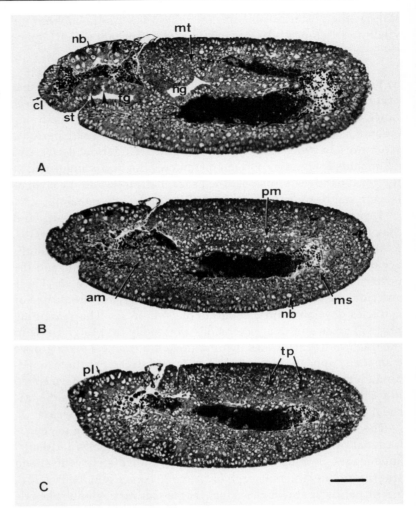

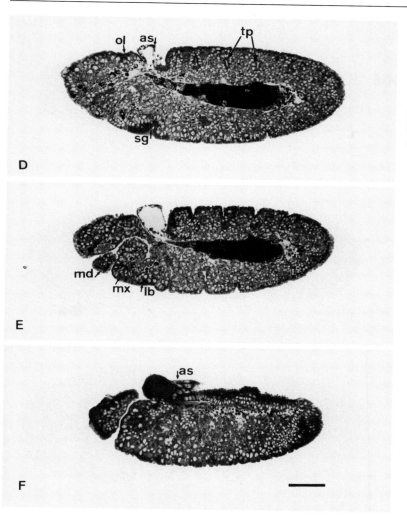

D

E

F

primordia of the optic lobes (*ol*) and of the salivary glands (*sg*) are invaginating. *as* amnioserosa; *cl* clypeolabrum; *hg* hindgut; *lb* labial bud; *md.* mandibular bud; *ms* mesoderm; *mt* primordium of Malpighian tubules; *mx.* maxillary bud; *nb* neuroblasts; *pl* procephalic lobe; *st* stomodeum. *Bar* 50 μm

Fig. 2.18. A to **F** are horizontal sections of a late stage 11 embryo of about the same age as that in Fig. 2.17. The regular arrangement of the neuroblasts (*nb*) in longitudinal rows separated by the mesectodermal cells (*me*) can be clearly distinguished in **A** and **F**. **B** shows that behind the last tracheal pit (*tp*), that of *a8*, there is a deep slit that corresponds to *a9*. This slit, however, does not contribute to the formation of the tracheal tree and will disappear later in development. Pole cells (*pc*) are arranged in rows dorsally on either side of the midline, extending from *a8* to *a5*. The orientation of sectioning, along the horizontal plane, permits one to observe the extent of the growing Malpighian tubules (*mt*). Two dorsal tubules have been

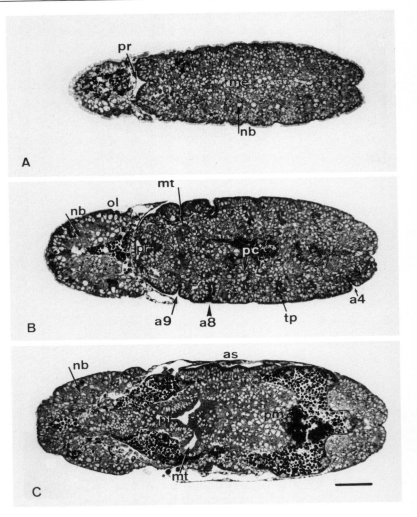

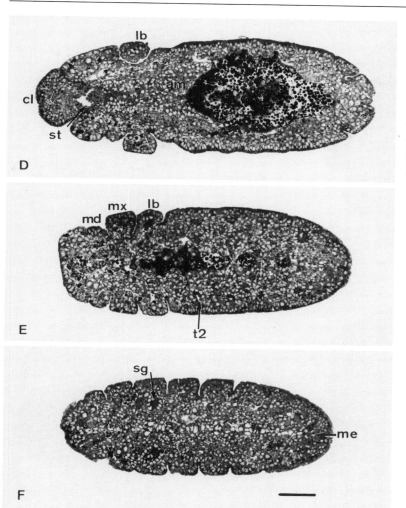

sectioned in **B**, two anterolateral, in association with the hindgut (*pr* proctodeum) in **C**. Cells of the anterior (*am* in **D**), and posterior midgut primordia show a Y-like arrangement. Abbreviations. *as* amnioserosa; *cl* clypeolabrum; *mb* mandibular bud; *mx* maxillary bud; *nb* neuroblasts; *ol* primordium of the optic lobe; *sg* salivary gland; *st* stomodeal opening

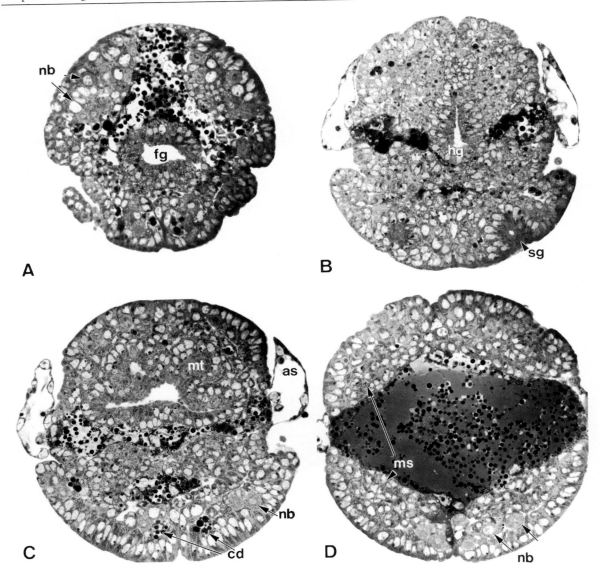

Fig. 2.19. A to D are transverse sections of a late stage 11 embryo. The primordia of the salivary glands (*sg*) are invaginating (**B**). Somatic and visceral mesoderm are distinct. *as* amnioserosa; *cd* cell death; *fg* foregut; *hg* hindgut; *ms* mesoderm; *mt* primordium of Malpighian tubules; *nb*. neuroblasts; *st* stomodeum. *Bar* 50 μm

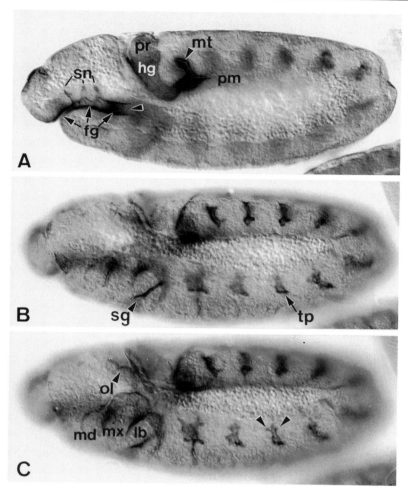

Fig. 2.20. A to C are midsagittal and lateral planes of focus through a late stage 11 embryo stained with an anti-Crumbs antibody. The three invaginations that will give rise to the stomatogastric nervous system (*sn*) can be seen at the roof of the foregut. Notice that the fundus of the foregut (*arrowhead* in **A**) now does not express Crumbs, its cells having lost their earlier epithelial organization. The primordium of the Malpighian tubules (*mt*) is visible at the boundary with the posterior midgut primordium (*pm*). The anlagen of the salivary glands (*sg*) and of the tracheal tree (ten tracheal pits, *tp*, in **B**) have invaginated. The tracheal pits elongate and form a dorsal and a ventral stem. Anterior and posterior branches of the dorsal stem (*arrowheads* in **C**) have started to differentiate. *md*, *mx* and *lb* refer to the gnathal buds, mandible, maxilla and labium, respectively. *ol* primordium of the optic lobe

Fig. 2.21. A to F are parasagittal sections of an early stage 12 embryo at the beginning of germ band shortening, with the tip of the germ band at about 60% EL. The general organization of this embryo is very similar to that of those shown in Figs. 2.17–2.20. Anterior (*am*) and posterior (*pm*) midgut have extended to reach 50% EL and the posterior pole, respectively. As the germ band shortens, the posterior midgut will come into contact with the anterior midgut, with which it will eventually fuse. Dorsal and anterolateral Malpighian tubules (*mt*) can be clearly seen in **C** in association with the hindgut (*hg*). The invagination of the

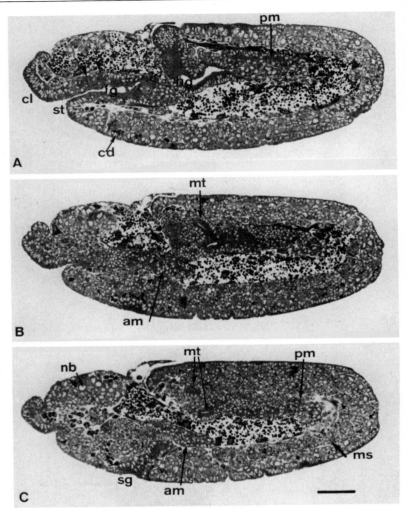

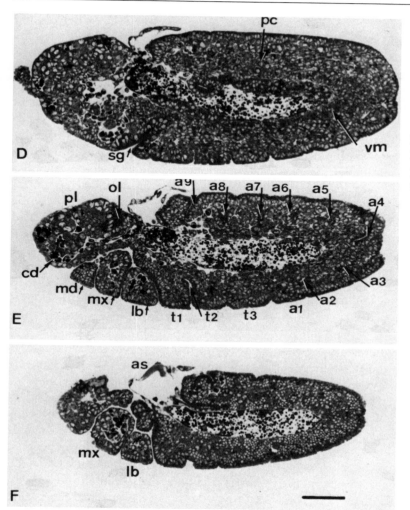

salivary gland (*sg*) has proceeded further, reaching deeper levels. All tracheal pits, as well as the slit at *a9*, and all segments are labelled in **E**. Abundant areas of cell death (*cd*) are visible. *as* amnioserosa; *cl* clypeolabrum; *fg* foregut; *lb* labial bud; *md* mandibular bud; *ms* mesodermal layer; *mx* maxillary bub; *nb* neuroblasts; *ol* optic lobe primordium; *pc* pole cells; *pl* procephalic lobe; *st* stomodeum. *Bar* 50 μm

death figures at this stage are located between the epidermis and the nervous system, forming large groups interpenetrated by macrophages. Macrophages are large, round cells that appear singly in the vicinity of dead cells and generally contain cell debris; thus, macrophages are involved in phagocytosing dead cells. The coincidence between the appearance of macrophages and cell deaths is very striking, suggesting that cell death itself elicits the differentiation of the macrophages. Macrophages will persist at least until the larva hatches from the egg; their ultimate fate is unknown. The majority of cell deaths found in stages 11 and 12 seem to derive from the epidermis, where pycnotic nuclei can frequently be seen (Figs. 2.19C, 2.21); a few are also present in the nervous system. The peculiar subepidermal arrangement of dead cells suggests that these cells may contribute to separate the nervous system from the epidermis.

By this stage both anterior and posterior midgut primordia have become morphologically distinguishable from foregut and hindgut, respectively, the midgut cells having extended considerably towards caudal levels (Figs. 2.17-2.22). During stage 11, the anterior midgut primordium will reach about 50% EL, where it will meet the cells of the posterior midgut primordium after germ band shortening; at the same time the posterior midgut primordium reaches the posterior tip of the germ band and bends ventralwards. At the end of stage 11, when mesodermal cells have undergone their third mitosis, the visceral mesoderm, or splanchnopleura, separates from the somatic mesoderm, or somatopleura (Fig. 2.19D), lateral to both anterior and posterior midgut primordia. At this stage the pole cells can be seen to be aligned dorsally on both sides of the posterior midgut at the level of the abdominal segments 8, 7 and 6, in contact with the mesodermal layer (Fig. 2.18). The hindgut is clearly individualized during stage 11, forming a tube connected to the cells of the posterior midgut primordium (Fig. 2.20). At the posterolateral side of the hindgut two pairs of Malpighian tubules have started to form as two posterolateral outgrowths of the hindgut (Figs. 2.19, 2.20). One of these pairs grows laterally, the other dorsally.

The pattern of neuroblast division during stage 11 is comparable to that described for previous stages. Large numbers of ganglion cells are already present in both prospective ventral cord and procephalic lobe. Towards the end of this stage two new neural primordia become apparent: the primordium of the optic lobes which forms a deep groove on both sides dorsocaudally in the procephalic lobe (Fig. 2.20C), and the three primordia of the sto-

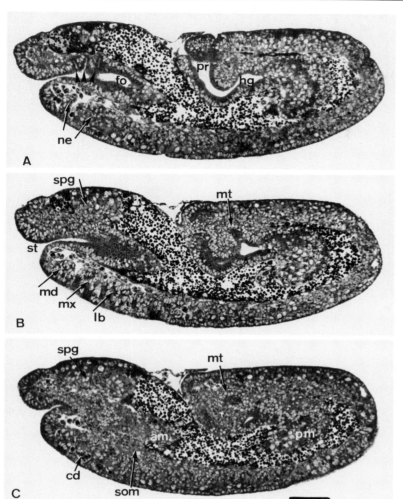

Fig. 2.22. **A to F** are parasagittal sections of a late stage 12 embryo during the fast period of germ band shortening (tip of the germ band at about 45% EL). In this embryo the three invaginations of the primordial stomatogastric nervous system can be seen in the dorsal wall of the foregut (*fg, arrowheads*). Neuromeric organization (*ne*) of the developing ventral cord is evident in **A**, an incipient supraoesophageal ganglion (*spg*) can be distinguished in the procephalic lobe (*pl*). The cells of the optic lobe primordium (*ol*) will join the developing supraoesophageal ganglion dorsolaterally. The posterior midgut (*pm*), which has been growing caudally, can be seen in this embryo to bend at the caudal pole (**B** and **C**), thereby reversing the direction of its movement. The cells

of the splanchnopleura (*spl*), organized in palisades, can be seen in **C** to join the posterior midgut laterally; medially the cells of the somatopleura (*som*) become apparent. The amnioserosa (*as*) starts to unfold and becomes distended as a consequence of shortening, to eventually cover the embryo dorsally. *cl* clypeolabrum; *dr* dorsal ridge; *hg* hindgut; *lb* labial bud; *mt* Malpighian tubules; *pr* proctodeum; *sg* salivary glands. *Bar* 50 µm

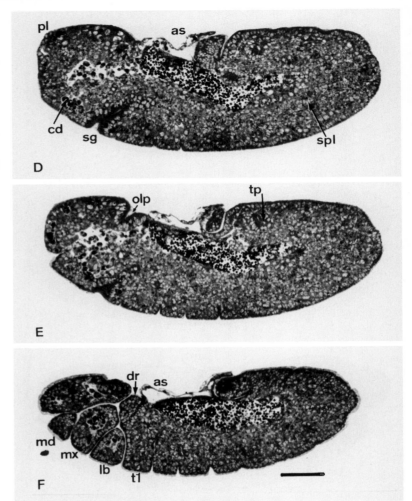

matogastric nervous system which invaginate from the roof of the stomodeum (Figs. 2.17, 2.20).

Stage 12

a) **Living embryos**. Corresponds to part of stages 10 and 11 of Bownes (1975). Stage 12 lasts for about 2 h (7:20-9:20 h). The major event in this stage is the shortening of the germ band (Figs. 2.1, 12 - 12c). The germ band retracts caudalwards in such a way that the opening of the hindgut comes to be located at the dorsal side of the posterior egg pole; at the same time the width (mediolateral extent) of the germ band increases approximately one and a half times. Anterior and posterior midgut rudiments, barely distinguishable in the living embryo in stages 10 and 11, can now be clearly seen as two distinct bilateral cell bands that move toward each other, eventually to meet at about 60% EL. One of the consequences of germ band shortening is that the yolk sac now protrudes on the dorsal egg surface, being covered only by the amnioserosa. Furthermore, the yolk sac retracts from cephalic regions, becoming restricted to the trunk, flanked by the posterior and anterior midgut anlagen. However, a ventral extension of the yolk sac is visible that extends beyond the limits of the posterior midgut, at its interface with the hindgut. Germ band segmentation becomes very prominent during shortening, most probably due to the increased cell density per unit volume of the segments and the depth of the intersegmental furrows after shortening.

b) **Fixed material**. Germ band shortening is accompanied by profound modifications of embryo morphology. The most obvious consequences of germ band shortening are: the fusion of the anterior and posterior midgut primordia to form a continuous midgut, the fusion of the tracheal pit extensions to form the tracheal tree and the definitive establishment of metamery. As described above, anterior and posterior midgut rudiments split medially into two cell groups each; these cell groups advance caudalwards during stage 11 and in early stage 12 (Fig. 2.21), partially due to growth by cell division and cell shape changes and, in the case of the posterior midgut, partly as a result of passive displacement by the retracting germ band. The visceral mesoderm has joined the midgut cells by the beginning of stage 12 (Fig. 2.22), and during shortening forms a palisade-like sheet of cells that overlies the cells of the midgut laterally (Figs. 2.22, 2.23, 2.24). The cells of the posterior midgut that are destined to join the anterior midgut pri-

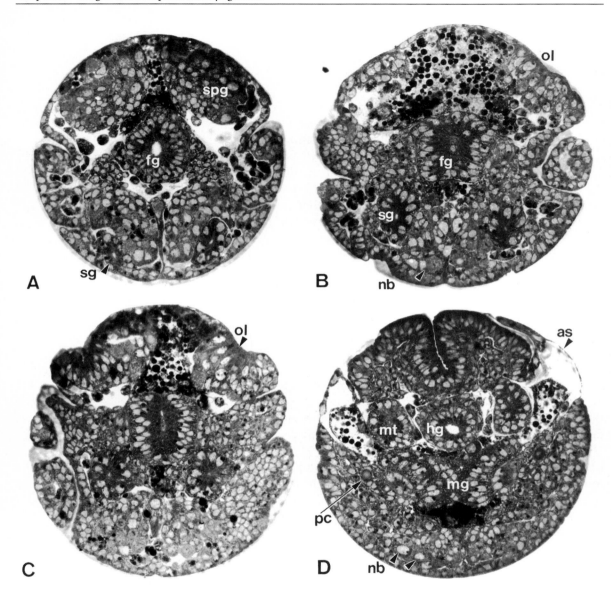

Fig. 2.23. A to D show transverse sections of a late stage 12 embryo in the fast period of germ band shortening; it is slightly younger than the embryo shown in Fig. 2.24. Somatic mesodermal cells (*sm*) have formed loose aggregates or clusters. Abbreviations. *as* amnioserosa; *fg* foregut; *hg* hindgut; *lb* labial bud; *md* mandibular bud; *mt* Malpighian tubules; *nb* neuroblasts; *ol*.primordium of the optic lobe; *ps* pole cells; *sg* salivary gland; *so* somatic musculature; *spg* supraoesophageal ganglion

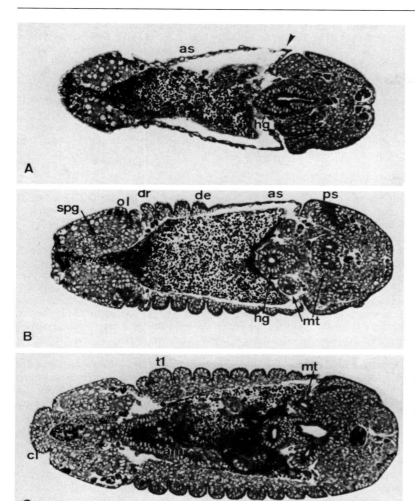

Fig. 2.24. **A to F** are horizontal sections of a late stage 12 embryo in the fast period of germ band shortening (tip of the germ band at about 30% EL). Except for its caudal portion (*arrowhead* in **A**) the amnioserosa (*as*) is unfolded and covers the yolk sac dorsally. The dorsal epidermis (*de*) slides over the amnioserosa to initiate dorsal closure of the embryo. Fusion of anterior and posterior midgut (*am/pm*) has already occurred (**D**), and a continuous midgut is surrounded by the circular and longitudinal visceral musculature (*vm*). The primordium of the optic lobe (*ol*) has joined the supraoesophageal ganglia (*spg*), though it still maintains an open connection to the outside (**B**). Pole cells are forming an incipient gonad (*go* in **D**). Cytodifferentiation in the posterior spiracles (*ps*) has begun. Definitive segmental metamery occurs gradually. Segments are forming within dorsal levels of the territory that has completed germ band shortening (**C**).

Abbreviations. *cl* clypeolabrum; *dr* dorsal ridge; *fg* foregut; *hg* hindgut; *lb* labial bud; *md* mandibular bud; *mt* Malpighian tubules; *mx* maxillary bud; *sg* salivary gland; *so* somatic musculature. *Bar* 50 μm

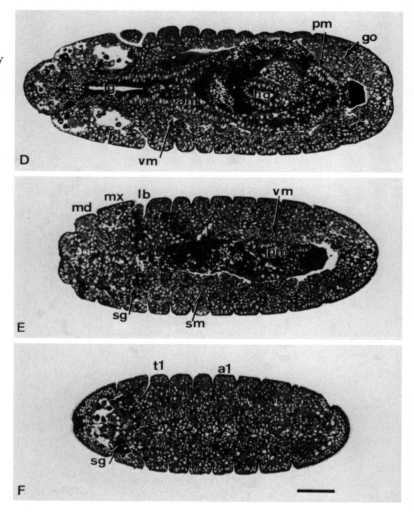

mordium exhibit distinct morphological characteristics (Fig. 2.22A-B), with large and irregular nuclei and abundant, basophilic cytoplasm, reminiscent of the pole cells (Poulson, 1950; see Chapter 7.3). Eventually, anterior and posterior midgut primordia meet at about 60% EL and fuse.

The advancing tracheal tubules initially remain open to the outside. This connection is eventually broken and at germ band shortening, as described by Poulson (1950), the tracheal sections (Fig. 2.25B) are brought into contact and will eventually fuse with each other, forming a continuous tube that extends from the anterior mesothoracic border up to the eighth abdominal segment

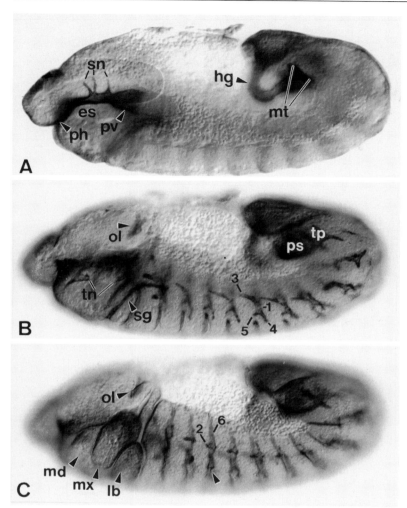

Fig. 2.25. A to C are midsagittal and lateral planes of focus through a late stage 12 embryo stained with an anti-Crumbs antibody. The various subdivisions of the foregut, pharynx (*ph*), oesophagus (*es*), and proventriculus (*pv*) are distinguishable. The tracheal pits have developed to form a ventral stem, which bifurcates into a posterior branch (*4*) and an anterior branch (*5*). The dorsal stem (*1*) gives off two side branches, one towards anterior (*2*), the other towards posterior and interior of the embryo (*3*). Finally, there is a dorsomedial branch (*6*). The tentorium (*tn*) differentiates from two invaginations, a posterior one at the boundary between the maxillary and the labial bud, and an anterior one in the intercalary segment. The primordium of the posterior spiracles (*ps*) has fused with the tracheal pit at *a8* (*tp*). Other abbreviations. *hg* hindgut; *mt* Malpighian tubules; *ol* primordium of the optic lobes; *sg* salivary glands; *sn* primordia of the stomatogastric nervous system

(Fig. 2.34). Finally, the pole cells, which were aligned along both sides of the posterior midgut primordium during the previous stage, form clusters at the level of *a5* during the retraction of the germ band (Fig. 2.23D), i.e. the position at which the gonads will form (Figs. 2.24, 2.26).

While the germ band is retracting, the definitive segmental furrows become apparent over their entire extent. Initially dorsal divisions of the intersegmental furrows appear, which are in register with the tracheal pits. During shortening, the dorsal segmental furrows join the pits and the ventral intersegmental furrows. Furthermore the amnioserosa unfolds to form the outer layer of

Fig. 2.26. A to F are parasagittal sections of a stage 13 embryo. Shortening of the germ band is complete. The amnioserosa (*as*) is fully unfolded, covering the embryo dorsally. Metamery is still apparent within medial levels of the developing ventral cord (*vc*), the various neuromeres are labelled. Salivary glands (*sg*) have a common salivary duct (*sd*). The palisade-like arrangement of the visceral mesoderm (*vm*) is clearly visible in both anterior (*am*) and posterior midgut (*pm*). Malpighian tubules arise from the junction between hindgut (*hg*) and midgut. Anterior (*asp*) and posterior spiracles (*ps*) are apparent. The gonads

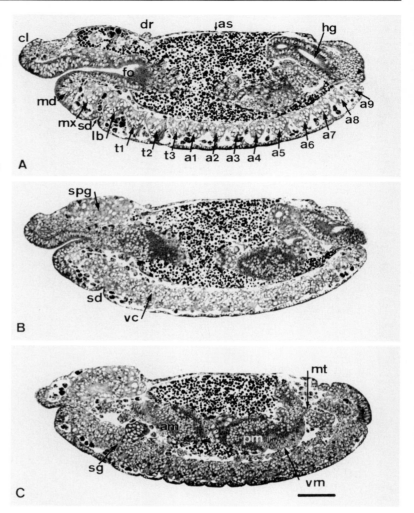

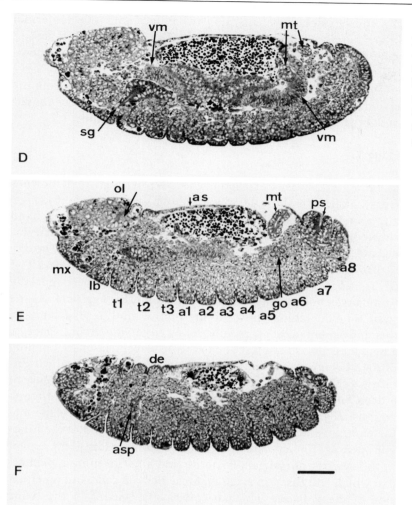

(go) have reached their definitive
position. Segmentation is complete.
Abbreviations. *cl* clypeolabrum; *fg*
foregut; *lb* labial neuromere and
labial bud; *md* mandibular neuromere; *mx* maxillary neuromere and
maxillary bud; *ol* optic lobe primordium; *spg* supraoesophageal ganglion. *Bar* 50 μm

the embryo dorsally, remaining in contact with the epidermal layer laterally, and with the hindgut. This relationship is lost in stage 14, when the epidermis overgrows the extraembryonic membrane at dorsal closure.

During stage 12 the ventral cord separates completely from the epidermis; the first neural processes and fibres appear. The remaining organs do not exhibit any significant modification.

Stage 13

a) **Living embryos.** Corresponds to part of stage 11 of Bownes (1975). Stage 13 is short, lasting for about 1 h (9:20-10:20 h) (Figs. 2.1, 13 - 13c). It is initiated by the completion of germ band shortening, when the prospective anal pads occupy the posterior egg pole, and ends with the beginning of head involution. The clypeolabrum becomes thinner and starts to retract (compare Fig. 2.1, 13 and 13c); retraction of the clypeolabrum gives rise ventrally to a conspicuous triangular gap at the anterior egg pole. At the same time the labium moves to the ventral midline (Fig. 2.1, 13c); this movement can be readily followed in the living embryo, where it becomes manifest due to the displacement of the opening of the salivary gland duct. The yolk sac protrudes dorsally, exhibiting a characteristic convex shape - apparently a consequence of a deep fold dorsally at the interface between cephalic and trunk regions. A gap appears there between the vitelline envelope and the embryo, and another is seen caudally at the level of the hindgut opening. The so-called dorsal fold (dorsal ridge) will appear in the anterior dorsal gap. Posterior and anterior midgut primordia, which had fused in stage 12, form a single cell band on either side of the yolk sac which becomes less evident in the living embryo. The caudal yolk sac extension mentioned above is still visible.

b) **Fixed material.** After completion of germ band shortening, the midgut is still open at both dorsal and ventral levels, merely consisting of two lateroventral cell bands between which the yolk sac lies (Fig. 2.26C-D, 2.27C). During stage 13, the midgut cells will become organized in a regular cylindrical, palisade-like epithelium. Apparently due to volumetric growth and stretching, cells of the midgut epithelia expand over the yolk sac to make contact along the ventral and dorsal midline and eventually fuse. This process takes quite a long time to complete, extending into stages 14 and 15. When the midgut has closed dorsally two medial holes

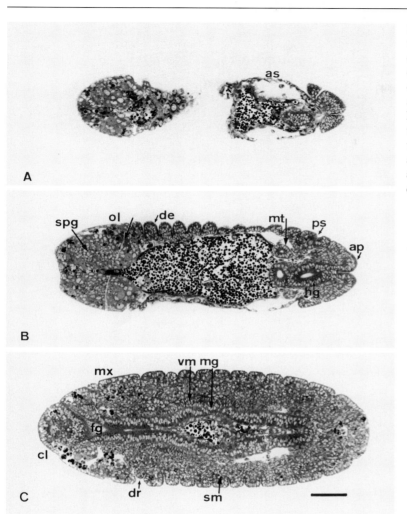

Fig. 2.27. A to F are horizontal sections of a stage 13 embryo, slightly older than that in Fig. 2.26. The cells of the amnioserosa (*as*) are being displaced inwards by the sliding dorsal epidermis (*de*), thereby becoming larger and thicker. The connection that linked the optic lobe primordium (*ol*) to the outside has been severed. The midgut is continuous and flanked by the visceral mesoderm (*vm*) along its entire length.

The anal pads (*ap*) begin cytodiffe-rentiation. The primordium of the somatic musculature (*sm*) becomes distinguishable. Abbreviations. *cl* clypeolabrum; *dr* dorsal ridge; *fg* foregut; *hg* hindgut; *lb* labial bud; *mx* maxillary bud; *ps* posterior spiracles; *sg* salivary gland. *Bar* 50 μm

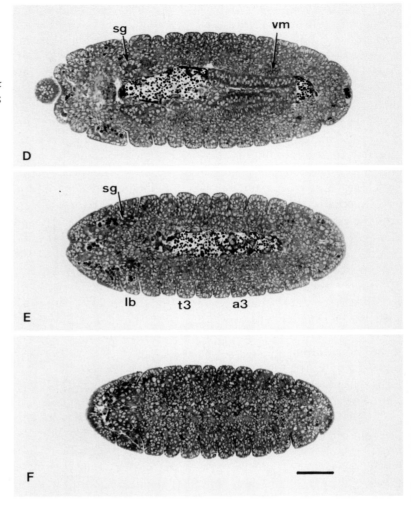

remain, one caudal and ventral and the other rostral and dorsal. In stage 13 processes of the yolk sac can be seen to protrude through these holes (Fig. 2.26A). One of these processes extends caudally between the midgut walls and contacts the hindgut wall, whereas the other extends anterodorsally midway into the clypeo-labrum. Both processes will eventually retract to be definitively absorbed into the yolk sac during stage 15.

The hindgut opens in the anus, which by now is located slight-ly dorsally at the posterior embryonic pole (Figs. 2.26B, 2.27B, 2.29A). At the beginning of stage 13 the hindgut consists of an empty longitudinal tube, contiguous with the midgut epithelia. Four thin Malpighian tubules can be distinguished sprouting from its ventral side (Figs. 2.26C-E, 2.29A). During stage 13, the hindgut grows and the tube adopts a sigmoidal shape oriented along the horizontal plane. The course of the foregut in this stage is also initially longitudinal and then sigmoidal; however, the foregut is oriented along the sagittal plane (see Fig. 2.31A). Two different regions can be distinguished in the foregut. The anterior region consists of the pharyngeal roof and hypopharyngeal lobes, with tall cylindrical cells and basal nuclei; the posterior region consists of the prospective oesophagus, with cuboidal cells. The three invaginations of the stomatogastric nervous system, which were visible in the foregut roof in stage 12, have pinched off the foregut epithelium and form closed vesicles (Figs. 2.19, 2.25, 2.26A, 2.29).

The two developing salivary gland lobes join to form a com-mon midsagittal duct behind the hypopharyngeal lobes at the beginning of stage 13 (Fig. 2.26B-C); the salivary gland duct will be displaced towards the foregut as the labial appendages move forward during mouth involution. The salivary glands diverge bilaterally from the common duct, extending between the epider-mis and the ventral cord beneath the midgut (Fig. 2.26D-E). Secretion products are now visible in the lumen of the salivary tubes. The wall of the clypeolabrum is continuous with the pharyngeal roof; its tip contains the anlagen of the labral sensil-lum, the so-called epiphysis (Hertweck 1931), which has already started to differentiate at this stage and is readily distinguishable as a slight depression surrounded by clypeolabral cells. The epi-dermis of the procephalic lobe and clypeolabrum is extremely thin and ends at the interface with the amnioserosa, contiguous with the dorsal ridge (Fig. 2.26). The dorsal ridge itself has alrea-dy become apparent dorsolaterally during stage 12, in the form of epidermal folds on both sides of the procephalic lobe in the vicinity of the labial bud; during stage 13, the dorsal ridge extends dorsalwards to fuse eventually across the dorsal midline with its complement on the other side. Both openings of the tracheal tree, the prospective anterior and posterior spiracles, have reached a considerably advanced stage of differentiation and by now they are very conspicuous structures, consisting of cylindrical cells with very basophilic cytoplasm, strikingly different from the surrounding epidermal cells (Figs. 2.26E-F, 2.27B).

After germ band shortening, the dorsal surface of the embryo is formed by the amnioserosa, the cells of which, at the beginning of stage 13, still show the same characteristic, extremely thin and flat shape seen in previous stages, and which abuts on the epidermis. During this stage the epidermal layer starts to extend towards the dorsal midline (Figs. 2.26F, 2.27B), ultimately leading to dorsal closure of the embryo.

The developing central nervous system consists of well differentiated ventral cord and supraoesophageal ganglion. Neuromere organization is clearly distinguishable within both the ventral nerve cord and the suboesophageal ganglion (Fig. 2.26A), although the neuromeres corresponding to the three gnathal segments on the one hand, and to abdominal segments $a8$ and $a9$ on the other hand, will fuse during this stage to form the suboesophageal ganglion and the caudalmost abdominal neuromere, respectively. Ventral cord and suboesophageal ganglion extend from the tip of the hypopharyngeal lobe to the region immediately ventral to the anus, and will maintain this length until stage 15, when ventral cord condensation sets in. The primordia of the optic lobes (Figs. 2.21E, 2.22E, 2.23) have become integrated into the posteroventral region of the supraoesophageal ganglion (Figs. 2.26E, 2.27B), although their cells can still be distinguished cytologically (Fig. 2.28A). Fibre connectives and commissures linking the different neuromeres are already visible (Fig. 2.27F). Progenitors of sensory cells, including the primordia of the antennomaxillary complex, appear in the epidermis of maxilla, mandible and procephalic lobe. At the end of stage 13 muscle cells become apparent, inserting at incipient apodemes of the lateral epidermis. During displacement of the posterior midgut primordia at germ band shortening, pole cells form two clusters; in stage 13 pole cells become intermingled with mesodermal cells, forming incipient gonads that already occupy their definitive position in abdominal segment 5 (Fig. 2.26E).

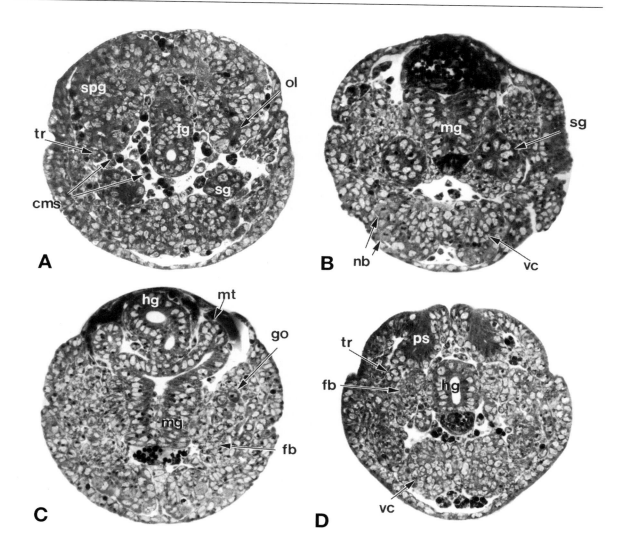

Fig. 2.28. A to D are transverse sections through a stage 13 embryo. Notice that although the connection between the optic lobe primordium (*ol*) and the outside has already disappeared, the primordium forms a vesicle in contact with the supraoesophageal ganglion (*spg*), with a central lumen. (**B**). The midgut (*mg*) is a continuous cell sheath. Abbreviations. *cms* cephalic mesoderm; *fb* fat body; *fg* foregut; *go* gonad; *hg* hindgut; *ps* posterior spiracles; *sg* salivary gland; *tr* trachea; *vc* ventral cord

Stage 14

a) **Living embryos**. Corresponds to part of stage 12 of Bownes (1975). Stage 14 lasts for 1 h (10:20-11:20 h) (Figs. 2.1, 14 - 14d). It begins with the onset of head involution, and in this stage three major morphogenetic events, i.e. closure of the midgut, dorsal closure and head involution , continue. At dorsal closure, the dorsolateral edges of the epidermis advance to fuse at the midline. By the end of stage 14, the epidermis already covers about 80% of the ventrodorsal perimeter. However, before dorsal closure is completed, both anterior and posterior dorsal gaps disappear and the dorsal surface of the living embryo of stage 14 briefly takes on a smooth texture (Fig. 2.1, 14c). Head development continues with the caudalward retraction of the clypeolabrum, and the inward movement of the labium, which is reflected in the corresponding displacement of the opening of the salivary glands into the atrium. The anal plate shows a slight ventral displacement from the posterior tip of the egg to its definitive position. The posterior spiracles become evident in the living embryo.

b) **Fixed material**. The midgut has closed ventrally during the previous stage and then, in stage 14, dorsal closure proceeds; in both cases the mechanism is the same, namely volumetric growth displaces the midgut walls towards the midsagittal plane, at both ventral and dorsal levels, where they fuse. The last portions of the the midgut to close correspond to a mid-dorsal, rostral opening and a midventral, caudal opening, through which processes of the yolk sac protrude (Figs. 2.31A-B, 2.32A,D); we have referred above to these openings. After closure, the midgut exhibits a pyramidal shape, being wider in anterior regions; however, no cytological differences can be discerned in the organization of the midgut wall. At the end of stage 14, the midgut constricts once in the middle, thereby acquiring a characteristic heart-like shape. Dorsal embryonic closure and head involution have been thoroughly depicted and described by Turner and Mahowald (1979) on the basis of scanning electron micrographs. Before germ band shortening the dorsal epidermis consists of tightly packed epithelial cells contiguous to the tracheal pits. After completion of germ band shortening dorsal epidermal cells flatten out. At the end of stage 14, dorsal closure has reached about 80% EL at mid-embryo levels (Fig. 2.30B-D). Intersegmental folds persist during epidermal spreading, so that segmental individuality is also maintained after closure.

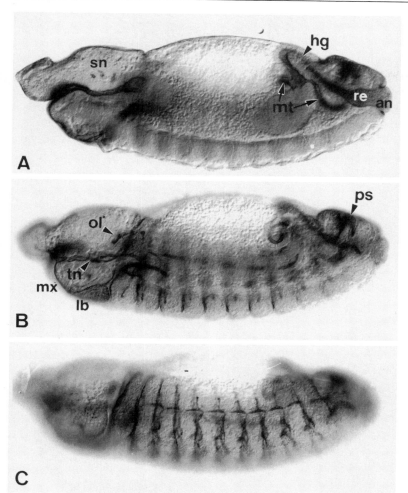

A

B

C

Fig. 2.29. A to C are midsagittal and lateral planes of focus through an early stage 13 embryo stained with an anti-Crumbs antibody. The various subdivisions of the hindgut (*hg*) have started differentiation (*re.* rectum; *an.* anus). For further details compare to Fig. 2.27. Other abbreviations. *lb* labial bud; *mt* Malpighian tubules; *mx* maxillary bud; *ol* primordium of the optic lobes; *ps* posterior spiracles; *sg* salivary glands; *sn* primordia of the stomatogastric nervous system; *tn* tentorium

Head involution progresses simultaneously with the dorsalward extension of the epidermis to achieve dorsal closure; head involution '*sensu stricto*' starts in stage 14 when the dorsal fold forms. On sagittal sections the dorsal ridge can be seen to progress cephalad to form the dorsal pouch; at the same time the clypeolabrum is pulled dorsally against the dorsal ridge. Ventrally the hypopharyngeal lobes have been displaced into the stomodeum, and accordingly the salivary duct can now be seen ending in the floor of the atrium (Fig. 2.31A). The gnathal appendages have moved anteromedially; whereas the labial appendages of both sides join at the midline and move further cephalad to form the most anterior part of the mouth floor, maxillary and mandibular

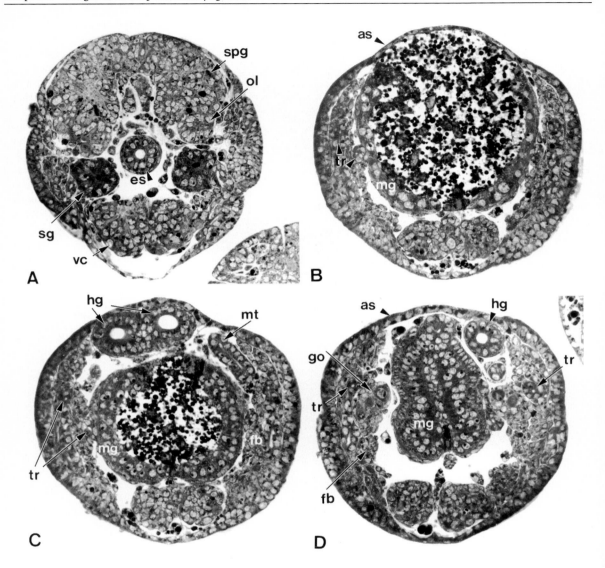

Fig. 2.30. A to D are transverse sections of a stage 14 embryo. The midgut (*mg*) has not yet completed dorsal closure. The primordium of the optic lobe (*ol*) is still distinguishable as a vesicle in contact with the supraoesophageal ganglion (*spg*). Abbreviations. *es* oesophagus; *fb* fat body; *go* gonads; *hg* hindgut; *mt* Malpighian tubules; *sg* salivary gland; *tr* trachea; *vc* ventral cord

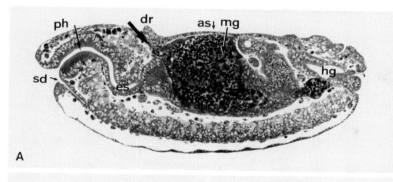

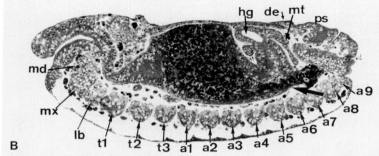

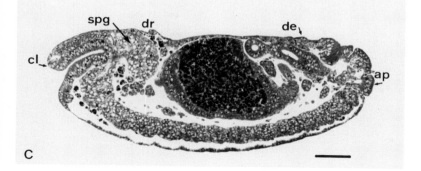

Fig. 2.31. A to F are parasagittal sections of an early stage 14 embryo. The dorsal ridge (*dr*) has reached middorsal levels. Changes in the foregut have brought about differentiation of the pharynx (*ph*) and the oesophagus (*es*), which is connected to the proventriculus, now almost closed dorsally. The yolk sac is almost completely included in the midgut; two large *arrows*, one dorsal and anterior in **A** and the other ventral and posterior in **B**, indicate processes of the yolk sac which still remain external to the midgut. Metamery is still apparent within medial levels of the ventral cord, mandibular (*md*), maxillary (*mx*) and labial (*lb*) neuromeres on the one hand, and *a8* and *a9* on the other hand have fused and are indistinguishable.

Mouth involution is in progress, the salivary duct (*sd*) having reached the level of the stomodeum. The clypeolabrum is clearly distinct in form from earlier stages. Cytodifferentiation in the anterior (*asp*) and posterior spiracles (*ps*) is in progress. Somatic musculature (*sm*) shows segmental organization. The tracheal tree (*tr*) exhibits a main trunk and segmental branches. Abbreviations. *amx* antenno-maxillary complex; *ap* anal pad; *as* amnioserosa; *de* dorsal epidermis; *go* gonads; *mt* Malpighian tubules; *sg* salivary gland; *spg* supraoesophageal ganglia. *Bar* 50 μm

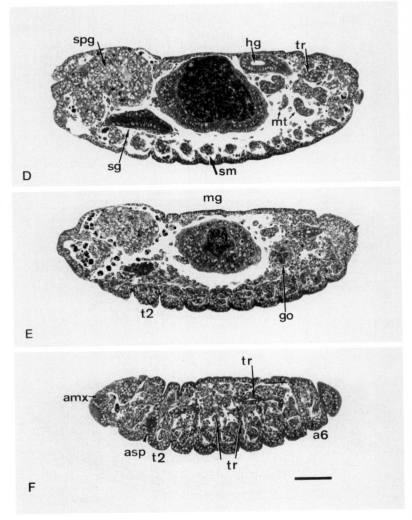

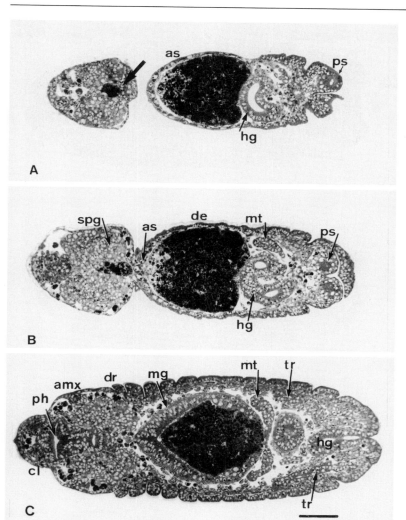

Fig. 2.32. **A to F** are horizontal sections of a stage 14 embryo. The midgut (*mg*) is not yet completely closed dorsally, nor is the entire yolk sac included in it. The two large *arrows* in **A** and **D** point to as yet unincorporated yolk sac processes, one separating the supraoesophageal ganglia (*spg*), the other attached ventrally to the anal pad (*ap*). Ventrally the midgut is closed and shows a transient constriction (see **D**), which will soon disappear. The amnioserosa (*as*) is being displaced into the embryo, apparently being pushed by the overlying dorsal epidermis (*de*). Supra- (*spg*) and suboesophageal (*sbg*) ganglia, and ventral cord (*vc*)

are very advanced in their development. Notice that two commissures (*arrowheads* in **F**) occur per segment; since ventral cord condensation has not yet taken place neuromeres and dermomeres (refer to the deep intersegmental furrows) still maintain isotopic relationships. Commissures are located within the posterior half of the neuromere. The salivary glands show a secretory region (*sg*) clearly separated from the excretory (*sd*, salivary duct) section. Abbreviations. *amx* antennomaxillary complex; *cl* clypeolabrum; *dr* dorsal ridge; *es* oesophagus; *go* gonads; *hg* hindgut; *mt* Malpighian tubules; *ph* pharynx; *ps* posterior spiracle; *tr* tracheal tree. *Bar* 50 μm

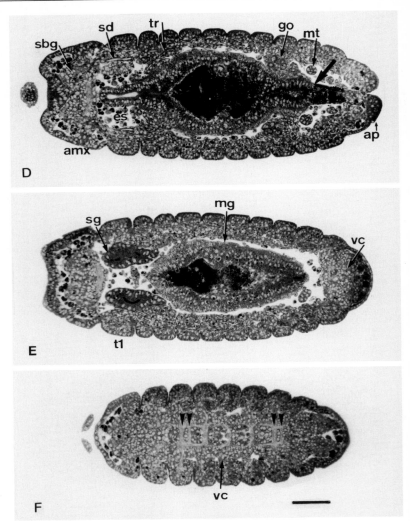

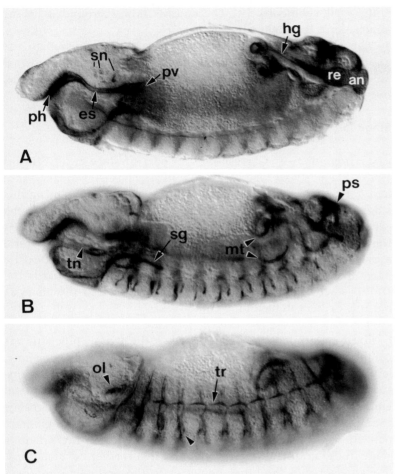

Fig. 2.33. A to C are midsagittal and lateral planes of focus through a stage 14 embryo stained with an anti-Crumbs antibody. Compare to Fig. 2.30. Abbreviations. *an* anus; *es* oesophagus; *hg* hindgut; *lb* labial bud; *mt* Malpighian tubules; *mx* maxillary bud; *ol* primordium of the optic lobes; *ph* pharynx; *ps* posterior spiracles; *re* rectum; *sg* salivary glands; *sn* primordia of the stomato-gastric nervous system; *tn* tentorium

appendages come to lie behind the lateral borders of the stomode-um and the lateral walls of the atrium, respectively.

Pharynx, oesophagus and proventriculus can be clearly distin-guished within the foregut (Figs. 2.31A, 2.32D, 2.33A). The hindgut grows considerably during stage 14, acquiring a hooked shape (Figs. 2.31, 2.32, 2.33). It consists of a tube that opens between the anal pads and extends longitudinally up to 50% EL; there it bends and courses further ventrocaudally to 30% EL, to connect up with the midgut. The origin of the Malpighian tubules lies within this most anterior part of the hindgut, immediately posterior to the junction with the midgut and shortly before the bend (Fig. 2.31B). At this stage Malpighian tubules form four thin tubules (Fig. 2.33B) with small cuboidal cells. The anus is now surrounded by the epidermis of the anal pads.

The somatic musculature, although already attached to the apodemes, is not yet completely stretched, nor can the normal larval pattern be recognized (Fig. 2.31F). Cytodifferentiation, i.e. outgrowth of axonal processes, begins in sensory organs.

Stage 15

a) **Living embryos.** Corresponds to part of stage 12 of Bownes (1975). Stage 15 lasts for 1 h and 40 min (11:20-13 h). Dorsal closure and dorsal epidermal segmentation are accomplished during this stage (Figs. 2.1, 15 - 15c). The gut forms a closed tube when the caudal yolk sac process is integrated into the posterior midgut. At this time the gut completely encloses the yolk sac. In lateral aspect, the gut shows a constriction that gives it its characteristic heart-like shape; two further constrictions appear towards the end of stage 15. The supraoesophageal ganglia and the pharynx become evident in the living embryo.

b) **Fixed material.** At the end of stage 14, the dorsal epidermis has completed fusion at anterior and posterior, but not at intermediate, levels (Figs. 2.34, 2.35C, 2.36A-C). At the end of stage 15, dorsal closure is complete. After being covered by the overgrowing dorsal epidermis, the cells of the amnioserosa enlarge considerably, round up and disaggregate (Fig. 2.36A-B), finally becoming internalized. A deep constriction has appeared during the previous stage in the middle of the midgut and two further constrictions, one anterior and the other posterior to the previous one, form during this stage (Fig. 2.34). Head involution continues and by the end of stage 15 the tip of the advancing dorsal ridge will reach approximately 85% EL (Fig. 2.23A-B). The different parts of the ring gland join posteriorly to the base of the dorsal pouch. Condensation of the ventral cord begins.

Stage 16

a) **Living embryos.** Corresponds to part of stage 12 of Bownes (1975). Stage 16 lasts for about 3 h (13-16 h) (Figs. 2.1, 16-16e). Stage 16 begins when the intersegmental grooves can be distinguished at mid-dorsal levels and ends when the dorsal ridge has completely overgrown the tip of the clypeolabrum, which is thereby inclu-

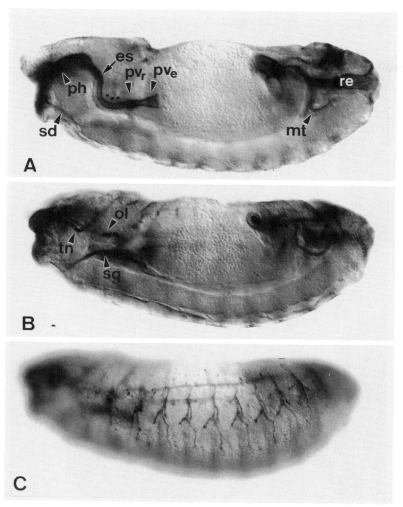

Fig. 2.34. A to C are midsagittal and lateral planes of focus through a late stage 14 embryo stained with an anti-Crumbs antibody. Notice in **A** the differentiation of the various prospective subdivisions of the proventriculus, and in **C** the ramifications of the tracheal tree. The three dots in **A** delimit the region that gives rise to the internal wall of the proventriculus, which is devoid of visceral musculature, pv_r will give rise to the recurrent portion and pv_e forms the external wall. Refer to the text for further details. Abbreviations. *es* oesophagus; *mt* Malpighian tubules; *ol.* primordium of the optic lobes; *ph* pharynx; *re* rectum; *sg* salivary glands; *tn* tentorium

ded in the atrium. Shortening of the ventral cord occurs to bring its posterior tip to about 60% EL.

b) **Fixed material.** The final stages of cytodifferentiation take place during stage 16 (Figs. 2.38, 2.39, 2.40). Secretion of cuticle has begun in the epidermis, in the tracheal tree, and in fore- and hindgut. The proventriculus and the gastric caeca appear. The proventriculus, or cardiac valve, forms at the junction of fore- and midgut with elements derived exclusively from the stomodeum, and the four gastric caeca evaginate from the midgut epithelium at the base of the proventriculus. In the anterior portion of the

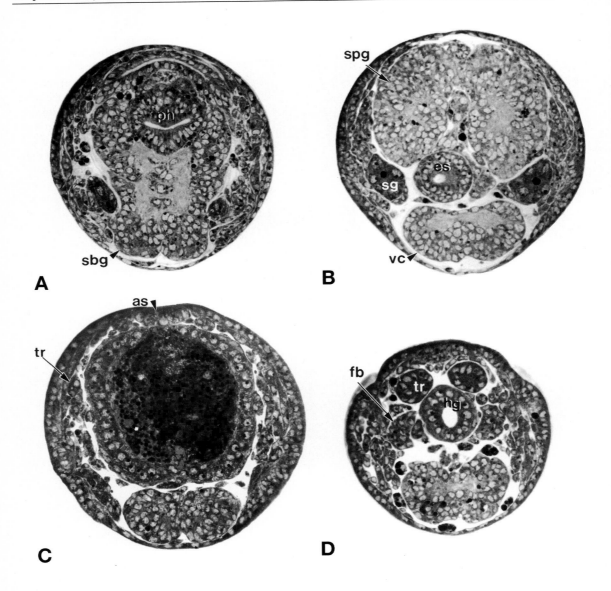

Fig. 2.35. A to D are transverse sections of an early stage 15 embryo. Dorsal epidermal closure is almost complete; midgut closure is accomplished. Abbreviations. *as* amnioserosa; *es* oesophagus; *fb* fat body; *hg* hindgut; *ph* pharynx; *sbg* suboesophageal ganglia; *sg* salivary glands; *spg* supraoesophageal ganglia; *tr* tracheal tree; *vc* ventral cord

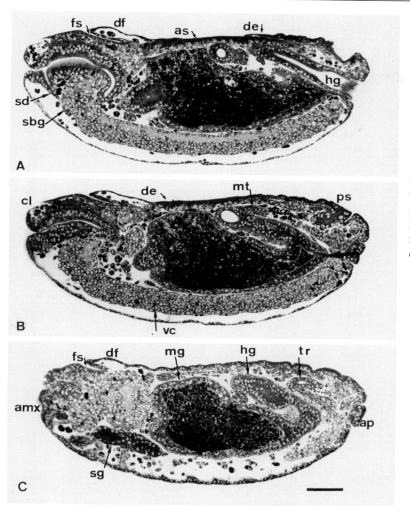

Fig. 2.36. A to F are parasagittal sections of an early stage 15 embryo. Midgut (*mg*) closure has now ended, though a caudoventral yolk sac process still has to be incorporated into the yolk sac (large *arrow* ventrally in **B**). The amnioserosa (*as*) is clearly separate from the midgut, as is evident in **A** and **B**; embryonic closure by stretching of the dorsal epidermis (*de*) is well advanced. Mouth involution is essentially complete, the salivary duct (*sd*) having already been incorporated into the atrium. Head involution is in progress, the dorsal fold (*df*) has originated from the dorsal ridge, and by this stage it

already covers about one-third of the clypeolabrum (*cl*); the frontal sac (*fs*) appears. Abbreviations. *amx* antennomaxillary complex; *ap* anal pads; *asp* anterior spiracle; *go* gonads; *hg* hindgut; *mt* Malpighian tubules; *ps* posterior spiracle; *sbg* suboesophageal ganglia; *sg* salivary glands; *spg* supraoesophageal ganglia; *tr* tracheal tree. *Bar* 50 μm

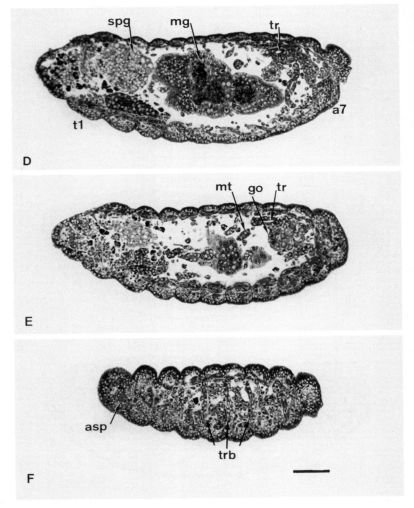

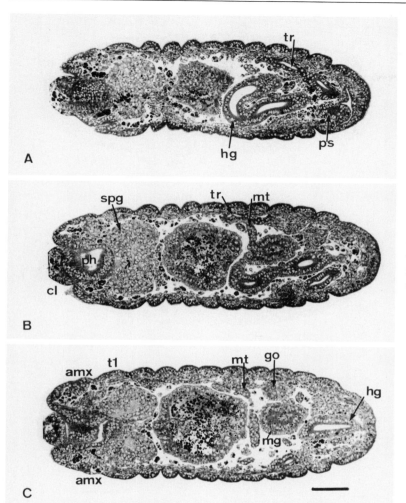

Fig. 2.37. **A to F** are horizontal sections of a stage 15 embryo. The plane of section is slightly oblique, so that the two sides are not completely in register. Horizontal sections permit one to distinguish the course of the hindgut (*hg*), with a distal, sigmoidal portion (see **A**), from which the Malpighian tubules originate (*mt*) and which is continued by the midgut (*mg*), and a proximal, longitudinal portion that ends in the anus at the anal pads (*ap*, see **B-D**). The degree of progression of head involution can be inferred from the rather lateral position of the antennomaxillary complex (*amx*); this sensillum will reach the dorsoanterior tip of the embryo after completion of head

involution. The definitive pattern of the segmental somatic musculature is not yet distinguishable, though well structured muscles (*sm*) are present. Abundant cell death figures (*cd*) as well as macrophages (*mf*) are visible. Abbreviations. *asp* anterior spiracle; *cl* clypeolabrum; *es* oesophagus; *go* gonads; *ph* pharynx; *ps* posterior spiracle; *sbg* suboesophageal ganglion; *sd* salivary duct; *sg* salivary gland; *spg* supraoesophageal ganglion; *t1* prothorax; *tr* tracheal tree; *vc* ventral cord. *Bar* 50 µm

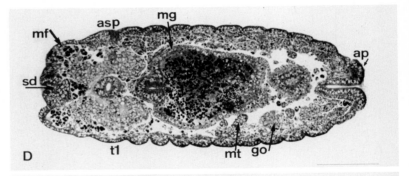

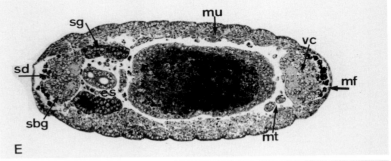

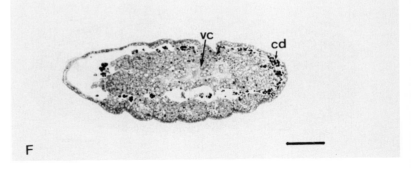

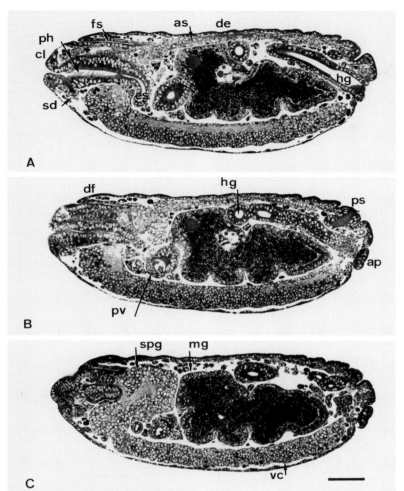

A

B

C

Fig. 2.38. A to F are parasagittal sections of a stage 16 embryo. Closure of the embryo by stretching of the dorsal epidermis (*de*) is complete. The cells of the amnioserosa (*as*) have been incorporated into the embryo, though they are still distinguishable middorsally, in association with the mesodermal crests. Formation of the frontal sac (*fs*) between dorsal fold (*df*) and clypeolabrum (*cl*) has progressed about halfway;

formation of the proventriculus (*pv*) has begun. The constrictions of the midgut (*mg*) that give rise to its various portions are distinguishable. Abbreviations. *amx* antennomaxillary complex; *ap* anal pads; *as* anterior spiracle; *es* oesophagus; *fb* fat body; *go* gonads; *hg* hindgut; *sm* somatic musculature; *mx* maxilla; *ph* pharynx; *ps* posterior spiracle; *sd* salivary duct; *sg* salivary gland; *tr* tracheal tree; *vc* ventral cord. *Bar* 50 µm

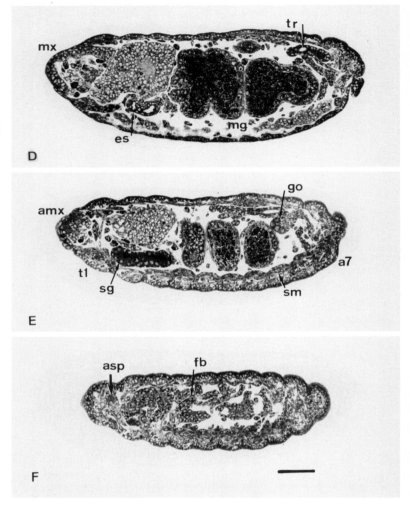

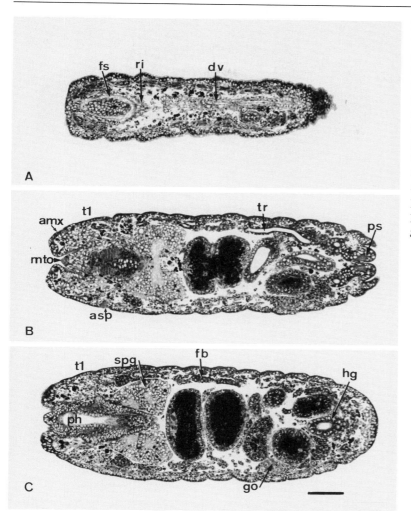

Fig. 2.39. A to F are horizontal sections of a late stage 16 embryo. **A** shows the dorsal vessel (*dv*) in almost its entire length. The median tooth (*mto*) can be seen at the tip of the labrum, whereas the antennomaxillary complex has almost reached its definitive location. The different parts of the midgut (*mg*) are evident, the invaginations that give rise to the midgut caeca (*gc*) are present. Abbreviations. *ans* anal slit; *ap* anal pads; *asp* anterior spiracle; *es* oesophagus; *fb* fat body; *fs* frontal sac; *go* gonads; *hg* hindgut; *ph* pharynx;

ps posterior spiracle; *sbg* suboesophageal ganglia; *sd* salivary duct; *sg*
salivary gland; *spg* supraoesophageal
ganglia; *t1* prothorax; *tr* tracheal
tree; *vc*. ventral cord. *Bar* 50 μm

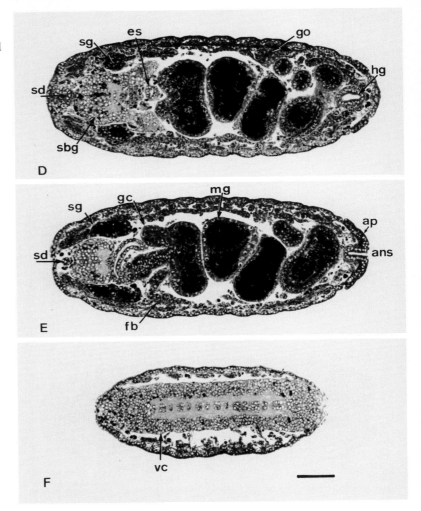

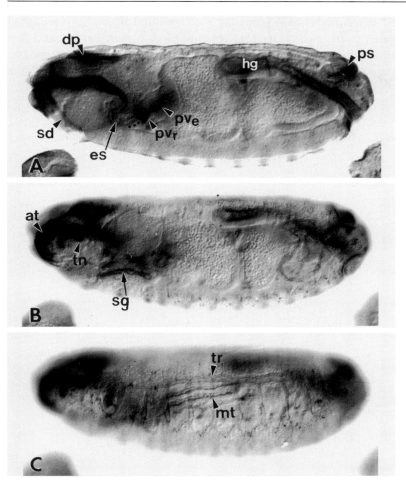

midgut, the so-called ventriculus, four evaginations of the midgut appear, which are called gastric caeca. The pharyngeal ridges appear. The fat body can be seen extending laterally from the gonads to anterior thoracic levels. The larval pattern of somatic musculature and sensory organs becomes distinguishable. The heart has also formed at a mid-dorsal position.

Stage 17

a) **Living embryos.** Corresponds to part of stages 13 and 14 of Bownes (1975). Stage 17 lasts until hatching of the embryo (Fig. 2.1, 17). The tracheal tree fills with air, thus becoming clearly visible. Retraction of the ventral cord continues.

b) **Fixed material.** Except for the involuted head, the shortened ventral cord, whose tip reaches 60% EL during this stage, and further elaboration of cuticle specializations, e.g. sensilla, no conspicuous differences can be discerned in the morphology of the embryo relative to that in stage 16 (refer to Figs. 2.41, 2.42, 2.43).

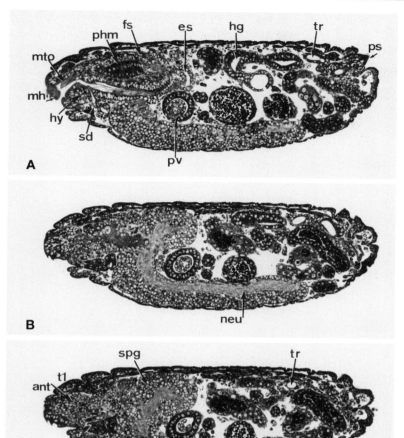

Fig. 2.41. A to F are parasagittal sections of a stage 17 embryo. Organogenesis is almost complete, so that the organization of the embryo at this stage practically corresponds to that of the first-instar larva. Only the ventral cord condenses further before hatching. Abbreviations. *ant* antennal ganglion; *es* oesophagus; *fb* fat body; *fs* frontal sac; *gc* gastric caeca; *go* gonads. *hg* hindgut; *hy* labial sensory complex (hypophysis);

mh mouth hook; *mt* Malpighian tubules; *mto* median tooth; *mu* somatic musculature; *neu* neuropile; *phm* pharyngeal musculature; *pv* proventriculus; *sbg* suboesophageal ganglia; *sd* salivary duct; *sg* salivary gland; *spg* supraoesophageal ganglia; *t1* prothorax; *tr* tracheal tree.
Bar 50 μm

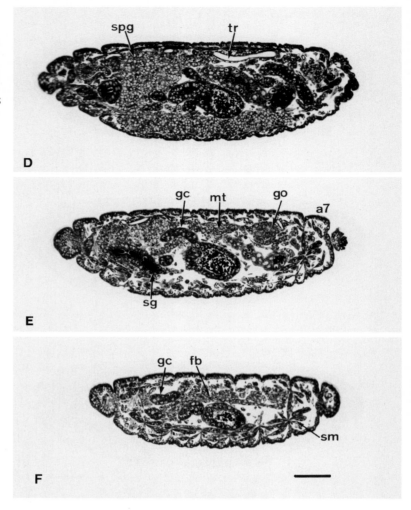

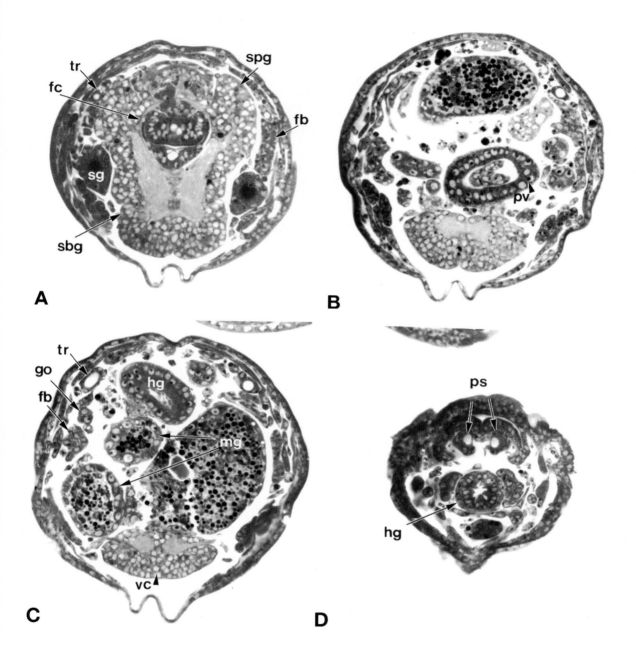

Fig. 2.42. A to D are transverse sections of a stage 17 embryo. Abbreviations. *fb* fat body; *fc* frontal commissure; *go* gonads. *hg* hindgut; *mg* midgut; *pv* proventriculus; *sbg* suboesophageal ganglia; *sg* salivary gland; *spg* supraoesophageal ganglia; *tr* tracheal tree; *vc* ventral cord

Fig. 2.43. A and C are midsagittal and parasagittal planes of focus, respectively, through a stage 17 embryo stained with an anti-Crumbs antibody. Abbreviations. *at* atrium; *dp* dorsal pouch; *es* oesophagus; *hg* hindgut; *ph* pharynx; *re* rectum; *sd* salivary duct; *tn* tentorium

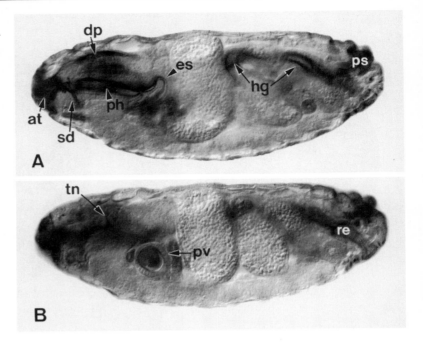

Chapter 3
Mesoderm Development

Whereas the ectoderm splits up into different organ primordia shortly after gastrulation, cells of the mesoderm remain together as a coherent germ layer for several hours, from stage 7 until late stage 11. At stage 12, various differentiations become distinguishable in the germ layer from which the mesodermal derivatives, i.e. somatic and visceral musculature, fat body, heart etc., will develop. In this chapter, we describe the pattern of proliferation and the morphogenetic movements of the mesoderm that occur during this period of time.

3.1
Gastrulation and Formation of the Mesodermal Germ Layer

After closure of the ventral furrow and consequent formation of the mesodermal tube, germ band elongation starts. This process coincides with the onset of mitotic activity within the cells of the mesodermal tube (first postblastodermal mitosis, refer to Chapter 14), which then disaggregates and loses its regular epithelial architecture (Figs. 2.7, 2.8). In stage 9 (proctodeal opening at 60% EL), during interphase 15, mesodermal cells reorganize themselves into a regular monolayer of small cuboidal cells lying on the ectoderm along the whole extent of the germ band (Figs. 2.12, 2.13, 2.16C-D, 11.1D). During this stage, the mesodermal layer of the germ band territory extends laterally from the midline up to the amnioserosa. At the midline, the mesodermal cells intermingle to form two regular cell rows between the lateral halves of the ectodermal layer; these constitute the so-called mesectoderm (Fig. 3.1). The mesodermal layer of the germ band displays in stage 9 clear signs of metamery , forming periodic bulges along the whole of the germ band (Figs. 2.12, 2.13). At this time neuroblast segregation begins and it appears that the delamination of the neuroblasts might in part be responsible for the appearance of these metameric bulges (see Chapter 11).

3.2
Splanchnopleura, Somatopleura, Somites

At the end of stage 10, mitotic activity reappears in the mesoderm (third postblastodermal mitosis), apparently affecting all the cells of the germ layer. This third mitosis brings about an increase in cell number and consequent dislocation of the mesodermal layer, which was previously organized as a monolayer. It is worth mentioning that, during this mitosis, the primordia of the various mesodermal derivatives (somatic and visceral musculature, dorsal vessel, fat body), are actually foreshadowed as specific mitotic domains (see Chapter 14). In the mesoderm lining the ectoderm, two different cell layers (Figs. 2.19D, 3.1E) become discernible in stage 11. One of these layers is internal, and in contact with the yolk sac, and will give rise to the visceral musculature; the other layer is external, contacting both the developing nervous system and the ectoderm, and will give rise to the somatic musculature and fat body. The two layers can be regarded as homologues of the splanchnopleura and the somatopleura in embryos of animals with well differentiated somites. However, strictly speaking only incipient somites can be distinguished in the territory of the metameric germ band in the *Drosophila* embryo, since a clearly differentiated coelomic cavity is missing.

Cells at the dorsalmost edge of the mesoderm, where outer and inner layers converge, form the so-called mesodermal crest (Poulson 1950), which gives rise to the dorsal vessel (Fig. 3.1E). Medially, the somatopleura thickens to form an irregular sheet 2-3 cell diameters thick. The lateral and ventral somatic musculature develops from the external cells within this sheet, whilst the internal cells give rise to the fat body.

3.3
The Procephalic Mesoderm

At the stage when the mesoderm of the germ band forms a monolayer, the mesoderm of the procephalon is also organized into two bilaterally symmetrical, vertical plates. It was previously propo-

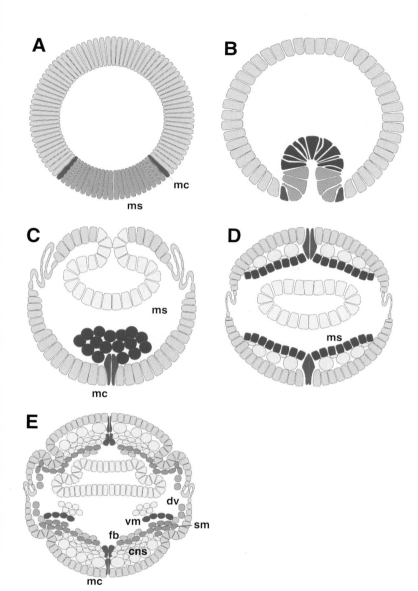

Fig. 3.1. Early mesodermal development. **A to E** show drawings of transverse sections through stage 6 (**A**), stage 7 (**B**), stage 8 (**C**), stage 9 (**D**) and early stage 12 (**E**) embryos. In the blastoderm (**A**), the prospective mesoderm occupies a midventral stripe of about 20 cell diameters in width, separated from the ectodermal anlage by two rows of cells, the mesectoderm (*mec*). This region invaginates during gastrulation (**B**) and forms the mesoderm (*ms*). In the stage 8 embryo (**C**), the mesoderm forms a dorsoventrally flattened tube connected to the ventral ectoderm by the mesectodermal cells (*mec*). At the transition to stage 9 (**D**), mesoderm cells rearrange and form a monolayer. From late stage 11 onward (**E**), the mesoderm splits up into separate cell masses that give rise to the somatic musculature (*sm*), visceral musculature (*vm*), dorsal vessel (*dv*), and fat body (*fb*). *cns.* developing central nervous system. Modified from Hartenstein (1993), with permission of Cold Spring Harbor Laboratory Press

sed that the procephalic mesoderm derives from cells of the ventral furrow located between the anterior midgut rudiment and the head fold (Poulson 1950). Following gastrulation, these cells would have to move anteriorly and laterally to end up as the two plates mentioned above. However, recent studies suggest that the origin of the head mesoderm is more complex. Thus, besides the anterior tip of the ventral furrow, those parts of the procephalon that flank the stomodeal plate anteriorly and laterally give rise to ingressing mesoderm cells (Tepass and Hartenstein 1994a). This of course implies that, in the head, the process of separation of the germ layers continues well into stage 10 and is not completed at gastrulation. A similar conclusion can be drawn from the observation that part of the stomodeal epithelium, which is part of the ectoderm, later contributes to the anterior midgut rudiment (see Chapter 7).

Suggestive evidence for a more anterior origin of the head mesoderm than previously thought is provided by the expression pattern of the *twist* gene, which is known to be expressed in the cells of the mesoderm and anterior midgut rudiment (Fig. 3.2). Thus, an anti-Twist antibody reveals Twist-expressing cells located on either side, and in front, of the anterior midgut rudiment from early stage 9 onward. Cells which ingress from the surface of the procephalon flanking the stomodeum can also be observed by BrdU labelling, due to the fact that they replicate DNA simultaneously. BrdU pulses during stage 9 reveal that the cells that will join the head mesoderm form a distinct domain different from those of the remaining mesoderm. Thus, whereas the cells of the trunk mesoderm parasynchronously undergo three rounds of mitosis at stage 8 (mitosis 14), stage 9 (mitosis 15), and stage 10 (mitosis 16), respectively, these three mitoses occur slightly later in the anlage of the head mesoderm. At the third round of mitosis (cycle 16), in late stage 10, BrdU labeling shows a clear continuity between the vertical plates formed by the head mesoderm and a ventral region of the surface of the procephalon, which is part of the same domain. These findings strongly suggest that cells from this particular ventral ectodermal region ingress to form part of the head mesoderm. It appears, therefore, that a process of recruitment of new cells into the vertical plates of head mesoderm continues until as late as the end of stage 10. In addition to their DNA replication behaviour, one other feature suggests that these cells form a special group, namely the fact that they do not show the regular columnar shape of their epithelial neighbours. Instead, they are round or bottle-shaped and give the strong impression of cells segregating from the epithelium.

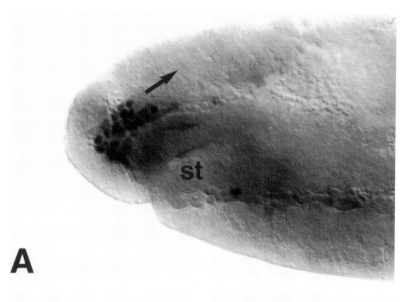

A

st

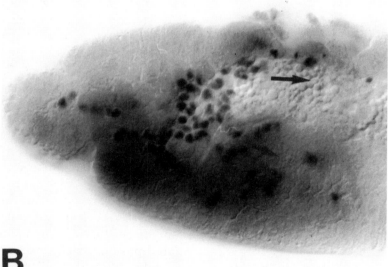

B

Fig. 3.2 Origin of hemocytes in the head mesoderm. Panels **A and B** show wholemounts of embryos from the enhancer detector line E2-3-18 expressed in hemocytes and their precursors. **A** shows a late stage 10, **B** a stage 11 embryo in lateral view. Hemocyte precursors (prohemocytes) appear in the mesoderm of the gnathal buds, the procephalon, the clypeolabrum, and the anterior portion of the median mesoderm. Some prohemocytes in **B** have already migrated beneath the amnioserosa toward the tail end of the germ band (*arrow* in **B** marks migration route)

Following the fourth mitosis, the number of procephalic mesodermal cells within the vertical plates has reached approximately
350-400 on each side of the embryo. The anteriormost of these
cells (about 70), which come to lie in the clypeolabrum, will differentiate into the myoblasts of the pharyngeal musculature, as
shown by using an enhancer trap expressed in these cells (Tepass
and Hartenstein 1994a). Whereas the myoblasts remain closely
packed, the remaining procephalic mesodermal cells (approximately 300), i.e. the precursors of the hemocytes (Fig. 3.2), dissociate
and begin to disperse throughout the embryo (see Chapter 6).

Chapter 4
Musculature

4.1 Visceral Musculature

The visceral musculature comprises circular and longitudinal fibres, which surround the entire intestinal tract, with the exception of the recurrent layer of the proventriculus (Fig. 4.1). The two systems of fibres differ in developmental origin: the circular fibres originate from a region of the mesodermal layer analogous to the splanchnopleura of other animals, i.e. a bilaterally symmetrical band of mesoderm cells extending continuously throughout most of the germ band (Fig. 4.2A). The longitudinal fibres derive from clusters of mesodermal cells which appear during stage 12 at the posterior end of the embryo and migrate anteriorly (Fig. 4.2A).

At the beginning of germ band shortening in stage 12 (Fig. 2.21), the precursors of the circular fibres are organized like a palisade, consisting of a dorsal and a ventral row of tightly packed, slender cells (Fig. 4.2B, C). These cells adhere to both anterior and posterior midgut rudiments as germ band shortening progresses (Fig. 2.22). Further anteriorly and posteriorly, cells of the visceral mesoderm also surround the primordia of the foregut and hindgut. These terminal parts of the visceral mesoderm, which do not exhibit a regular, layered structure, are separated by wide gaps from the mesodermal cells associated with the midgut; these gaps will remain unchanged throughout development. The segment of the foregut that lacks visceral mesodermal cells, and thus visceral musculature, will become the recurrent layer of the proventriculus; the corresponding segment of the hindgut will develop into the region where the Malpighian tubules open into the hindgut (Figs. 4.1, 4.3).

During closure of the midgut the precursor cells of the circular visceral muscles move dorsally and ventrally in association with the endodermal cells (Figs. 4.2B-D). This movement follows a highly stereotyped pattern. The visceral muscle precursors

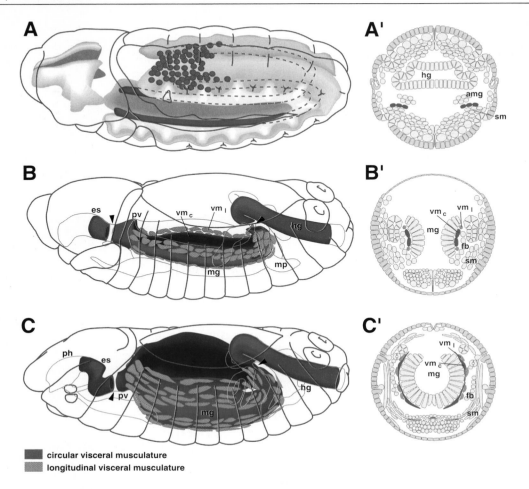

Fig. 4.1. Development of the visceral musculature. Diagrams show dorso-lateral views of late stage 11 (**A**), stage 13 (**B**) and late stage 14 (**C**) embryos. **A' - C'** present transverse sections of corresponding stages. The circular visceral muscle fibres (*vm$_c$*) are coloured turquoise, longitudinal fibres (*vm$_l$*) are shown in purple. In **A**, the mesoderm is shaded grey. During late stage 11, the primordia of the circular visceral muscles appear as longitudinal cell bands at the lateral edge of the mesoderm in each segment (visceral mesoderm or splanchnopleura). Precursors of the longitudinal fibres are located in the tail end of the germ band at this stage. After germ band retraction, the visceral mesoderm forms a plate flanking the fused midgut rudiments (*mg*); further anteriorly and posteriorly, visceral mesoderm cells already form continuous sheaths around the primordia of the foregut and hindgut, respectively. Narrow segments of fore- and hindgut will remain devoid of visceral mesoderm (*arrowheads*) and will give rise to the recurrent layer of the proventriculus (*pv*) and the junction between Malpighian tubules and hindgut (*hg*), respectively. Precursors of the longitudinal fibres migr-ate dorsally and ventrally along the plate flanking the midgut and thereby spread into anterior parts of the embryo. As the midgut closes (**C**), the circular visceral muscle fibres elongate. Since visceral myoblasts do not fuse, four individual cells surround the perimeter of the midgut at any given level (**C'**). Other abbreviations. *amg* anterior midgut rudiment; *es* oesophagus; *fb* fat body; *mp* Malpighian tubules; *ph* pharynx; *sm* somatic musculature. Modified from Hartenstein (1993) with permission of Cold Spring Harbor Laboratory Press

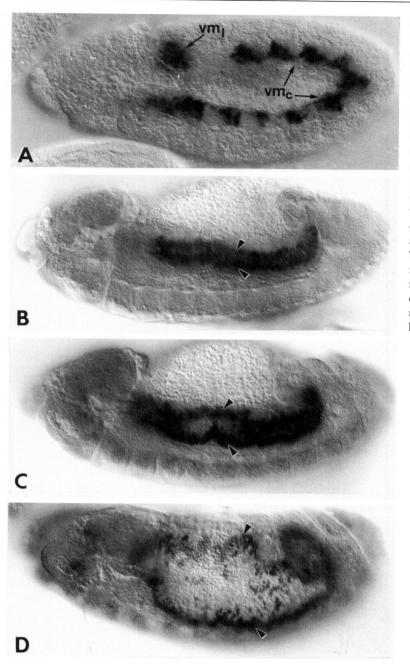

Fig. 4.2. Development of the circular visceral musculature. **A** is a stage 11 embryo stained with an anti-β3 tubulin antibody (kindly provided by Andrea Wolk), **B-D** carry an enhancer trap that highlights the developing circular musculature. During stage 11 (**A**), the primordia of the circular visceral mesoderm (vm_c) consist of ten distinct clusters of dorsal mesodermal cells. After fusion of the midgut primordia, the circular muscle cells form palisades, which will extend eventually to ensheath the midgut. Their nuclei separate into two rows (*arrowheads*). The longitudinal visceral musculature (vm_l) originates from a group of mesodermal cells at the caudal end of the embryo (**A**). These cells later migrate along the developing circular muscle cells

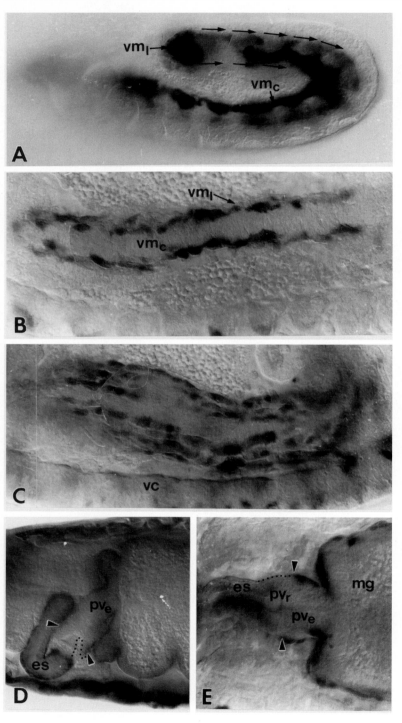

Fig. 4.3. Development of the visceral musculature. Longitudinal muscle fibres. **A.** Stage 11 embryo stained with an anti-β3 tubulin antibody. Stage 13 (**B**) and stage 14 (**C**) and stage 15 (**E**) embryos, carrying an enhancer trap insertion in the *couch-potato* gene that highlights the longitudinal visceral musculature. **D** Stage 16 embryo stained with an anti-myosin antibody. In *A* the primordia of the circular visceral muscles (vm_c) are organized as longitudinal cell bands at the lateral edge of the mesoderm in each segment. Precursors of the longitudinal fibres (vm_l) are located in the tail end of the germ band at this stage. From this location, longitudinal muscle fibre precursors will migrate during germ band shortening along the pathways indicated by *arrows* in **A**. After germ band retraction (**B, C**), the longitudinal fibre precursors are arranged in dorsal and ventral rows along the circular muscle fibre precursors (vm_c, unstained) at the wall of midgut (*mg*). In **D** and **E**, longitudinal and circular muscle fibres form continuous sheaths around the developing foregut and hindgut. Note segment of the foregut devoid of visceral muscles (*arrowheads*) that gives rise to the recurrent layer of the proventriculus (pv_r) and the junction to the oesophagus (*es*, **D, E**). pv_e external portion of the proventriculus

derived from the former dorsal row of splanchnopleura cells move dorsally; those derived from the ventral row move ventrally. At the same time, precursor cells spread out in the transverse axis to encircle the midgut tube. Unlike somatic myoblasts, the precursors of the visceral musculature do not fuse with other precursor cells. Consequently, visceral muscle fibres are formed by mononucleate cells.

The precursors of the longitudinal visceral muscles migrate from the caudal abdominal region, where they originate, throughout the entire length of the germ band during germ band retraction (Figs. 4.1 - 4.3). Following this stage, these cells form two irregular, elongated clusters of fusiform cells which flank the splanchnopleura dorsally and ventrally. During midgut closure, these cells spread out dorsally and ventrally, to form approximately 10 rows of 6-10 serially arranged fibres each.

4.2
Somatic Musculature

Following the fate of the somatic mesodermal cells that will form individual somatic muscles is more difficult. The difficulty is caused mainly by the disaggregation of the external mesodermal layer into loose, segmental, cellular clusters that occurs at germ band shortening (Figs. 2.22, 2.23). Formation of muscles occurs during stages 13 to 15, by fusion of single mesodermal muscle founder cells to fusion-competent cells, to form syncytial cells, visible in stage 14, and further growth of these syncytia (Figs. 4.4, 4.5). Bate (1990) describes fused doublets and triplets as early as 7.75 h (onset of germ band shortening) in the ventral mesoderm, overlying the developing CNS. The distribution of these muscle precursors appears to be constant in different embryos, so that their identification is possible. Similar small syncytia with two to three nuclei appear at lateral and dorsal levels, which grow by fusion with other neighbouring myoblasts. In stage 16 the final pattern of the somatic musculature can be distinguished.

The organization of the somatic musculature in thoracic and abdominal segments of the fully mature embryo is illustrated in Figs. 4.6-4.11 and is adapted from Bate (1993), who has provided the most complete description of the complement of embryonic somatic muscles. Segmental muscles in *Drosophila* are organized in five different patterns, corresponding to those of thoracic segments *t1*, *t2* and *t3*, abdominal segments *a1-a7*, which all exhibit

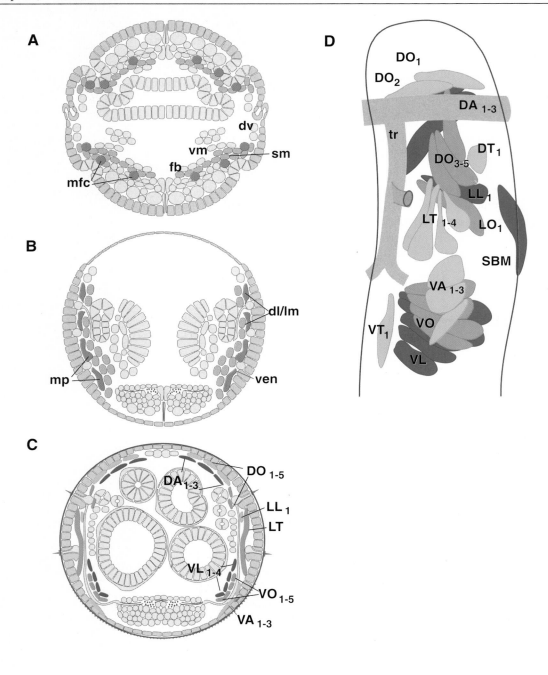

Fig. 4.4. Development of the somatic musculature. **A to C** show transverse sections of stage 11 (**A**), 13 (**B**) and 17 (**C**) embryos, **D** shows the pattern of the so-called muscle precursors in an abdominal hemisegment at stage 13 (after Bate 1990). In **C-D**, different muscle groups are shown in different shades of green (light. superficial muscles; dark. deep muscles). In **D**, segment boundaries and trachea (*tr*) are given as points of reference. During stage 11 (**A**), the mesodermal germ layer splits up into different organ primordia (*dv* dorsal vessel; *fb* fat body; *vm* visceral musculature). The progenitors of the somatic musculature (*sm*, light green) are found superficially in the entire mesoderm. From these, individual cells (dark green) segregate to become the muscle founder cells (*mfc*). By stage 13 (**B, D**), founder cells have fused into muscle precursors which foreshadow the mature pattern of muscle fibres (compare **D** with Fig. 4.6 and 4.7 showing this pattern). By stage 17 (**C**) all muscle fibres have attached to the epidermis and acquired their final pattern. For nomenclature of individual muscles, see Figs. 4.6 and 4.7. Other abbreviations. *d/lm* primordium of dorsal and lateral musculature; *vc* ventral cord; *vm* primordium of ventral musculature. Modified from Hartenstein (1993) with permission of Cold Spring Harbor Laboratory Press

essentially the same pattern, and *a8-a9*. As a rule, thoracic and abdominal muscles do not encompass more than one segment, inserting in prominent apodemes located either at intersegmental boundaries or within each segment (see Figs. 8.5, 8.6). However, the following are exceptions: massive ventral intersegmental muscles that extend from the anterior apodeme of the prothorax to that of the first abdominal segment, therefore passing over the meso- and the metathorax (*VIS5*); the slender ventral intersegmental *VIS1* which extends from the anterior apodeme of the prothorax to the anterior apodeme of the metathorax; furthermore, within each of the abdominal segments, two of the ventral external oblique muscles (*VO4-5*) extend beyond the posterior segmental border to terminate within the next segment caudally.

4.2.1
The Muscle Pattern of *a1-a7*

The basic pattern of segmental musculature in the fully developed embryo is identical to that in the third-instar larva, as worked out by Hertweck (1931, his Fig. 12) and Crossley (1978; see also Szabad et al. 1979). In our description, we essentially follow the terminology proposed by Crossley (1978) and extended by Bate (1993), who classified the muscles in the embryo into three different groups: dorsal, lateral (pleural) and ventral. The muscle pattern of *a1-a7* is considered to be the basic pattern (ref. to Figs. 4.6, 4.7), to which the patterns of the remaining segments will be referred. The dorsal muscles insert a short distance lateral to the dorsal midline, which is occupied by the dorsal vessel, and extend over about 30% of the transverse diameter. They are organized in two different layers, one external (the dorsal oblique or *DO* muscles)

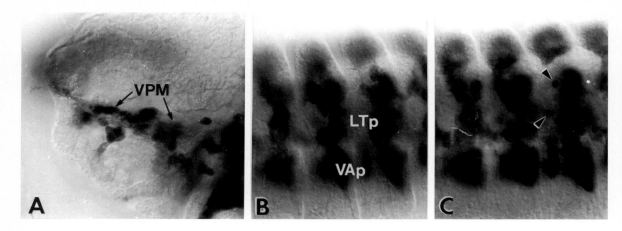

Fig. 4.5. A to C are lateral views of a stage 13 embryo stained with an anti-β3 tubulin antibody (slide kindly provided by Andrea Wolk; Leiss et al. 1988). They show that the myoblasts that will form the internal muscles (*arrowheads* in **C**) as well as those of the head (*VPM* designates fusion-competent myoblasts which will give rise to the ventral pharyngeal muscles), have not yet fused, whereas the external muscle precursors (*LTp, VAp*) are already established

and one internal (the dorsal acute or *DA* muscles). These two layers overlap each other completely. The dorsal oblique muscles extend from ventrocaudal to dorsorostral, whilst the dorsal acute muscles follow the complementary course. The ventralmost fibres of the dorsal group (*DO3-5; DA3*) insert below the level of the dorsal hair sensilla; the intermediate fibres (*DO2, DA2*) cover the main tracheal trunk, and the dorsal fibres (*DO1, DA1*) insert immediately adjacent to the dorsal vessel.

The lateral muscles can be subdivided into three components, lateral transverse muscles, lateral longitudinal muscles, and lateral oblique muscles. The lateral transverse muscles have a dorsoventral course and insert at the lateral epidermis of each segment. The internalmost of these muscles (segment border muscle, *SBM*) is located along the transverse apodeme between neighbouring segments. Three lateral transverse muscles (*LT1-3*) extend within the anterior one-third of a given segment. As a whole, they form a conspicuous vertical plate stretching from 35% to 70% of the dorsoventral extent, covering a subepidermal cleft which contains the lateral groups of sensilla. Two more caudally located transverse muscles (*LT4, DT1*) insert at a somewhat more dorsal level than their anterior counterparts. The lateral longitudinal muscle (*LL1*) extends perpendicularly to the lateral transverse muscles. The

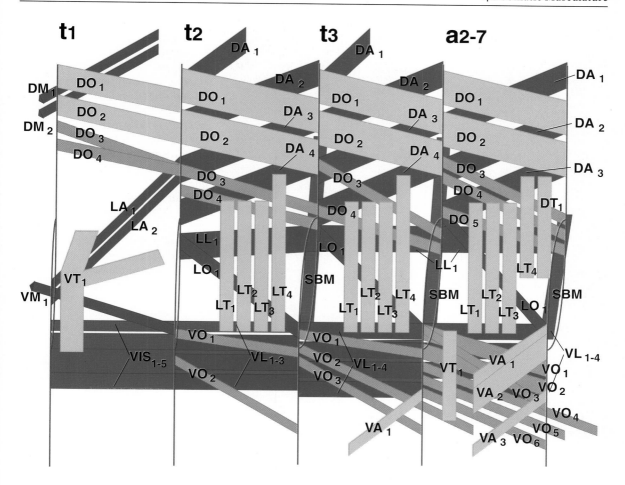

Fig. 4.6. Pattern of muscles of the thoracic (*t1-3*) and abdominal segments (*a2-7*). Diagrams show flattened views of hemisegments; dorsal midline is at the top, ventral midline at the bottom. Muscle groups are shown in different shades of green, with light green indicating the superficial stratum of fibres, and darker shades of green the deeper strata. The ventral intersegmental muscles are shown in turquoise. The A1 pattern is very similar to the *a2-7* pattern shown; two muscles, *VO6* and *VT1*, are missing in *a1*. Terminology follows that of Bate (1993). Abbreviations. *DM* dorsal mouthpart muscles; *DA* dorsal acute muscles; *DO* dorsal oblique muscles; *DT* dorsal transverse muscles; *LA* lateral acute muscles; *LL* lateral longitudinal muscle; *LO* lateral oblique muscle; *LT* lateral transverse muscles; *SBM* segment border muscle; *VA* ventral acute muscles; *VIS* ventral intersegmental muscle; *VM* ventral mouthpart muscles; *VO* ventral oblique muscles; *VT* ventral transverse muscle

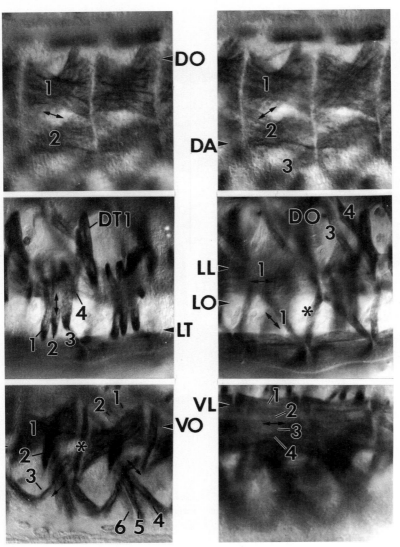

Fig. 4.7. The pattern of somatic muscles in *a2-a7*. Two segments of a late stage 16 embryo stained with an anti-β3 tubulin antibody are shown at (from top to bottom) dorsal, lateral and ventral levels; the plane of focus in the panels on the left is exterior, in the panels on the right interior. Anterior to the left. Terminology is that of Bate (1993). *DA 1-3* dorsal acute 1-3; *DO 1-4* dorsal oblique 1-4; *DT 1* dorsal transverse 1; *LL 1* lateral longitudinal; *LO 1* lateral oblique; *LT 1-4* lateral transverse 1-4; *VA 1-3* ventral acute 1-3; *VL 1-4* ventral longitudinal 1-4; *VO 1-6* ventral oblique 1-6. The *asterisks* in the middle right and lower left panels indicate the segment border muscle (slide kindly provided by Andrea Wolk; Leiss et al. 1988)

lateral oblique muscle (*LO1*) extends obliquely from posteroventral to anterodorsal, approximating the diagonal of the rectangle formed by the anterior and posterior segmental borders and the lateral longitudinal muscles.

The ventral muscles cover the area between the ventral midline and the lateral transverse muscles. They comprise intrasegmental and intersegmental muscles, which form three overlapping muscle layers. The internal muscle layer (ventral longitudinal muscles, *VL*) extends longitudinally from one intersegmental apodeme to the next. It comprises four closely spaced fibres (*VL1-4*). The intermediate layer (ventral oblique muscles, *VO*) comprises six slender fibres (five in *a1*) with a dorso-anterior to ventro-posterior orientation. The superficial layer comprises fibres with a ventro-anterior to dorso-posterior orientation (ventral acute muscles, *VA*), as well as one transverse fibre (*VT1*).

4.2.2
The Muscle Pattern in *t1-t3*

The pattern in *t1-t3* consistently differs from the basic pattern described above, although homologies between groups of muscles in thorax and abdomen clearly exist. In *t1* (Fig. 4.6), though grossly modified, dorsal, lateral and ventral groups of muscles are present. Dorsally four (external) oblique muscles (*DO1-4*) insert at the rostral prothoracic apodeme and run caudally. The lateral group is represented by two acute muscle fibres (*LA1-2*) and a transverse fibre (*VT1*) located slightly more ventrally. Ventrally, a massive bundle of five muscle fibres (ventral intersegmental muscles, *VIS1-5*) extends along the ventral prothoracic wall to cover approximately 30-40% of the prothorax, thereby constituting the largest of the larval muscles. Some of its fibres (*VIS2, 3, 4*) terminate at the *t1-t2* apodeme; one fibre (*VIS1*) continues caudally to reach the *t2-t3* apodeme; *VIS5* inserts at the apodeme between *t3* and *a1*. Laterally to the salivary duct, there is an oblique muscle that originates from the lateral pharyngeal wall (*VM1*) which joins the ventral intersegmental muscles.

In *t2* and *t3* (Fig. 4.6), all groups of muscles of the basic pattern are present, deviations from this pattern only concern the number of individual fibres within the different groups. The dorsalmost of the dorsal acute muscles (*DA1*) displays a mode of insertion which differs greatly from that encountered within the abdominal segments. The posterior part of this muscle is split up into two (*t3*) or three (*t2*) strands which become interwoven with their contralateral counterparts and terminate within the territory of the con-

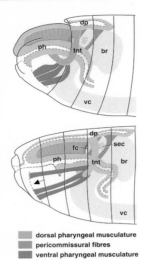

dorsal pharyngeal musculature
pericommissural fibres
ventral pharyngeal musculature

Fig. 4.8. Pattern of muscles in the head. Diagrams show side views of the cephalic region of stage 16 (top) and 17 embryos. Muscle groups are shown in different shades of green. Muscles are sparse in the head, relative to the trunk and telson. They comprise mainly two groups, the ventral pharyngeal muscles (*VPM*), derived from the gnathal segments, and the dorsal pharyngeal muscles (*DPM*), which originate in the labrum. Anteriorly, the ventral pharyngeal muscles insert at an apodeme located adjacent to the labial sensory complex (*arrow*). Posteriorly, they are attached to the ventral limb of the tentorium (*tn*). The dorsal pharyngeal muscles form a massive bundle of short, vertically oriented fibres attached to the roof of the pharynx (*ph*) ventrally, and the internal layer of the dorsal pouch (*dp*) dorsally. Three slender fibres (pericommissural muscles, *PCM*), which flank the frontal commissure (*fc*) and supraesophageal commissure (*sec*), form a posterior continuation of the dorsal pharyngeal muscles. Other abbreviations. *br* brain (supraoesophageal ganglion); *es* oesophagus; *vc* ventral nerve cord

tralateral segmental half. This peculiar structure is evident only during larval stages; in the late embryo, *DA1* is unbranched.

4.2.3
The Pattern of Cephalic Muscles

In the cephalic region of the fully developed embryo, three groups of muscles can be distinguished (Fig. 4.8, 4.9). The first (dorsal mouthpart muscles *DM1, 2*) comprises fibres that extend dorsolaterally, from the wall of the pharynx to the epidermis of the first thoracic segment. The second group (ventral pharyngeal muscles *VPM1-3*) is located ventrolaterally. Its fibres insert at the tentorium, an apodeme which develops by invagination from the maxillary and intercalary segment (Figs. 2.25B, 2.29B, 2.33B, 2.34B, 2.40B; see Chapter 16). The third group corresponds to the dorsal pharyngeal musculature (*DPM*), formed by numerous vertically oriented fibres that insert ventrally at the dorsal pharyngeal epithelium and dorsally at the bottom of the dorsal pouch. Three slender fibres (pericommissural muscles, *PCM*), which flank the frontal commissure and supraoesophageal commissure, form a posterior continuation of the dorsal pharyngeal muscles.

4.2.4
The Muscle Pattern of the Telson

The pattern of muscles in the telson (Fig. 4.10, 4.11) differs due to the presence of the posterior spiracles and the rudimentary nature of *a9* and *a10*. After shortening of the germ band, *a9* and the so-

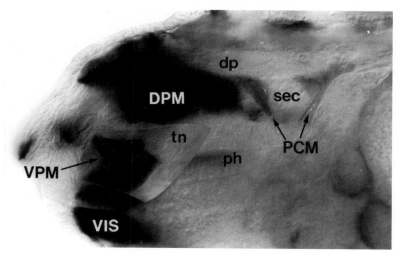

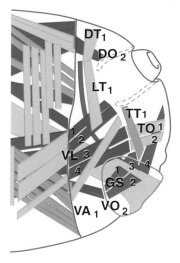

Fig. 4.9. The pattern of cephalic somatic muscles. A stage 17 embryo stained with an anti-β3 tubulin antibody is shown (slide kindly provided by Andrea Wolk). Anterior to the left. *dp* dorsal pouch; *DPM* dorsal pharyngeal muscles; *PCM* pericommissursal muscles; *ph* pharynx; *sec* supraoesophageal commissure; *tn* tentorium; *VIS* ventral intersegmental muscles; *VPM* ventral pharyngeal muscles

called *a10* fuse along the midline to form the blunt posterior end of the body wall. Since the caudal borders of *a9* fuse caudomedially, no longitudinal muscles can be expected. Instead, muscles homologous to the longitudinals in other segments should here exhibit an oblique or transverse course; this expectation turns out to be correct. Dorsal to the distal hindgut two muscle fibres extend on either side between the lateral epidermis and a vertical median septum at the *a8-a9* apodeme. These terminal oblique muscle fibres (*TO1,2*) are continuous with the ventral longitudinals of more anterior abdominal segments. Another group of terminal (*a9*) muscles which are located more ventrally and insert on either side of the anal pads are the gut suspension muscles (*GS1, 2, 3, 4*). These may be homologues of the ventral oblique and ventral acute muscles of more anterior segments. In *a9* no homologues of the lateral and dorsal group could be identified.

In *a8* all muscle groups of the basic pattern are clearly identifiable, but are profoundly modified due to the presence of the posterior spiracles. The dorsal group is reduced both in length and width. It consists of two flattened fibres on either side of the dorsal midline, which overlap (*DT1, DO1*). These fibres insert caudally in the epidermis anterior to, and above, the spiracles. Rostrally, *DO1* reaches the *a7-a8* apodeme where it is continuous with *DO/DA* muscles of *a7*. The *DT1* muscle terminates close to the *a7-a8* apodeme in the dorso-anterior region of *a8*. The lateral group of *a8* comprises one large transverse muscle (*LT1*). All other muscle fibres of *a8* (*VL1-4; VO1-3; VA1*) have to be related to the ventral group, since all insert at the *a7-a8* apodeme at the level of

Fig. 4.10. Pattern of somatic muscles of the telson. In the telson, ventral muscles predominate. Serial homologues of the ventral longitudinal (*VL1-4*), ventral oblique (*VO*) and ventral acute fibres (*VA*) can be found in *a8*. The posterior telson (*a9/10*) contains the terminal transverse (*TT1*) and terminal oblique muscles (*TO*); in addition, a number of fibres serve to suspend the distal hindgut from the body wall (*GS 1-5*). There are only a few lateral (*LT1*) and dorsal muscles (*DT1, DO1*) derived from *a8*. Other abbreviations. *ap* anal pads; *hg* hindgut; *ps* posterior spiracle

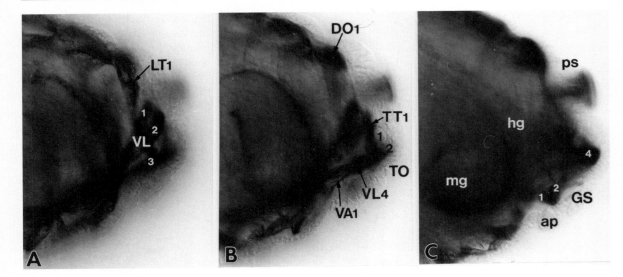

Fig. 4.11. The pattern of somatic muscles in the caudal terminal region. Three different planes of focus (lateral views) of a late stage 16 embryo stained with an anti-β3 tubulin antibody (slide kindly provided by Andrea Wolk; Leiss et al. 1988) are shown. *ap* anal pad; *DO 1* dorsal oblique 1; *GS 1-4* gut suspensor 1-4; *hg* hindgut; *mg* midgut; *ps* posterior spiracles; *TO 1-2* transverse oblique 1-2; *VA 1* ventral acute 1; *VL 1-4* ventral longitudinal 1-4

the *VL* and *VO* muscles of anteriorly adjacent segments. The prominent *VL1-4* and *VA1* fibres follow a dorso-posterior course and insert at the base of the posterior spiracle; by contracting they pull the posterior spiracle inside. *VO1-3* show a ventro-posterior course and insert close to the anal pads.

4.2.5
The Progenitor Cells of the Imaginal Muscles

Imaginal somatic muscles develop during pupal metamorphosis from progenitor cells associated with the imaginal discs, the so-called adepithelial cells, and histoblast nests. The progenitors of these adepithelial cells are of embryonic origin. Bate et al. (1991) showed that they can be easily identified due to the fact that they express the Twist protein. Fig. 4.12 shows the distribution of the progenitor cells of thoracic and abdominal imaginal somatic muscles in a stage 16 embryo. There is a distinct, highly reproducible pattern of progenitors in thoracic and abdominal segments. Their development and further behaviour is discussed in Bate (1993).

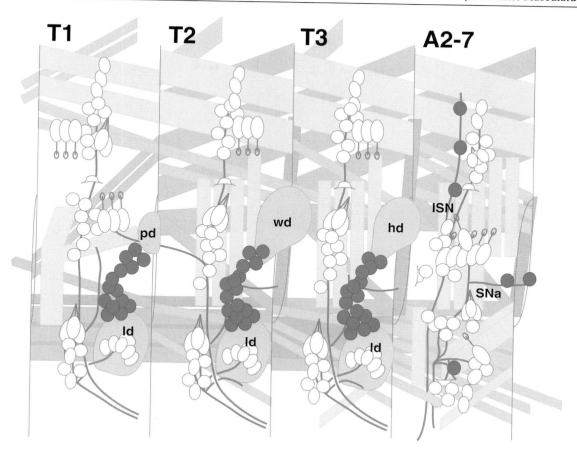

Fig.4.12 Location of adult muscle progenitor cells (AMPs; turquoise) in the late embryo, as revealed by an anti-Twist antibody. The drawing shows flattened views of thoracic hemisegments *t1-t3* and an abdominal hemisegment representing *a1-a7* (dorsal midline at the top) of a stage 16 embryo. Sensory neurons (grey circles), peripheral nerves (grey lines), muscle fibres (light green) and imaginal discs (light orange) are shown as points of reference. In abdominal segments, there are three groups of adult muscle progenitors which are all attached to branches of the peripheral nerves. Three dorsal AMPs are located adjacent to the intersegmental nerve (*ISN*), two lateral AMPs are on a posteriorly directed branch of the segmental nerve (*SNa*), and one ventral AMP sits on the SNd branch. In the thorax, there are large clusters of AMPs located in between the imaginal discs (*hd* haltere disc; *ld* leg discs; *pd* dorsal prothoracic disc; *wd* wing disc). In addition, there exist scattered AMPs along the peripheral nerves; the pattern of these cells is not well understood

Chapter 5
Circulatory System and Fat Body

5.1
Structure of the Dorsal Vessel in the Mature Embryo

The organization of the dorsal vessel in the mature embryo is comparable to that in the third instar larva (see Fig. 1 of Rizki 1978a; Rugendorff et al. 1994). In the embryo, the dorsal vessel consists of a tube that extends from the region immediately behind the supraesophageal commissure up to the arthrodial membrane in *a7*, where it inserts by means of two caudally directed muscles, which connect the posterior tip of the dorsal vessel to an attachment site in *a8* (Figs. 5.1, 5.2F). One distinguishes two main sections of the dorsal vessel; one is anterior, extending up to *a4*, corresponding to the so-called aorta, while the other is posterior, extending throughout *a5* and *a6*, and called the heart. At any given cross-sectional level, the lumen of the tube is formed by only two cells, one on each side; this lumen is wider in its caudal, cardiac portion than in its anterior, aortic portion. The tube formed by the cardial cells is surrounded by smaller, round, pericardial cells that extend continuously along both aorta and heart; cardial and pericardial cells are both of mesodermal origin (Bodmer et al. 1989; Hartenstein and Jan 1992), and develop from the two longitudinal rows of cells at the dorsal edge of the main mass of somatic mesodermal cells, called mesodermal crests (Poulson 1950), which meet and fuse at the midline with the homologous cells of the other side (Fig. 5.2).

Within anterior regions, the dorsal vessel is surrounded by the lymph glands, which in the embryo seem to consist of only two cell masses, rather than three or more as in the third-instar larva (Rizki 1978a; Fig. 5.1). In front of the lymph glands, the dorsal vessel runs through the ring gland to terminate by means of filamentous insertions immediately behind the supraoesophageal commissure (Rizki 1978a). Enhancer-trap lines show that the lymph glands develop from the mesodermal crests corresponding to the

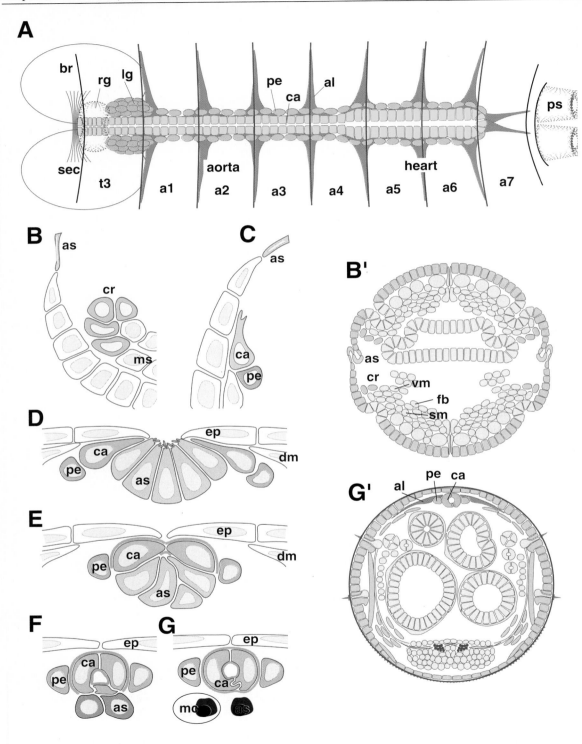

Fig. 5.1 Structure of the dorsal vessel of a stage 17 embryo, dorsal view (**A**). The dorsal vessel extends in the dorsal midline from segment *t3* to *a6*. Alary muscles (*al*) attach the dorsal vessel to the apodemes. The dorsal vessel, consisting of the heart posteriorly and the aorta anteriorly, is formed by the cardioblasts (*ca*), in association with a bilateral row of pericardial cells (*pc*). In its anterior portion, the dorsal vessel is flanked by the lymph glands (*lg*) and ring gland (*rg*). **B-G, B'** and **G'** depict the development of the dorsal vessel in transverse sections. During stage 12 (**B, B'**), precursors of the dorsal vessel appear in the lateral part of the mesoderm (*ms*), called the mesodermal crest (Poulson 1950) or cardiogenic region (*cr*, Hartenstein et al. 1992). In stage 13 (**C**), precursors of cardioblasts (*ca*) and pericardial cells (*pe*) separate from the neighbouring cells (*sm*) and move dorsally. Shortly before dorsal closure (**D**), the amnioserosa (*as*) becomes internalized. Cardioblasts move between amnioserosa cells and epidermal cells (*ep*). After dorsal closure (**E**), the leading edges of the cardioblasts meet along the dorsal midline, whilst amnioserosa cells remain attached to the cardioblasts. During late stage 16 (**F**), the trailing edges of the cardioblasts curl around to form a lumen. Finally, in stage 17 (**G, G'**), formation of the lumen of the dorsal vessel is complete. Amnioserosa cells die and are taken up by macrophages (*mc*). Abbreviations. *br* brain; *dm* dorsal somatic musculature; *fb* fatbody primordium; *ps* posterior spiracle; *sec* supraoesophageal commissure; *sm* somatic mesoderm; *vm* visceral mesoderm

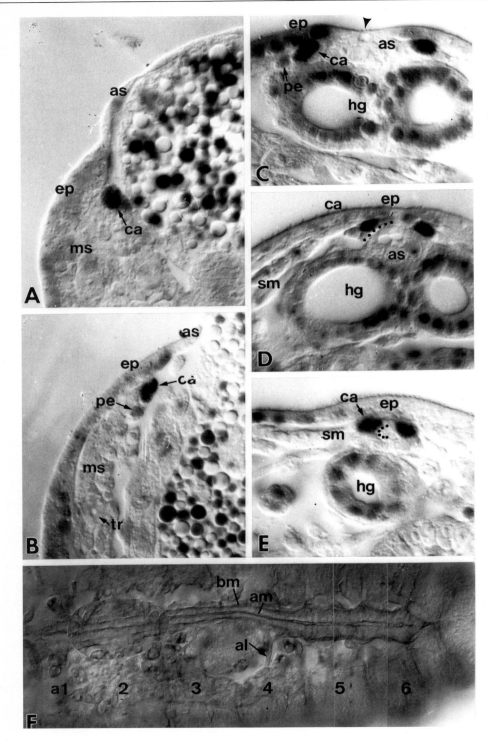

Fig. 5.2. Development (**A to E**) and organization (**F**) of the dorsal vessel. In the stage 12 embryo (**A**), the precursors of cardioblasts (*ca*) can be distinguished by their expression of enhancer trap B2-3-20 in the lateral extension of the mesoderm. Precursors of cardioblasts and pericardial cells (*pe*) move along the overlying epidermis (stage 14, **B**; stage 15, **C**).

In the stage 16 embryo (**D**), the leading processes of the cardioblasts meet at the midline. The cells of the amnioserosa become internalized at dorsal closure and transiently form a tube with a central lumen, attached to the cardioblasts. Amnioserosa cells will degenerate soon thereafter. In the stage 17 embryo (**E**), morphogenesis of the heart is complete.

F shows anti-laminin antibody staining of a stage 17 embryo. Note the size difference between the anterior (aorta) and posterior (heart) divisions of the dorsal vessel. *1-6* refer to *a1-a6*. *al* alary muscles; *as*. amnioserosa; *bm* basement membrane; *ep* epidermis; *hg* hindgut; *lm* luminal membrane; *ms* mesoderm; *sm* somatic musculature; *tr* trachea

t3 segment, where no pericardial cells normally occur. It seems, thus, as if the pericardial cells of *t3* have differentiated into lymph glands. As in the third instar larva, the dorsal vessel of the embryo is suspended from the body wall by seven pairs of alary muscles, inserting between the dorsal vessel and lateral apodemes. The alary muscles are among the first muscles to differentiate, and can be clearly seen with an anti-myosin antibody, or with an antibody against β3-tubulin as early as stage 13.

5.2
Development of the Dorsal Vessel

As mentioned above, the dorsal vessel develops from two longitudinal rows of mesodermal cells, the mesodermal crests, that meet at the dorsal midline after dorsal closure of the embryo; no event directly related to cell proliferation is involved in morphogenesis of the dorsal vessel (refer to Fig. 5.2). Apart from their topology, the cells that give rise to the dorsal vessel do not exhibit any morphological characteristic that would allow them to be distinguished at earlier stages; Poulson (1950) referred to them as 'inconspicuous cardioblasts'. However, these cells are clearly distinguishable from early developmental stages on by virtue of their *lacZ* expression in various enhancer trap lines, and this permits one to follow their development. The heart precursor cells develop from a cardiogenic region at the lateral edge of the mesodermal layer, where splanchnopleura and somatopleura fuse, and can be distinguished as early as stage 12 (Hartenstein et al. 1992; Rugendorff et al. 1994). These cells are segmentally organized; in each of segments *t2* to *a6* there are about six cardioblasts, whereas *t1* and *a7* give rise to three or four, and no cardioblasts are produced in *a8*, *a9* or the head segments.

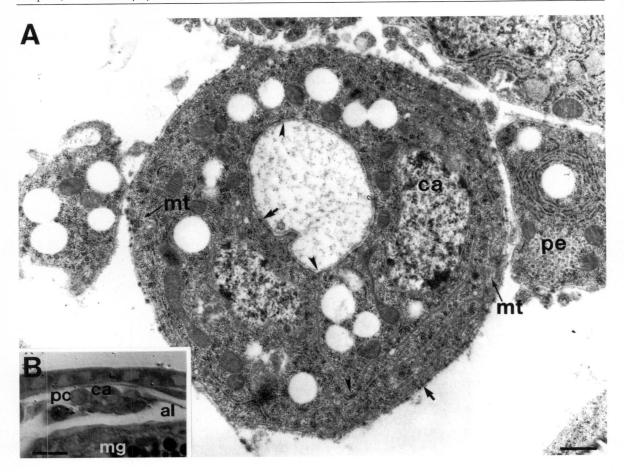

Fig. 5.3. A. Electron micrograph of a transverse section of the dorsal vessel of a stage 17 embryo. Notice that two cardioblasts (*ca*) delimit a central lumen. The small *arrows* indicate hemiadherens junctions between the cell membrane of the cardioblasts and the extracellular matrix. *Arrowheads* point to focal adherens junctions between cardioblasts. *mt.* microtubules. **B.** Transverse section of a stage 17 embryo to illustrate the relationship of the heart to surrounding organs (from Rugendorff et al. 1994)

Towards dorsal closure, in stages 14 and 15, prospective cardio-blasts and pericardial cells can be seen in histological sections, without special staining procedures, to derive from the medial margin of the mesodermal crests of each side: the cardioblasts stretch and flatten and develop dorsally directed processes, while the pericardial cells retain their rounded appearance. Based on the study of serial sections and cell transplantations, it was propo-sed (Campos-Ortega and Hartenstein 1985; Technau and Campos-Ortega 1986a) that the cells of the amnioserosa meet the cardio-blasts as they join mid-sagittally, becoming integrated into the dorsal vessel to form the pericardial cells. However, the use of enhancer trap lines that specifically label the amnioserosa cells has shown that the association of the amnioserosa with the dorsal vessel is actually transitory (Bodmer et al. 1989; Hartenstein and Jan 1992) and that the amnioserosa cells die after encountering the dorsal vessel (Fig. 5.2). At the end of stage 15, the epidermal primordia of both sides meet at the dorsal midline and the pros-pective cardioblasts squeeze between the amnioserosa and the epidermis. As a consequence of dorsal closure, the shape of the amnioserosa cells becomes columnar and the surface occupied by them is reduced to a narrow strip of about five cells, which invagi-nate as the epidermis closes, coming to lie ventrally to the cardio-blasts. Shortly thereafter, in stage 17, the internalized amnioserosa cells can be seen to degenerate and are phagocytosed by macro-phages.

After separation of the epidermis and amnioserosa has occur-red, leading processes of the cardioblast precursors meet at the midline with their counterparts from the other side (Figs. 5.1, 5.2); subsequently, prominent interdigitations appear on the ventral surface of these cells and the area of contact between the cardiob-lasts increases. Part of the surface of the medial cardioblast does not form this contact and becomes the lumen of the dorsal vessel (Fig. 5.3; Rugendorff et al. 1994).

5.3
The Fat Body

In the fully developed embryo, the fat body consists of a dorsal and a ventral division on either side of the embryo, located bet-ween the gut and the somatic musculature (Figs. 5.4, 5.5). These cells develop from segmentally arranged groups of precursors which, in stage 12, split from the most interior part of the meso-

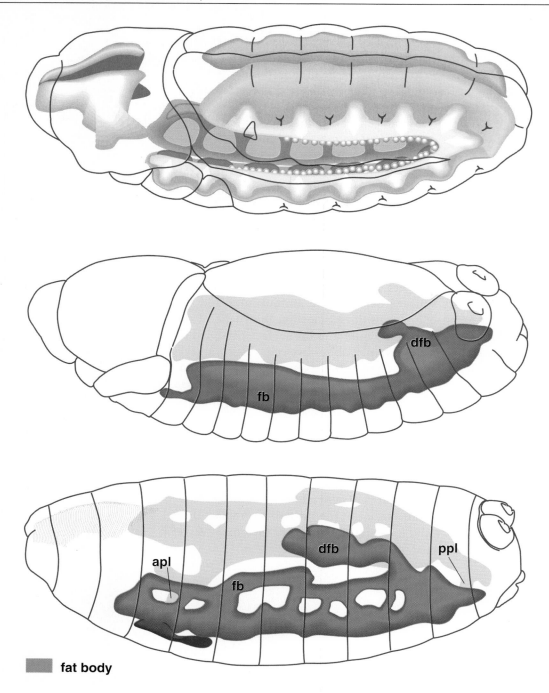

fat body

Fig. 5.4. Development of the fat body. Diagrams show dorsolateral views of late stage 11 (**A**), stage 13 (**B**) and stage 17 (**C**) embryos. The fat body is in green. In **A**, the mesoderm is shaded grey. During stage 11, the primordium of the fat body appears as a series of segmental cell clusters of in the ventral part of the mesoderm (**A**). After germ band retraction, these clusters fuse into a continuous band (*fb*) extending laterally throughout the segmented germ band (**B**). In later stages, the band expands dorsally and ventrally to form a loose layer. Prominent structures within the fat body are the dorsal process (*dfb*), and the anterior (*apl*) and posterior (*ppl*) plates, which form transverse rows of fat body cells located in the cleft between gut and CNS. Modified from Hartenstein (1993), with permission of Cold Spring Harbor Laboratory Press

derm and become sandwiched between the developing visceral and somatic musculature. The segmental clusters fuse into a continuous column, which splits into a dorsal and a ventral division. The cells of the dorsal division extend all the way from the gonads, which they surround anteriorly (Fig. 5.5), to the level of the brain hemispheres; the ventral division is even longer, extending from the telson to the mesothoracic segment.

The cells of the fat body are thin and do not show any signs of fat differentiation during the entire period of embryogenesis (Poulson 1950; Rizki 1987b). The two divisions of the fat body are initially fairly compact; however, holes and clefts appear during late embryonic stages, probably due to the movements of the other internal organs of the *Drosophila* larva.

Fig. 5.5. A shows a stage 17 embryo carrying an enhancer trap insertion that highlights the fat body (*fb*) and the oenocytes (*oe*). The various subdivisions of the fat body, i.e. the dorsal process (*dfb*), and the anterior (*apl*) and posterior (*ppl*) plate can be clearly distinguished. **B** shows that the fat body completely surrounds the gonads (*go*). **C-D** show the oenocytes in relation to the somatic musculature (*sm*)

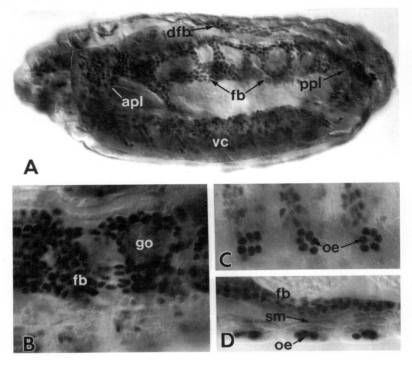

Chapter 6
Macrophages

6.1.
The Pattern of Embryonic Cell Death

Many cells within the developing epidermis and CNS die during embryogenesis, the overwhelming majority during the second half of stage 11 and first half of stage 12 (see Figs. 2.19C, 2.21). Restriction of cell death to this particular time period is quite striking, since at this time mitotic activity has not yet completely ceased in the epidermal primordium. Shortly after the appearance of cell death figures, the first macrophages become apparent in the developing embryo, suggesting a causal relationship between the onset of cell death and the appearance of macrophages. Whereas cell death is comparably intense in all regions of the developing central nervous system, in the epidermis it chiefly affects the ventrolateral and procephalic region, and is less prominent in the dorsal epidermis. However, the process follows a similar course in all regions.

Necrotic cells are clearly distinguishable with the light microscope on plastic sections by their density and pronounced shrinkage. Only a few cell death figures are found within the epidermis itself, most of them being located subepidermally, in the space left free by the retracting central nervous system, which we call paraneural; this suggests that the affected cells are forced by healthy ones to leave the epithelial context very soon after the necrotizing process has begun. In fact, cell death figures suggestive of pycnotic elements being pushed by normal cells are frequently seen in the epidermal epithelium (Figs. 2.19C, 2.21). During stage 12, dead cells start accumulating in large groups in the subepidermal, paraneural space of the gnathal segments and the dorsal tip of the extended germ band, where they are intermingled with macrophages (Figs. 2.21, 2.26, 2.27). In stage 13, cell death figures are more widespread, being visible in the procephalic lobe, posterior abdominal segments and dorsal ridge, besides the gnathal segments and the dorsal tip of the germ band (Abrams et al. 1993). Although most dead cells are phagocytosed by macrophages, normal epidermal cells can apparently also act as phagocytes (Tepass et al. 1994).

The peculiar accumulation of dead cells in the subepidermal space coincides with the separation of the developing CNS from the epidermis (Fig. 2.22). Therefore, cell death within the epidermal layer might be causally related to this phenomenon. In addition, the cell death that occurs in late embryogenesis may play a role in the reshaping of the dorsal head region, as suggested by various features of the pattern of *engrailed* expression in the head (Younossi-Hartenstein et al. 1993). Particularly interesting is the behaviour of the three *engrailed* stripes in the gnathal segments. Whereas the expression domains within the gnathal protuberances persist, the dorsal halves of the stripes fuse into a single strip the cells which give rise to the dorsal ridge. The basis of this process of fusion is cell death (see Fig. 16.4, Chapter 16). Another example is the single stripe marking the antennal segment prior to head involution, which covers a broad region between the clypeolabrum and the dorsal ridge. Subsequently, most of these *engrailed*-expressing cells disappear from the surface. While a few of them become incorporated into the brain, the majority degenerate and are taken up by macrophages (Younossi-Hartenstein et al. 1993).

6.2.
Origin and Dispersal of Macrophages

Macrophages develop from a scattered population of hemolymph cells, or hemocytes. These cells have recently been shown to derive exclusively from the mesoderm of the procephalon (Tepass et al. 1994). Using an antibody against the extracellularly deposited peroxidasin produced by the hemocytes, the dispersal and differentiation of these cells, which begins in stage 11, has been analyzed (Abrams et al. 1993; Tepass et al. 1994; Nelson et al. 1994). After mitosis 17 (stage 11), hemocytes remain mitotically quiescent during the rest of embyogenesis. Thus, the number of hemocytes remains constant throughout development and cannot be increased by experimentally raising the level of cell death in the embryo (Tepass et al. 1994). During stages 11 and 12, hemocytes slowly move out from the head, following various routes (Figs. 6.1, 6.2). Moving anteriorly and ventrally, they come to populate the

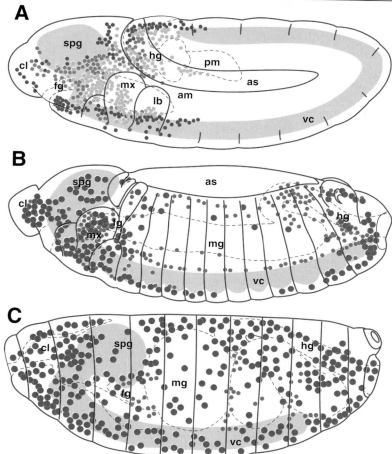

Fig. 6.1. Development of the hemocytes. **A to C** are drawings of late stage 11 (**A**), stage 13 (**B**), and late stage 16 (**C**) embryos stained with an antibody against peroxidasin (Nelson et al. 1994). Hemocytes derive from the cephalic mesoderm and follow four main pathways to populate the anteroposterior extent of the embryo (**C**). One lies ventrolateral to the supraoesophageal ganglion (*spg*, **A**), underneath the amnioserosa. Two other pathways are at the dorsal and ventral surface of the developing ventral cord (*vc*, **A to B**). Notice that hemocytes are also located midsagittally in the ventral cord. The fourth pathway is lateral to the anterior (*am*) and posterior (*pm*) midgut primordia and the hindgut (*hg*). *cl* clypeolabrum; *fg* foregut; *lb*. labium; *mg* midgut; *mx* maxillary bud; *pv* proventriculus

clypeolabrum and gnathal buds. Posteriorly directed migration brings them directly into the tail end of the germ band, which is folded over the anterior part of the germ band during this stage and abuts the head region. A large number of hemocyte precursors remain in the dorsal head region. Germ band retraction in stage 12 carries the group of hemocyte precursors that had previously reached the tail end posteriorly. In the following stages (12-14), hemocyte precursors migrate from both ends of the embryo towards the middle. During this migration hemocytes follow four main routes. The first one is superficial, between the ventral epidermis and the ventral nerve cord; the second stream is located between the dorsal surface of the ventral nerve cord and the mesoderm; the third is deep inside the embryo, surrounding the gut, while the fourth follows the dorsal edge of the epidermis. From late stage 14 on, all segments of the trunk are rather evenly populated with hemocytes (Figs. 6.1, 6.2).

During migration, hemocytes start to differentiate. They develop long processes which radiate out from the cell bodies in all directions. Many hemocytes, in particular those of the head and those surrounding the ventral nerve cord, show phagocytotic activity and contain one or more dark inclusions which represent the pycnotic nuclei of ingested cells that have undergone apoptotic cell death (Abrams et al. 1993; Tepass et al. 1994). These inclusions, easily visible with the light microscope, as well as the larger size of the macrophages, permit one to conclude from wholemount preparations of anti-exoperoxidase stained embryos that the macrophages derive from hemocytes.

By late stage 13, approximately half of the entire population of hemocytes have differentiated into macrophages. Most of them are located around the CNS and in the head, where most cell death occurs. Hemocytes located deeply around the oesophagus, midgut and hindgut, are small and do not yet exhibit phagocytic activity. However, during later stages, the percentage of macrophages, i.e. of cells containing phagocytosed debris, among the population of peroxidasin positive cells, increases, and by the end of embryogenesis 80-90% of these cells are macrophages (Tepass et al. 1994).

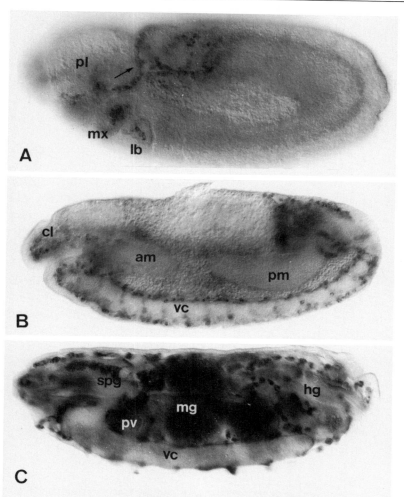

Fig. 6.2. Development of the hemocytes. **A to C** are early stage 12 (**A**), stage 13 (**B**), and late stage 16 (**C**) embryos stained with an antibody against peroxidasin (Nelson et al. 1994). Refer to Fig. 6.1. *am* anterior midgut; *cl* clypeolabrum; *hg* hindgut; *lb* labium; *mg* midgut; *mx* maxillary bud; *pl* procephalic lobe; *pm* posterior midgut; *pv* proventriculus; *spg* supraoesophageal ganglion; *vc* ventral cord

Chapter 7
The Gut and its Annexes

The three parts of the alimentary tract of *Drosophila*, foregut, midgut and hindgut, differ in origin. Fore- and hindgut are derivatives of the ectoderm, and as such are lined with cuticle; the midgut derives chiefly from the endoderm, although parts of the anterior midgut are of ectodermal origin (Technau and Campos-Ortega 1985; Hartenstein et al. 1985). In the fully developed embryo, the foregut consists of atrium, pharynx, oesophagus and proventriculus, with the salivary glands ending in the floor of the atrium, at its border with the pharynx (Fig. 7.1). The proventriculus is a complex structure, comprising an inner epithelial layer which essentially represents a continuation of the oesophagus; a recurrent layer which is folded back over the inner layer; and an outer epithelial layer which is continuous with the midgut (Fig. 7.2). The midgut is a convoluted tube that still contains yolk granules at hatching (see Fig. 2.41), and several parts may be distinguished from both topological and morphological points of view (Bodenstein 1950; see Fig. 7.3); additionally, four blind tubules called gastric caeca project from the initial portion of the midgut, at the base of the proventriculus (Figs. 7.2, 7.3). The hindgut has a dorsally directed ascendant portion, from which the Malpighian tubules originate, and a corresponding descendant portion that ends in the anus (Figs. 7.3, 7.6). A thorough description of the anatomical organization of the *Drosophila* larval gut, in particular of the intricate foregut, can be found in Strasburger (1932).

7.1.
The Foregut

Details of the anatomical organization of the foregut in the fully developed embryo can be seen in Figs. 7.1, 16.1, 16.2. The foregut derives entirely from the ectoderm and has a complex origin. It develops on the one hand from stomodeal cells that invaginate

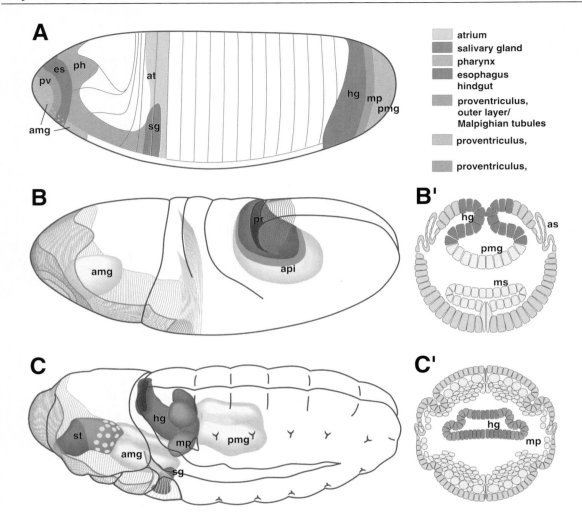

Legend:
- atrium
- salivary gland
- pharynx
- esophagus hindgut
- proventriculus, outer layer/ Malpighian tubules
- proventriculus,
- proventriculus,

Fig. 7.1. Development of the gut tube. Foregut and hindgut. **A to F** are dorsolateral views of stage 5 (**A**, fate map), stage 8 (**B**), stage 9 (**B**), stage 11 (**C**), stage 13 (**D**), stage 15 (**E**) and stage 17 (**F**) embryos; **B'-F'** are transverse sections through the middle of embryos in the corresponding stages. The foregut and hindgut are exclusively ectodermal (stomodeal, proctodeal). In the blastoderm (**A**), the anlagen of the midgut rudiments are located at the embryonic poles. The

ectodermal anlagen surrounding the anterior and posterior midgut rudiments will form the foregut and hindgut, respectively. Foregut and hindgut are divided up into structurally and functionally different parts. The spatial arrangement of their blastoderm anlagen is indicated in **A** (*ph* pharynx; *es* oesophagus; *pv* proventriculus). The foregut primordium begins to invaginate in stage 10, when the stomodeum forms; however, internalization of foregut structures continues

until stage 16. The hindgut primordium invaginates immediately after the posterior midgut rudiment (stages 7-9). In stage 8 (**B, B'**), the posterior midgut rudiment (*pmg*) and the invaginated part of the hindgut primordium (*hg*) form a pouch called the amnioproctodeal invagination (*api*). In stage 11 (**C, C'**), four spherical outgrowths, which will form the Malpighian tubules (*mp*), appear in the hindgut primordium close to its boundary with the posterior midgut

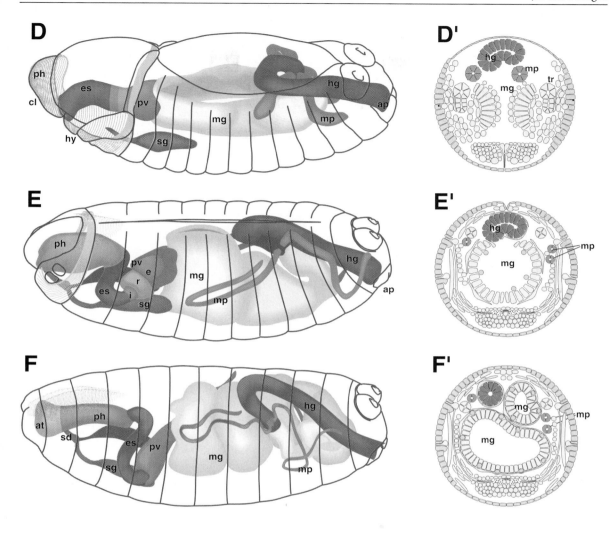

rudiment. In stage 13 (**D, D'**), the foregut has elongated considerably; the
prospective proventriculus (*pv*) and
oesophagus (*es*) become distinct. The
midgut (*mg*) consists of two longitudinal plates of epithelial cells flanking
the yolk sac, which abut on the prospective proventriculus and hindgut
(*hg*), respectively. The hindgut has
adopted its characteristic sigmoid
shape; the Malpighian tubules (*mp*)
have elongated. The pharynx (*ph*) differentiates from the oesophagus. *sm*.

somatic mesoderm; *vm*. visceral
mesoderm. By stage 15 (**E, E'**), the
midgut lumen has become continuous with the lumen of the prospective proventriculus (*pv*) and hindgut
(*hg*), respectively. The different parts
of the foregut are now further advanced in their differentiation. The
pharynx (*ph*) forms a wide, vertically
flattened structure; the oesophagus
(*es*) is a round tube bent like an S. A
circular constriction separates the
presumptive proventriculus into an

anterior and posterior chamber. The
posterior chamber opens into the
midgut. During later development
(stage 16), the anterior chamber invaginates into the posterior chamber.
The salivary duct (*sd*) has formed. Its
external opening has reached the ventral lip of the stomodeal opening. *sg*
salivary gland. By stage 16 (**F, F'**),
Malpighian tubules (*mp*) have reached their final size. Modified from
Hartenstein (1993), with permission of
Cold Spring Harbor Laboratory Press

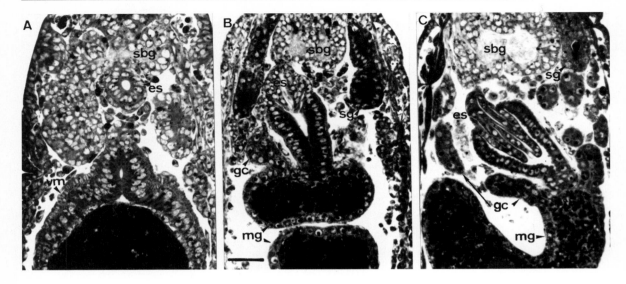

Fig. 7.2. Horizontal sections to illustrate three stages of proventriculus and midgut caeca development. **A** is a stage 14 embryo, in which no trace of regionalization is visible in the midgut. *es* oesophagus; *sbg* suboesophageal ganglion; *vm* visceral mesoderm. **B** is a stage 16 embryo in which both gastric caeca (*gc*) and proventriculus can already be distinguished. *mg* midgut; *sg* salivary glands. **C** is a stage 17 embryo. Gastric caeca and proventriculus are almost fully developed. *Bar* 50 μm

during stages 10 - 12, and on the other hand from cells of the clypeolabrum, hypopharyngeal lobe and gnathal segments that become incorporated into the stomodeum during head involution in stages 13, 14 and 15 (Fig. 7.1).

At the end of stage 9, the anlage of the stomodeum forms a plate of tall cylindrical, strongly basophilic cells, located ventromedially at the anterior embryonic pole, which differ markedly from all neighbouring ectodermal cells. At invagination during stage 10, cells surrounding the stomodeal plate (domain $\partial 15_1$ of Foe 1989) divide synchronously (Figs. 2.13A, 2.14A), the invaginated cells divide immediately afterward, and thereafter become organized into a monolayered epithelium (Figs. 2.14B, 2.15D). During the initial period of its development, the stomodeal invagination is oriented strictly perpendicular to the horizontal plane; however, it bends progressively towards the ventral plane.

Subsequently, up to stage 13, the stomodeum grows further caudalwards, partially by means of mitotic divisions during the third postblastodermal mitosis, and partially by incorporating additional cells from the neighbouring, paraoral ectoderm in the first instance and, later, from the gnathal segments. During

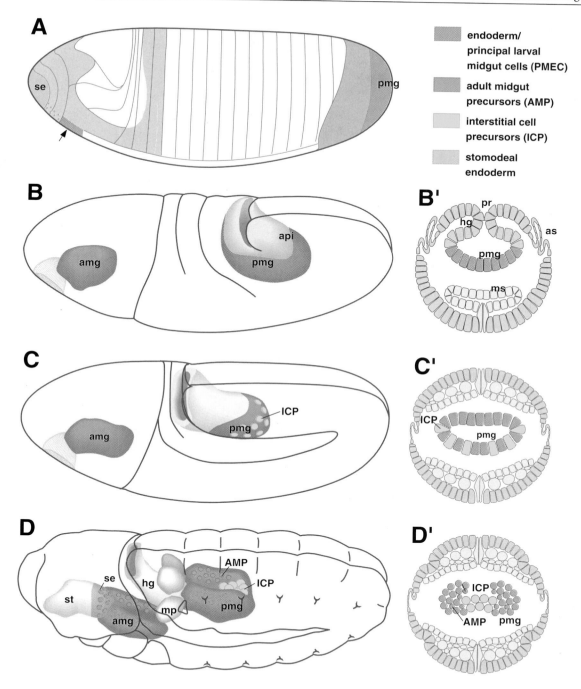

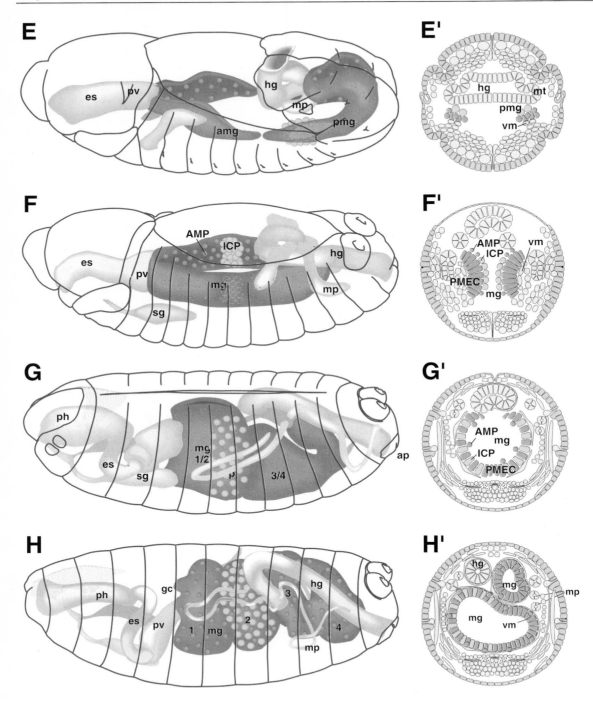

Fig. 7.3. Development of the gut tube. The midgut. **A to H** are dorsolateral views of stage 5 (**A**, fate map), stage 8 (**B**), stage 9 (**C**), stage 11 (**D**), stage 12 (**E**), stage 13 (**F**), stage 16 (**G and H**) embryos; **B'-H'** are transverse sections through the middle of embryos in the corresponding stages. The midgut originates mainly from endodermal (*am, pm*) derivatives; however, parts of the anterior midgut derive from ectodermal cells that invaginate with the stomodeum. In the blastoderm (**A**), the anlagen of the midgut rudiments are located at the embryo poles. During gastrulation in stage 7, anterior and posterior midgut rudiments invaginate. In stage 8 (**B, B'**), both the posterior midgut rudiment (*pm*) and the invaginated part of the hindgut primordium (*hg*) form a pouch called the amnioproctodeal invagination (*api*). During stage 9 (**C, C'**), parts of the hindgut primordium become incorporated into the proctodeum and, consequently, the length of the amnioproctodeal invagination increases. The interstitial cell progenitors (*ICP*) are intermingled with the midgut epithelial cells. In stage 11 (**D, D'**), anterior (*am*) and posterior midgut rudiments (*pm*) have lost their epithelial organization and form solid clusters of rounded cells. *mt* primordia of the Malpighian tubules. During stage 12 (**E, E'**), as germ band shortening sets in, anterior and posterior midgut rudiments stretch in the longitudinal axis, become bilobular and fuse at about 50% egg length. The midgut of a stage 13 (**F, F'**) embryo (*mg*) consists of two longitudinal plates of epithelial cells flanking the yolk sac. These plates will spread in the transverse axis and eventually (stage 15) fuse at the ventral and dorsal midline, thereby forming a closed chamber which encloses the yolk. By stage 16 (**G, G'**), the presumptive midgut (*mg*) has closed ventrally and dorsally. A constriction in its posterior third gives the midgut a heart-like shape. *1/2* and *3/4* indicate the prospective midgut chambers. Anteriorly and posteriorly, the midgut lumen has become continuous with the lumen of the prospective proventriculus (*pv*) and hindgut (*hg*), respectively. In stage 16 (**H, H'**), four midgut chambers (*1 - 4*) can be seen. Four evaginations, the gastric caeca (*gc*), develop from the first midgut chamber. *as* amnioserosa; *AMP* adult (imaginal) midgut progenitors; *es* oesophagus; *hg* hindgut; *ICP* interstitial cell progenitors; *mp* Malpighian tubules; *ms* mesoderm; *ph* pharynx; *pv* proventriculus. *sm* somatic mesoderm; *vm* visceral musculature. Modified from Hartenstein (1993), with permission of Cold Spring Harbor Laboratory Press

growth, the stomodeum maintains its tubular architecture, except for its innermost portion or fundus, from which the proventriculus, the gastric caeca and adjacent portions of the midgut will develop (Technau and Campos-Ortega 1985). These cells establish contact with the endodermal anterior midgut primordium (Poulson 1950) and thereafter lose their epithelial organization (Fig. 2.20A) and intermingle with the oralmost cells of the (endodermal) anterior midgut primordium.

No conspicuous regional differences can be recognized in the stomodeal epithelium during this initial growth phase. However, the pattern of mitoses of stomodeal cells reveals the anlagen of the individual subdivisions of the foregut (see Chapter 9). Mitotic domain $\partial 14_9$ (see Chapter 14), occupying the fundus, includes the presumptive proventriculus, as well as those stomodeal cells which will become the anteriormost part of the midgut. Mitotic domain $\partial 14_{15}$, which surrounds $\partial 14_9$ and invaginates during stage 11, represents the oesophagus. Mitotic domain $\partial 14_3$, corresponding to the presumptive clypeolabrum that will invaginate during stage 13, will form the pharynx roof (epipharynx); mitotic domain $\partial 14_2$, the presumptive hypopharyngeal lobe, will form the pharynx floor (hypopharynx).

During stages 12 and 13, the subdivisions of the foregut become morphologically distinct (Fig. 7.1D-F). The fundus of the stomodeum widens to acquire a somewhat bell-like shape. Cells in the roof of the part of the stomodeum that will become the oesophagus form three invaginations, which represent the primordia of the stomatogastric nervous system (Figs. 2.17, 2.21, 2.22; see Chapter 10). From stage 13 onwards, further regional differences become evident in the cellular architecture of the foregut, beginning when the hypopharyngeal lobes ventrally and the tip of the clypeolabrum dorsally form the stomodeal opening (Figs. 2.31, 2.36). By this stage, four morphologically distinct subdivisions of the stomodeum can be recognized. These comprise (from anterior to posterior) the primordia of the pharynx, oesophagus, recurrent layer of the proventriculus, and outer layer of the proventriculus (Figs. 7.1, 7.2).

The pharynx is characterized by tall cylindrical cells, larger in the floor than in the roof. Anteriorly, the pharynx is delimited by the opening of the salivary glands ventrally and by the median tooth dorsally. The oesophageal and proventricular primordia during stages 13 -15 consist of small cuboidal cells.

During later stages (14-16) the anteriormost part of the foregut, represented by the atrium, invaginates and head involution occurs (see Chapter 16). During these stages the foregut lengthens considerably. The oesophagus moves posteriorly between the developing brain hemispheres and adopts a sigmoid shape (Fig. 7.1). Morphogenesis of the proventriculus proceeds with the invagination of the inner layer into the recurrent layer which, at the same time, folds into the outer layer (Fig. 7.2).

7.2
The Salivary Glands

The salivary glands develop from the sternal portions of the labium. However, it appears that the maxillary sternum contributes to formation of the salivary gland ducts: maxillary cells that express *engrailed* become incorporated into the salivary duct (Younossi-Hartenstein et al. 1993). Two epidermal placodes become evident at the medial surface of the labial buds during stage 11; they invaginate to give rise to the distal portion of the secretory epithelium. The cells of the placodes are cylindrical, with the nucleus in a basal position, and have basophilic cytoplasm, and are thus clearly distinguishable from the neighbouring cells (Figs. 2.17, 2.19). By means of double staining with anti-Forkhead and anti-Engrailed antibodies, Panzer et al. (1992) have determined the position of the placodes; they extend from the border between parasegments 1 and 2 to the border between parasegments 2 and 3, and are therefore contained entirely within parasegment 2, which spans the posterior region of the maxillary and the anterior of the labial segment. The cells of the salivary gland anlage divide twice before the placodes become evident; they invaginate at the end of stage 11 (Figs. 2.17, 2.19, 7.1), without further division, and form two tubes of cylindrical cells that diverge laterocaudally and run into a common, wide opening at the ventromedial surface of the labial epidermis (Figs. 2.19, 2.22, 2.26, 2.31, 2.36). Poulson (1937b, 1950) and Sonnenblick (1950) had already noticed that no cell divisions are associated with the embryonic development of the salivary glands. Therefore the number of salivary gland cells must be determined at least as early as the time of invagination of the placodes. Staining with anti-Forkhead antibody (Weigel et al. 1989) in stage 11 (Panzer et al. 1992) shows that the placodes contain about 180 cells each (H. Saumweber, personal communication). Growth of salivary gland epithelial cells occurs by volumetric increase. The cells of the salivary glands enter the first round of endomitosis as the germ band starts to retract in stage 12 (Smith and Orr-Weaver 1991).

All cells in the tubes formed by the invaginated placodes remain basophilic during subsequent development, with the exception of those that form their external portions. As the labial appendages move medio-orally during stage 13, the openings of the salivary tubes join at the embryonic midline and fuse into a common duct (Fig. 2.31). Prior to the fusion of the openings, the distalmost cells in each of the salivary tubes change their size and shape from large and cuboidal into small and flat in order to form the duct. Thus, three different regions can be distinguished in the developing salivary glands from stage 13 onwards: the common salivary duct, the duct of each salivary gland and the secretory epithelium (Figs. 2.32, 2.39, 7.1). In late embryogenesis the secretory epithelium is located dorsolaterally to the ventral cord (Figs. 2.33, 7.1), ending in the duct at the level of the anterior border of the prothoracic neuromere; the excretory part extends past the suboesophageal ganglion on either side, fusing in front of the ganglion with that of the other side to form the common portion of the duct. From stage 13 onwards, the salivary glands display signs of secretory activity in their secretory portions (see, for example, Fig. 2.30A). The cells have small secretory granules at their apical poles, distinguishable at high magnification with the light microscope, and the lumen of the tubes contains a substance that stains very darkly with methylene blue. A description of the larval salivary glands can be found in Berendes and Ashburner (1978).

7.3
The Midgut

7.3.1.
Origin and Early Development of the Midgut Primordia

Based on various kinds of observations, the midgut has been proposed to be of dual origin (Poulson 1950; Technau and Campos-Ortega 1985). On the one hand, the midgut develops from two endodermal primordia that were located by Poulson (1950), one anterior and the other posterior to the mesodermal anlage in the blastoderm. We shall see below that the posterior midgut primordium is not contiguous with the mesodermal primordium, as Poulson (1950) thought, but rather is separated from the latter by parts of the proctodeal anlage. On the other hand, the anterior-most midgut region, including the gastric caeca, has been found to derive from the stomodeum and is therefore formally of ectodermal origin (Technau and Campos-Ortega 1985).

The conventional view of how the anterior midgut develops is that the cells of the endodermal portion of the anterior midgut primordium invaginate, and/or delaminate (see below) ventrally during stage 7, at the anterior end of the ventral furrow, corresponding to the crossbar of the T whose stem is the ventral furrow itself (see Figs. 2.7, 7.3). Once the ventral furrow has closed, the so-called midgut invagination is pushed anterolaterally, now forming a thin slit. After ingression, all cells of the primordium divide once (first postblastodermal mitosis in stage 8; see Fig. 2.8), slightly later than the mesodermal cells, and lose their epithelial organization, arranging themselves into an amorphic mass located dorsal to the anlage of the cephalic mesoderm. During stage 9, all cells of the midgut primordium divide a second time (second postblastodermal mitosis). Once the stomodeum invaginates during stage 10, the cells of Poulson's anterior midgut primordium attach to its internal end and begin to spread caudalwards (Fig. 2.15D-E), flattening laterally and thereby producing deformations of the yolk sac. At this stage, the anterior midgut primordium has two lobes, with one mass of cells extending on either side of the midline (Figs. 2.18D). Poulson (1950) describes this arrangement as Y-like. Besides this, no other obvious signs of patterning can be distinguished in the anterior midgut primordium, which during the period of its caudal extension consists of rather large, fairly round cells. Progression of the anterior midgut primordium towards caudal is due partially to mitotic divisions of the midgut cells (third postblastodermal mitosis) and partially to passive displacement caused by the continuously ingrowing foregut. Technau and Campos-Ortega (1985) fate-mapped the anteriormost part of the midgut to the fundus of the stomodeum. These cells start out as regular epithelial cells integrated in the stomodeal epithelium. However, during stage 11 they lose their epithelial morphology, as evidenced by their loss of expression of the epithelial cell marker Crumbs (Figs. 2.14, 2.19), and become mesenchymal. Subsequently, cells of the stomodeal fundus mingle with, and can no longer be distinguished from, the remainder of the anterior midgut rudiment. They form most of the vertical stem of the Y-shaped structure.

This view of the development of the anterior midgut from two different regions, the cells in front of the cross-bar of the T of the ventral furrow (Poulson's anterior midgut primordium) and the cells of the fundus of the stomodeum (Technau and Campos-Ortega 1985), has recently been challenged by Foe (1989). According to

Foe, the entire anterior midgut rudiment derives from the cells in mitotic domain $\partial 14\delta$ (corresponding to the stomodeal plate). Furthermore, according to this view, the presumptive anterior midgut cells delaminate from this domain at an early stage, before invagination of the stomodeum has even begun. Thus, $\partial 14\delta$ is peculiar in that the orientation of the mitotic spindle is perpendicular to the surface of the embryo, in such a way that one of the daughter cells of each mitosis within this domain becomes internalized. These internalized cells, according to Foe (1989), represent the entire anterior midgut rudiment.

Due to the lack of early markers for anterior midgut, no definitive answer can presently be given to the question of the origin of the midgut. We consider it very likely that the cells internalized during mitosis of domain $\partial 14\delta$ do indeed contribute to the anterior midgut rudiment. However, several arguments indicate that additional cells, namely those of the so-called stomodeal endoderm (Reuter 1994), remain in domain $\partial 14\delta$ until long after it has invaginated as stomodeum, and only separate from it during stage 11 to join the anterior midgut primordium. Subsets of presumptive midgut cells are indeed labelled by their expression of the gene *asense* (see below). In these preparations, one can directly observe that, as described by Technau and Campos-Ortega (1985), some of these cells dissociate from the stomodeum during stage 11. Finally, the early pattern of expression of the gene *serpent*, which later labels the entire midgut, indicates that at least part of the T of the ventral furrow, as postulated by Poulson (1950), contributes to the anterior midgut rudiment (Reuter 1994). *serpent* expression becomes concentrated in the anterior lip of the crossbar of the T, and in cells lying further anterior that delaminate from domain 8. With respect to the head mesoderm, our view of its origin is discussed in Chapter 3.

The posterior midgut primordium derives from the cells of the posterior pole of the blastoderm, where the anlage has a circular shape (Fig. 7.3). During stage 7, these cells shift their position from the embryonic pole to the horizontal plane, forming a round plate comprising about 150 cells, on which the germ line progenitors lie (Figs. 2.6F-G, 2.7A). Thereafter the cell plate sinks ventrally into the embryo to form a hollow, dorsally open space, called the amnioproctodeal invagination, the posterior rim of which is oval in profile (Figs. 2.9D). Invagination of the posterior tip is due to the isotopical constriction of all of its cells at their apical poles, which thereby change their shape from columnar to cone-like (Costa et al. 1993). As it sinks, the posterior edge of the midgut primordium and the surrounding hindgut primordium are transiently continuous with the tube formed by the ventral furrow, the lumen of which ends in the cavity formed by the posterior midgut plate (Figs. 2.6I, 2.7A-B). However, this connection between the posterior midgut primordium and the tube disappears when the cells of the tube disaggregate to undergo their first mitosis at the beginning of germ band extension (Fig. 2.8). Then, the lateral walls of the amnioproctodeal invagination, of ectodermal origin, approach to meet each other at the dorsal midline, without fusing. Their further behaviour will be considered below, because these cells will later form parts of the hindgut. The segregation of the posterior midgut primordium coincides with the beginning of germ band extension; further development of the midgut is bound up with this phenomenon. One of the consequences of germ band extension is that the hollow open space formed by the posterior midgut primordium is covered dorsally by the expanding germ band (Fig. 2.8), and thereby transformed into a wide flat pocket that moves cephalad as germ band extension progresses (Figs. 2.9, 2.11, 2.12).

Starting from the neighbouring proctodeum, the cells of the posterior midgut primordium perform the second postblastodermal division at the beginning of stage 10 (Figs. 2.13, 7.6). The cells of the dorsal wall of the posterior midgut are pushed apart by the pole cells, which leave the midgut pocket in stage 10. Like the anterior midgut, the posterior midgut primordium, at the onset of caudal expansion, adopts a bilobular shape (Figs. 2.18C, 7.6). Both lobes consist of many small, mitotically active cells (third postblastodermal mitosis) that are joined ventrally by a plate of conspicuously large and basophilic, although less abundant, cells (Figs. 7.3 - 7.6). This ventral plate consists of cells that have segregated from the posterior midgut epithelium before its disaggregation and constitute the interstitial cell progenitors (Tepass and Hartenstein 1995; see below). Obvious morphological differences between dorsal and ventral cells permit easy distinction of the existence of a lumen in the mass of posterior midgut cells.

As it advances caudally, the posterior midgut primordium compresses the yolk sac, just as the anterior primordium does during its expansion. A thin yolk sac process extends up to the tip of the germ band, separating the midgut primordium from the overlying mesoderm (Fig. 7.6). The posterior midgut lobes reach the posterior pole of the extended germ band at the end of stage 11, and bend ventrally (Figs. 2.22, 7.4, 7.6). Germ band shortening then begins, and this serves to bring the posterior midgut primordium into contact with the corresponding lobes of the anterior midgut at about 50% EL. At this stage, most cells of the lobes have

Fig. 7.4. Development of the midgut. **A to D** carry an enhancer trap that highlights the developing anterior (*am*) and posterior (*pm*) midgut primordia. The *arrowheads* in **B-D** indicate the point of fusion of the two primordia. *1-3* refer to the prospective midgut constrictions. *hg* hindgut; *ol* primordium of the optic lobe; *es* ; *st* stomodeum

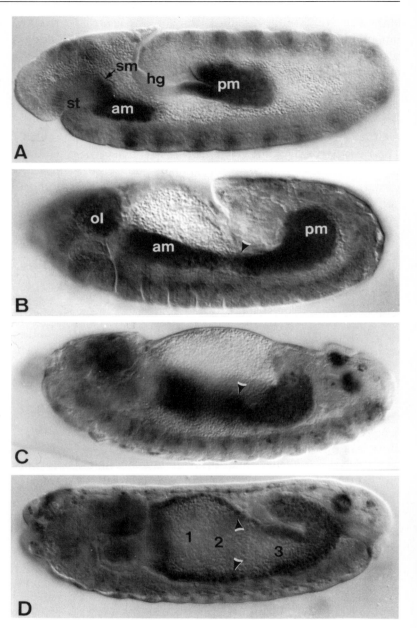

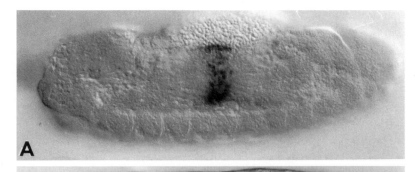

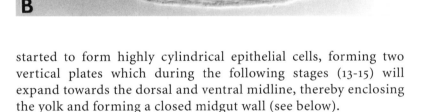

Fig. 7.5. Stage 13 (**A**) and 16 (**B**) embryos stained with an antibody against the Labial protein (provided by Marian Bienz). *1, 2* and *3* indicate the midgut chambers

started to form highly cylindrical epithelial cells, forming two vertical plates which during the following stages (13-15) will expand towards the dorsal and ventral midline, thereby enclosing the yolk and forming a closed midgut wall (see below).

As they elongate and move toward each other, the lobes of the midgut primordia follow the track formed by the adjacent visceral mesoderm. Experimental data show that the visceral mesoderm is instrumental in both migration and epithelial reorganization of the midgut primordia (Reuter and Leptin 1994; Tepass and Hartenstein 1994a).

7.3.2.
Segregation of Interstitial and Adult Midgut Progenitors

Long before the midgut rudiments fuse and form the midgut epithelium, they split up into three separate cell types. The majority of cells, called principal midgut epithelial cells (Tepass and Hartenstein 1995; see Fig. 7.3) undergo the epithelial transition mentioned above, to form the larval midgut. Two minor cell populations, called interstitial progenitor cells, and adult midgut progenitor cells, respectively, become morphologically distinct from the principal midgut cells and initially do not form part of

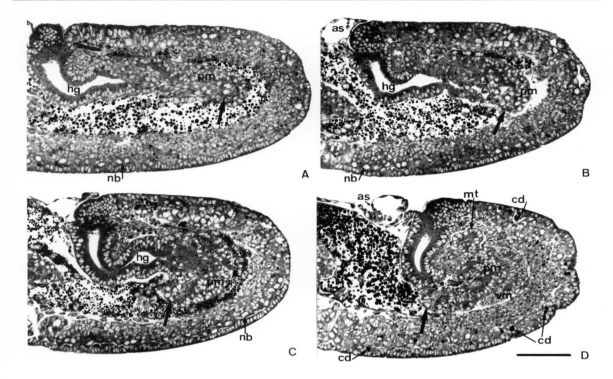

Fig. 7.6. A to D are parasagittal sections of increasingly older embryos to illustrate modifications of the hindgut (*hg*) and posterior midgut (*pm*) accompanying germ band shortening. In **A** the posterior midgut can be seen to consist of a thick dorsal layer of small cells and a thin ventral layer of larger cells, which correspond to the interstitial cell progenitors (ICP, see text for further details), separated by a lumen which becomes clearly visible as shortening progresses (**B**). During shortening the large ventral ICPs lose their association with the small-celled portion, and come to lie medially at the tip of the posterior midgut (large *arrows*). *as* amnioserosa; *cd* cell death; *mt* Malpighian tubules; *nb* neuroblasts; *vm* visceral mesoderm. *Bar* 50 μm

the developing midgut epithelium. The interstitial cell progenitors comprise 30-40 cells, which delaminate from the tip of the posterior midgut rudiment during late stage 10-early 11 and come to lie in the lumen of the posterior midgut primordium. When the midgut primordium loses its epithelial structure and becomes bilobed, the interstitial cell progenitors move postero-ventrally and form a ventral plate of basophilic cells wedged between the lobes of the midgut rudiment (Fig. 7.6). The ICPs undergo the third postblastoderm division later than the remainder of the midgut precursors, to produce 60-80 cells. During germ band retraction, due to the bending of the posterior midgut primordium, the ICPs become situated anterodorsally at the advancing edge (Fig. 7.6).

The interstitial cell progenitors remain distinct for a relatively long time after fusion of anterior and posterior midgut rudiments and after the epithelial reorganization of their cells into the midgut walls (Figs. 2.26C, 2.27D). They can also be specifically stained in an enhancer trap line from stage 11 on (Bier et al. 1989; Skaer 1993), as well as with an anti-Asense antibody in the wild type (Tepass and Hartenstein 1995), allowing their subsequent behaviour to be followed (ref. to Fig. 11.6). During stages 13 and 14, the interstitial cell progenitors are clustered at the inner surface of the developing midgut epithelium (Fig. 11.6). Smith and Orr-Weaver (1991) report on endomitotic replication of these cells at this stage, prior to the onset of endoreplication in the surrounding principal midgut cells. Towards the end of stage 14, the interstitial cell progenitors intercalate into the midgut epithelium. When the midgut becomes constricted into four portions during stage 16 (see below), the interstitial cells all populate the second. They stand out because they are much larger than the remainder of the (principal) midgut cells. In shape and size the interstitial cell progenitors actually resemble pole cells. This similarity explains why Poulson (1950) failed to distinguish between the two cell types and reported that pole cells, and vitellophages, become integrated into the midgut epithelium. However, cell transplantations (Underwood et al. 1980a; Technau and Campos-Ortega 1986b) and antibody staining (Hay et al. 1988a, b; Lasko and Ashburner 1988, 1990) demonstrate that pole cells do not contribute to the midgut. Furthermore, the present observations clearly indicate that vitellophages play no part in midgut formation.

Progenitor cells of the imaginal midgut appear in both anterior and posterior midgut primordia during late stage 11. These cells, which can be readily followed both with enhancer-trap lines and an anti-Asense antibody (Hartenstein and Jan 1992; Hartenstein et al. 1992; Tepass and Hartenstein 1994a), predominantly appear close to the inner surface of the bilobulate midgut primordia (Fig. 7.3D-H). Shortly after they have separated from the larval midgut progenitor cells, they divide synchronously. This division, which occurs later than that of the larval midgut and interstitial progenitor cells, probably corresponds to the third postblastoderm mitosis. Division brings the number of progenitor cells to approximately 80 (anterior midgut rudiment) and 150 (posterior midgut rudiment), respectively. When, during stage 12, the larval progenitors reorganize themselves into epithelial plates that continue to grow into the larval midgut, the adult progenitors remain on the inner (luminal) surface of the midgut. Here they persist until stage 15, scattered rather evenly over the apical midgut surface (Fig. 7.3G-H). Subsequently, they intercalate again into the larval midgut epithelium, a movement which in the embryo is unique to this cell type, such that in the late embryo, the adult progenitor cells occupy a basal position in the midgut epithelium. They are readily distinguishable from the neighbouring larval midgut cells because they remain small and do not flatten (Tepass and Hartenstein 1995). The adult midgut progenitors resume proliferation during the larval period.

7.3.3
Late Stages of Midgut Development

After the fusion of anterior and posterior midgut primordia, the principal midgut cells are converted into two epithelial plates of cylindrical cells. These plates stretch ventrally and dorsally to surround the yolk sac completely (Fig. 7.4). Consequently, the cell shape changes from cylindrical through cuboidal to squamous. It is noteworthy that no further mitotic divisions can be discerned in the midgut epithelium after fusion of the primordia. Thus, all modifications undergone by the midgut thereafter must be due to other factors. During stages 15 and 16, three constrictions (Figs. 2.37, 2.38, 2.39, 7.4, 7.5) appear consecutively in the wall of the midgut, which lead to the formation of the four intestinal convolutions present in the fully developed embryo (Fig. 2.41). In stage 16, four evaginations appear at the base of the anterior-most of these convolutions, the so-called ventriculus, from which the gastric caeca develop (Figs. 7.2, 7.3). Reuter and Scott (1990) have provided a detailed description of the development of midgut convolutions and gastric caeca, to which the interested reader is referred. The order of appearance of these midgut constrictions is invariant and has been used to stage the embryo (Broadie et al. 1992).

7.4
The Hindgut and the Malpighian Tubules

The hindgut is a derivative of the ectoderm (Poulson 1950; Hartenstein et al. 1985; Jürgens 1987; Jürgens and Hartenstein 1993). Its anlage is arranged annularly around the posterior pole of the blastoderm, extending from medioventral to mediodorsal levels, caudal to the mesodermal anlage, between it and the anlage of the posterior midgut (Fig. 7.1). When the posterior midgut anlage sinks inwards, the mediodorsal division of the hindgut anlage comes to form the anterior wall of the amnioproctodeal invagination, still open dorsally, whereas the remaining cells of the ring form its lateral and posterior walls (Figs. 2.8, 2.9, 7.1). As germ band extension begins, the caudal tip of the germ band moves over the midgut cavity and covers its dorsal opening; the hindgut primordium comes to lie on top of its floor due to the anterior displacement of the lateral and posterior walls of the amnioproctodeal cavity. Simultaneously the lateral walls approach each other to join at the embryonic dorsal midline, defining in this way a thin slit (Fig. 2.9). Therefore, due to the movements during germ band extension, the hindgut anlage disappears from the embryonic surface to be included in the midgut pocket, forming the proctodeum (Fig. 2.11).

While the proctodeal opening onto the ectodermal surface is at this time (second half of stage 8) very narrow in the anterolateral direction, where anterior and posterior borders are very near to each other, it extends widely in the mediolateral axis. As germ band extension progresses, the mediolateral extent of the proctodeal opening becomes reduced to a tube of triangular cross-section in stage 11. Within the hindgut primordium, the first postblastodermal mitoses (14th division) are found in stage 8; all cells in both the anteroventral and the posterodorsal walls undergo mitosis. These mitoses contribute to an increase in the size of the hindgut primordium during germ band extension (Figs. 2.11, 2.12, 2.13, 7.1).

At the beginning of stage 10, once the initial, fast phase of germ band extension has been accomplished, the first regional differentiation can be observed in the developing hindgut. This consists of a transverse depression in the dorsal wall which appears at the junction between hindgut and posterior midgut primordia, from which the Malpighian tubules will develop (Fig. 2.15). At this time, the cells of the hindgut go through their second postblastoderm mitosis. Embryonic development of the Malpighian tubules can be subdivided into two phases (Figs. 7.1, 7.7). In the first, the lateral corners of the initial proctodeal depression form blind tubes, one on either side, which will subsequently split to form four different processes, two anterodorsal and two lateroventral (Figs. 2.17, 2.19, 2.28, 7.1, 7.7). All cells in these processes divide (third postblastodermal mitosis), initiating the elongation of the buds. In a second phase, the buds elongate further by means of cell rearrangement and volumetric growth; mitotic divisions are then observed only rarely (Figs. 2.21, 2.22). Janning et al. (1986) have carried out a detailed clonal analysis of Malpighian tubule development and shown that, although the majority of the cells of the tubules divide three times, some cells perform up to five mitotic divisions. Mitoses continue for longer in the prospective anterior pair of tubules (until the end of stage 13) than in the prospective posterior pair; hence, the anterior pair of tubules contains 30 cells more than the posterior pair (Janning et al. 1986). The two ureters form during stage 12 and the two pairs of tubules become linked (Skaer 1993). The ureters may develop either by recruitment of cells from the proctodeum or by fusion of the proximal regions of the Malpighian tubules (Harbecke and Janning 1989). During stage 11, single cells become distinguishable at the distal tip of each of the tubules. Skaer (1989) has presented experimental evidence to show that, although these tip cells do not themselves divide, they are required for division of the cells in the tubules. Hoch et al. (1994) have shown that the tip cells segregate from clusters of equipotent cells, in a process somewhat similar to the segregation of neuroblasts. Tip cells differentiate various neural markers, which are normally expressed in nerve cells. By dye-filling the differentiated tip cells, Hoch et al. (1994) have shown that each cell has a long neck-like process extending into the corresponding tube, and that the cells of the posterior pair establish contact with the hindgut branch of nerve a9. It is, however, not clear whether these contacts are synaptic and whether the tip cells actually have neural functions.

During shortening of the germ band, the hindgut is carried caudalwards to open eventually at the dorsal surface of the posterior embryonic pole (Fig. 2.26). While this process is taking place, the hindgut changes its shape, bending transiently. However, by the end of germ band shortening, the definitive organization of the hindgut has in principle become established and the hindgut forms a loop that extends longitudinally from the proctodeal ope-

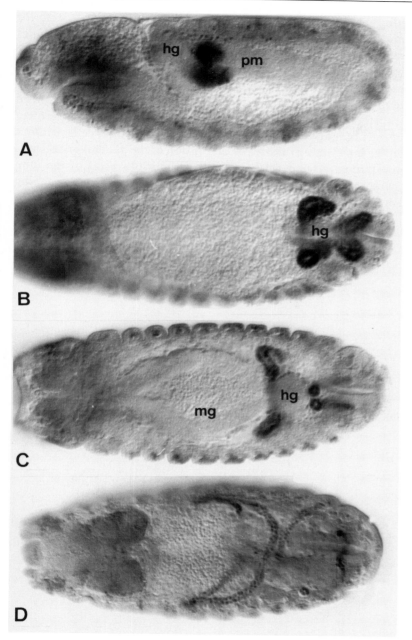

Fig. 7.7. Development of the Malpighian tubules. **A to D** carry an enhancer trap that highlights the developing Malpighian tubules. **A** is a lateral view of a late stage 11 embryo to show two of the four spherical evaginations at the boundary of the hindgut primordium to the posterior midgut rudiment. **B-D** show dorsal views in stages 13, 14 and 16, to illustrate the growth of the four tubules. Refer to text for further details. *hg* hindgut; *mg* midgut; *pm* posterior midgut

ning up to 30% EL; here it bends laterally to join the posterior midgut. It is at this point that the developing Malpighian tubules insert at the hindgut epithelium (Figs. 2.31, 2.32, 2.36). Obviously, the orientation of the hindgut walls has been inverted due to the caudal displacement of the proctodeum during germ band shortening, so that the formerly dorsal wall of the hindgut has been transformed into its ventral wall, and vice versa. Further growth of the hindgut occurs through stretching of hindgut cells, without either intervening mitoses or disruption of the general organization attained after germ band shortening. Both arms of the loop increase considerably in length, the longitudinal arm reaching about 45-50% EL in stage 16 (Figs. 2.38, 2.40, 2.41, 2.43, 7.1).

Chapter 8
Epidermis

The epidermis develops from those cells of the original ectodermal germ layer that remain at the embryo surface after segregation of the neuroblasts and formation of tracheal pits. These cells were called dermoblasts by Poulson (1950). However, we use the denomination epidermoblasts, since, by analogy to the vertebrates, dermis designates cells underlying the epidermis, i.e. cells of the connective tissue and, thus, of mesodermal origin. Clear regional differentiations in cell shape and size can be seen within the ectodermal layer in the epidermal primordium from very early stages on. For example, neurogenic and non-neurogenic divisions of the ectoderm are distinguishable before neuroblast delamination; also the various epidermal placodes (tracheal, salivary) become apparent during stages 10 and 11. Diversification of cells within the developing epidermis becomes even more prominent during later periods of embryogenesis, from germ band shortening onwards, and eventually leads to a rather complex morphological pattern.

Cell diversification within the epidermis is eventually manifested externally in a large variety of pattern elements displayed by the cuticular sheath, which confer morphological identity on the various segments; internally, diversification is manifested in the organization of the cells that have excreted the various pattern elements of the cuticular sheath. Epidermal pattern elements include setae, hairs and sensory organs, which have a characteristic topological distribution (Figs. 8.1-8.3). The cuticular pattern of the first-instar larva has been carefully described by Lohs-Schardin et al. (1979) (see also Hertweck 1931, Kankel et al. 1980, and Dambly-Chaudiére and Ghysen 1986, especially for epidermal sensory organs of the third-instar larva), and their terminology will be used in the present account. In the following the development of the epidermis of the head and the trunk is considered.

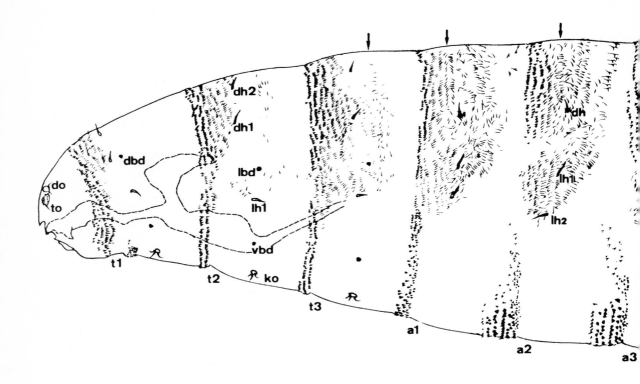

Fig. 8.1. Drawing of the cuticle of a first-instar larva. Care was taken to draw every single cuticular structure of this particular larva. *t1-a8* and *te* (telson) label the segments. Segments are subdivided at the level of the ventral sensory hair (*lh2*) into a dorsal and a ventral half, which exhibit different patterns. Dorsally there is a complex pattern of tiny hairs in all segments, with the exception of the prothorax which is devoid of them.

Dorsal hairs occur in three types, which are distributed in three regions according to a well defined pattern (labelled *a*, *b* and *c* in *a6*). The hair type anteriorly (*a*) is small and broad, the intermediate (*b*) is long and thin, and the posterior type is broad and larger than the anterior type. Irrespective of the type, dorsal hairs are oriented towards posterior within the anterior part of each segment, and towards anterior within

the posterior part. The *arrows* dorsally point to the plane of polarity inversion in each of the segments *t3* to *a7*. Notice that the plane of mirror-image symmetry is in the middle of *t3* but it becomes displaced to more anterior levels in the more posterior segments, until in *a8* the orientation of dorsal hairs is predominantly towards anterior. Ventral setae, or denticles, form prominent belts which vary slightly in arrangement from segment

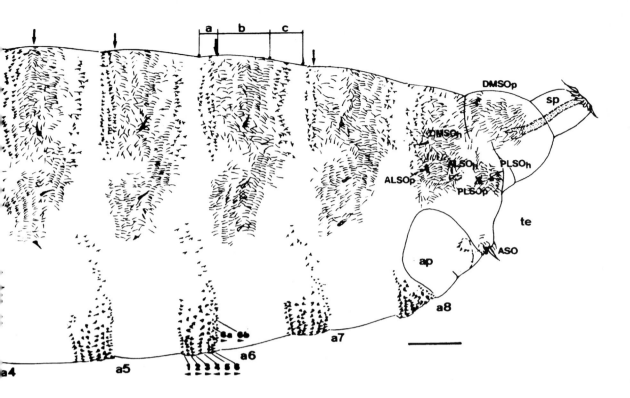

to segment. The muscle attachment sites (apodemes) lie between the first and the second denticle row. Denticles are also polarized as indicated in *a6*. Additionally, in the dorsal half of abdominal segments there are three long sensory hairs (labelled *dh*, dorsal sensory hair, *lh*, lateral sensory hair and *vh*, ventral sensory hair), and within dorsal and ventral halves of thoracic segments three long sensory hairs (*dh2, dh1* and *lh*), two large

basiconical sensilla called black dots (*lbd* and *vbd* lateral and ventral black dots) and Keilin's organs (*ko*). Notice that prothorax has a dorsal black dot (*dbd*) instead of a lateral one. The dorsal half of *a8* and the entire telson (derived from *a9* and *a10*) are highly modified due to the formation of the terminal organs, e.g. anal pads (*ap*) and posterior spiracles (*sp*), and to the appearance of sensory cones (*ALSOh, ALSOp* hair and peg sensilla

of anterior dorsolateral sensory cone; *ASO* caudal or anal sensory cone; *DMSOh, DMSOp* hair and peg sensilla of dorsomedial sensory cone; *PLSOh, PLSOp* hair and peg sensilla of posterior dorsolateral sensory cone). *do* and *to* are the dorsal and terminal organs of the antenno-maxillary complex. *Bar* 50 μm

8.1
The Pattern of Early Ectodermal Mitoses

Mitotic activity within the ectoderm is organized according to a characteristic spatio-temporal pattern, that shows little or no variation. Since the pattern of ectodermal mitoses is extensively discussed in Chapter 14, we shall deal here only with the essentials of this pattern. Characteristically, mitotic activity starts in the ectoderm from several different mitotic domains. Three consecutive mitotic cycles, that involve all ectodermal cells, can be unequivocally observed. This means that all cells of the ectoderm, excluding the neuroblasts, divide at least three times. Additionally, a certain number of epidermal cell progenitors divide two more times, and the progenitors of epidermal sensilla divide even more frequently.

8.2
Morphogenesis of the Epidermal Primordium

Preceding and during germ band elongation, the cells of the ectodermal layer and, after neuroblast segregation, those of the epidermal primordium exhibit conspicuous regional differences. A careful description of these cellular differences during morphogenetic movements has recently been provided by Martinez-Arias (1993), mainly based on stainings with an antibody against spectrin (Pesacreta et al. 1989), and the following mainly relies on this description (see also Tepass and Hartenstein 1994b).

At the onset of germ band extension, in stage 7, differences can be seen in the arrangement of ectodermal cells within ventral and dorsal regions. These differences arise before the major modifications in the shape and size of the cells that take place during stage 8, in the rapid phase of germ band extension, when the ventral cells enlarge relative to the dorsal cells. This process delimits the mediolateral extent of the neurogenic region, from which neuroblasts will start to delaminate at the end of stage 8 (Hartenstein and Campos-Ortega 1984). Such differences in cell size between dorsal and ventral regions persist as long as neuroblasts delaminate, throughout stages 9-11 (Fig. 11.1). However, once the segregation of the S1 and S2 neuroblasts is complete, only groups of large cells can be observed within the ventral region, probably corresponding to those areas from which SIII and subsequent subpopulations of neuroblasts (Doe 1992) will segregate. At the onset of sto-

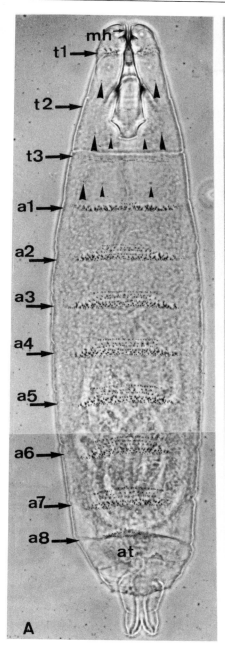

Fig. 8.2. Cuticular sheath of the first instar larva. **A** shows a ventral plane of focus, **B** a lateral plane. The various segments are bounded anteriorly by a conspicuous denticle belt, the shape of which confers on the segments a characteristic appearance that enables most of the segments to be distinguished from each other. *t1, t2, t3, a1, a3-a7* and *a8* can be so distinguished. In addition, segment identification is facilitated by a characteristic pattern of epidermal sensory organs. Some of these sensilla, the Keilin's organs (small *arrowheads* in metathorax and mesothorax), and the ventral and dorsal black dots (large *arrowheads* in metathorax, mesothorax and prothorax), can be clearly seen at this low magnification. Abbreviations. *amx* antenno-maxillary complex; *at* anal tuft; *mh* mouth hook; *ps* posterior spiracle. *Bar* 100 μm

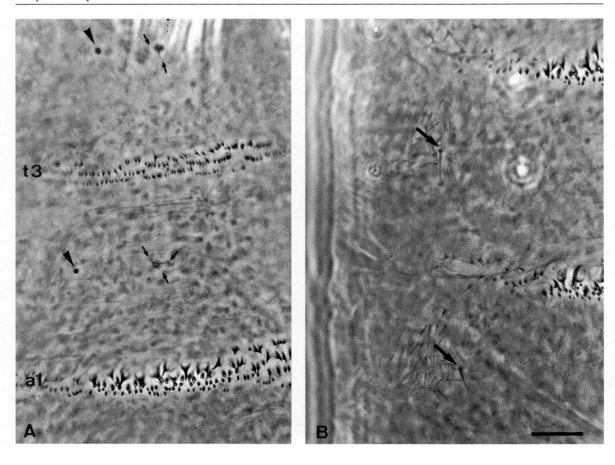

Fig. 8.3. A shows part of the denticle belts in *t3* and *a1*. In *t3* there are two to three rows of small denticles, whereas in *a1* three anterior rows of large and two of small denticles are present. In addition the Keilin's organ can be seen, small *arrows* point to the sensory hairs and the lateral black dot (*arrowhead*) of mesothorax (top) and metathorax (centre). **B** shows the lateral sensory hair (*arrowheads*) in two consecutive abdominal segments. *Bar* 20 μm

modeal invagination, in stage 10, the dorsalmost cells of the epidermal primordium - those in contact with the amnioserosa - enlarge and become arranged in a palisade (Fig. 2.16C). This organization persists during the entire period of extended germ band and, at dorsal closure (see Fig. 5.1F-G), these cells meet their homologues from the other side.

A number of important differentiations of the developing epidermal sheath and its cells become apparent during stages 10-14. They are related to the subdivision of the germ band into metameres and include the formation of the parasegmental and segmental furrows as well as the appearance of tracheal placodes and tracheal pits; these are discussed in Chapters 17 and 12, respectively.

8.3
Cell Differentiation in the Epidermis

After completion of the main period of mitotic activity, at the end of germ band shortening, the embryo is surrounded by the primordial epidermal sheath and by the amnioserosa. The epidermis exhibits obvious regional differences, with cuboidal cells in laterodorsal regions and thin, elongated cells ventrally, in the neighbourhood of the ventral cord (Fig. 11.1F). This is also the time at which developing sensory organs become apparent. The dorsolateral epidermal cells then become more basophilic and larger than the remaining cells, subsequently flattening in order to accomplish dorsal closure of the embryo. Once closure is complete, the apical epidermal cell surface changes depending on the location of the epidermal cells considered, i.e. cytoplasmic protrusions are formed which reflect the variety of cuticular differentiations characteristic of the different epidermal regions. Secretion of the cuticular sheath is initiated in stage 16 (for reviews on cuticle formation see Poodry 1980, and Martinez-Arias 1993).

The cuticle pattern of the first instar larva is illustrated in Figs. 8.1 - 8.4. Apart from epidermal sensilla, to be discussed in the next chapter, the cuticle shows distinctive features in thoracic and abdominal segments (Lohs-Schardin et al. 1979). Ventrally the cuticle of thoracic segments shows a belt of small denticles at the anterior border of each segment, the prothoracic belt being broader than that of the meso- and metathoracic segments (4-5 rows in the former vs. 2-3 in the latter). Dorsally the prothoracic cuticle is naked; in *t2-t3* there are several rows of thin hairs at dorsal levels, the anterior rows being thicker and the posterior ones longer; the

Fig. 8.4. A shows the differences between the denticle belt of *a8* and that of *a7*. In *a7*, as in *a2-a6*, the denticle belt comprises roughly seven rows of denticles; the denticles in the five anterior rows are larger than those in the two posterior rows; the latter two are not continuous over the ventral midline, where generally only one row is visible. In *a8* only the five rows of large denticles are present, which correspond to the five anterior rows of larger denticles in *a2-a7*. Denticles exhibit a specific polarity, whereby the denticles of the first and fourth rows point anteriorly, and the remainder posteriorly. Further differences between *a8* and the other abdominal segments concern the shape of the belts, which is trapezoidal in *a2-a7* and rectangular in *a8*. The anal plate (*ap*) is located caudal to the *a8* belt. It consists of two laterally symmetrical halves, the anal pads, which are in fact markedly convex, each exhibiting a central depression, and separated from each other by the anal slit (*ans*). **B** shows the posterior spiracles, with a crown of large radiating hairs and the 'Filzkörper' (*fk*). *Bar* 20 µm

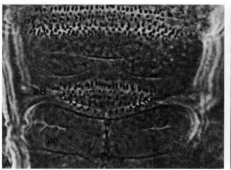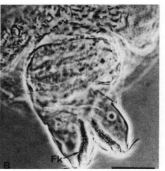

former are continuous with the ventral setae. Abdominal segments carry more complex, though similarly organized, denticle belts, that permit one unequivocally to distinguish abdominal from thoracic segments. Furthermore, shape of denticle belts and distribution of setae allow distinction of abdominal segments *a1* and *a8* (Figs. 8.1 - 8.4) from the remaining abdominal segments, whereas abdominal segments *a2-a7* cannot be reliably distinguished from each other. The denticle belt of *a1* is narrower (four rows), all its denticles being oriented posteriorly, continuous with the dorsal hairs. In abdominal segments *a2-a7* belts consist of seven rows of denticles, of which the five anterior rows are larger than the posterior two rows. Furthermore, the posterior rows exhibit interruptions in their course, varying in extent between segments. Denticles of abdominal segments exhibit a clear polarity in that the first and fourth rows are oriented anteriorly and all others posteriorly. Finally the denticle belt in *a8* only contains five rows of setae, which on the basis of polarity and size seem to correspond to the five anterior rows of larger setae of abdominal segments *a2-a7*. The shape of denticle belts is roughly trapezoidal in abdominal segments *a2-a7* and rectangular in abdominal segment *a8*. Dorsally the cuticle shows irregular rows of thin hairs, similar to those of thoracic segments; a single dorsal row of setae is present in the posterior half of segments *a2-a7*.

8.4
The Oenocytes

Bodenstein (1950) describes the oenocytes as bilateral, spherical clusters of conspicuously large cells located beneath the lateral epidermis of abdominal segments. An ectodermal origin for larval

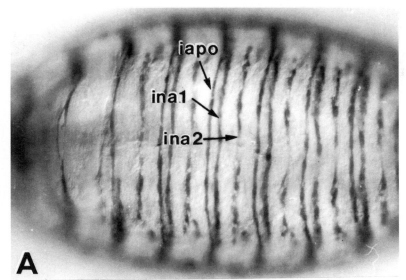

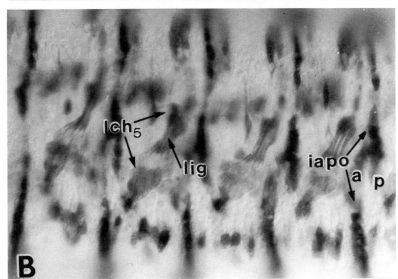

Fig. 8.5. The apodemes. Embryos in **A and B** carry a molecular construct comprising part of the β3 tubulin promoter fused to *lacz* that highlights, among other structures, the apodemes (slide kindly provided by D. Buttgereit and R. Renkawitz-Pohl). **A** is a ventral view of a stage 16 embryo; intersegmental (*iapo*) and intrasegmental (*ina1* and *ina2*) apodemes are clearly distinguishable. **B** shows a lateral view of a stage 17 embryo. Note that the anterior and posterior halves (*a* and *p*) of the *iapo* are out of register. Other abbreviations. *lch5* pentascolopidial chordotonal organ; *lig* ligament cell

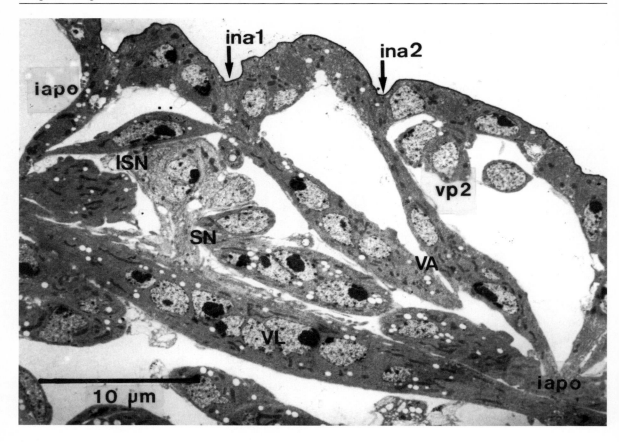

Fig. 8.6. Electron micrograph to illustrate the apodemes of an abdominal segment. There is one intersegmental apodeme (*iapo*) per segment, at which ventral longitudinals and ventral obliques insert, and two intrasegmental apodemes per segment, one anterior (*ina1*), on which the ventral acute muscle (*VA*) inserts, and one posterior (*ina2*) for the insertion of the ventral acute muscles 2-3 (*VA2-3*). *vp* designates the cells of a ventral papilla sensillum, *SN* and *ISN* the segmental and intersegmental nerves. *Bar* 10 μm

oenocytes has been proposed by Lawrence and Johnston (1986), based on a clonal analysis. *LacZ* expression in particular enhancer trap lines (see Figs. 5.5, 8.8F) permits one to follow the origin of the oenocytes back to the posterolateral epidermis within *a1-a7* of the stage 11 embryo, from which groups of 4-7, generally 6, cells segregate (Hartenstein and Jan 1992).

8.5
The Imaginal Discs and Histoblasts

The adult epidermis develops from the imaginal discs and the histoblasts, groups of epidermal cells that become obvious towards the end of embryogenesis. The embryonic origin of some of the imaginal discs and histoblasts has been studied recently using a variety of markers (refer to Bate and Martinez-Arias 1991; Cohen et al. 1991, 1993; Hartenstein and Jan 1992; Cohen 1993, for detailed descriptions). The primordia of the imaginal discs invaginate during stages 15-17 (Figs. 8.7, 8.8) from the epidermis into the interior of the embryo, to form small cell pouches connected with the outside, whereas the histoblasts remain as specialized cell nests integrated in the abdominal epidermis.

The ventral thoracic, or leg, discs develop in association with Keilin's organs, sensory organs comprising five neurons (Dambly-Chaudiére and Ghysen 1986; Ghysen et al. 1986; Hartenstein and Campos-Ortega 1986; Tix et al. 1989a) and about 15 accessory cells each (Bodmer et al. 1989). All of these cells appear to invaginate with the leg disc, the cells of which can be stained with a variety of antibodies (Cohen 1993). Tix et al. (1989a) demonstrated that the sensory neurons of Keilin's organs persist during larval life, being situated in the centre of the invaginated leg disc (Jan et al. 1985). Approximately 10 additional cells surround the 15-20 cells of each of Keilin's organs, forming the leg discs. The primordia of the dorsal thoracic, wing and haltere discs, become detectable in stage 13-14, as pouches of 24 and 12 cells, respectively (Bate and Martinez-Arias 1991), close to the posterior boundary of the second and third thoracic segments, in close proximity to the ventral, leg discs (Bate et al. 1991; Cohen 1993). Anti-Vestigial and anti-Distal-less antibody staining, as well as enhancer trap lines, show that the wing and haltere disc cells originate in association with the leg discs in stage 12, but become displaced dorsally prior to invagination in stage 15 (Cohen et al. 1993; Cohen 1993). The eye-antenna disc develops from the posteriormost extension of the dorsal

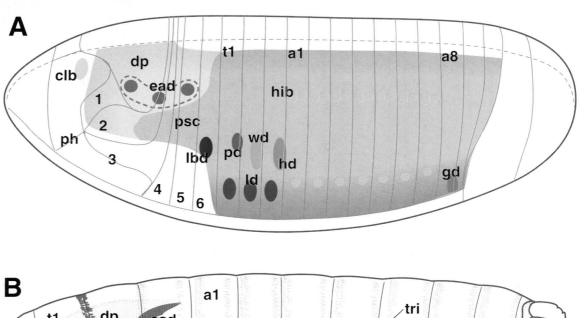

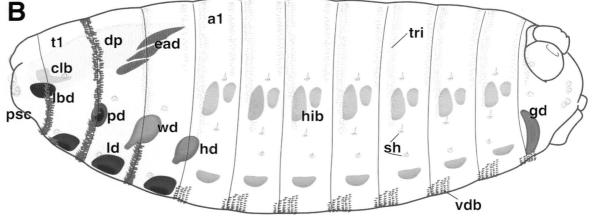

Fig. 8.7. A shows the anlage of the larval epidermis in the blastoderm (shaded blue). The major part corresponds to the ventral neurogenic region. The primordia of the adult epidermis are represented in the fate map as differently coloured ovals. The drawing of the stage 17 embryo (**B**) shows the main cuticle specializations formed by the larval cuticle. the ventral denticle belts (*vdb*), the dorsal trichomes (*tri*) and sensory structures (*sh*). Most of the larval head of the *Drosophila* involutes to form the rostral part of the foregut comprising the pharynx (*ph*) and the atrium, and the dorsal pouch (*dp*, shaded). Only small parts of the head segments, which bear the sensory antennomaxillary complex, are exposed to the outside at the anterior tip of the larva (pseudocephalon; *psc*). By early stage 17, some of these primordia have invaginated as imaginal discs and remain connected to the larval epidermis only by a thin peripodial stalk (leg discs *ld*; wing disc *wd*; haltere disc *hd*). Some other primordia (genital disc *gd*; labial disc *lbd*; eye-antenna disc *ead*) invaginate late in embryogenesis, shortly before hatching. The primordia of the abdomen (abdominal histoblasts *hib*), prothorax (*pd*), and labrum (clypeolabral disc *cld*) remain integrated in the larval epidermis. Modified from Hartenstein (1993), with permission of Cold Spring Harbor Laboratory Press

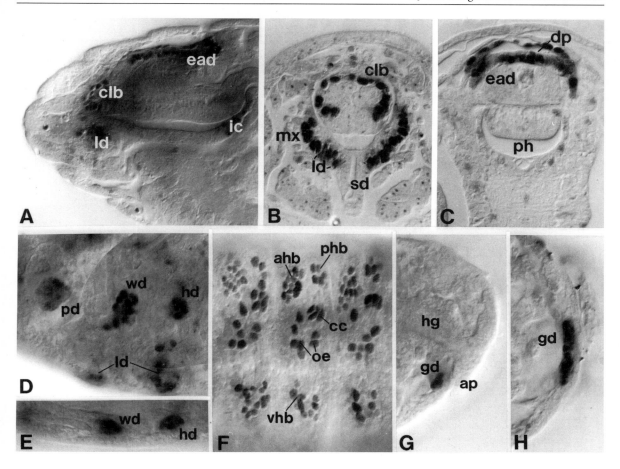

Fig. 8.8. Imaginal discs and histoblasts, as revealed by *lacZ* expression in enhancer trap line *l(2)4B7*. **A** shows a midsagittal view of a whole-mount stage 17 embryo, **B** and **C** two cross sections through the anterior part of an embyros of the same age, to illustrate the location of the eye-antenna disc (*ead*), which grows out of the dorsal pouch (*dp*), the labial and the clypeolabral discs (*ld* and *clb*). **D** shows a stage 14 embryo in lateral view. Label becomes detecta-ble at this stage in the histoblast nests. **E** shows a stage 16 embryo in dorsal view (anterior is to the left). The wing disc and haltere disc have invaginated. **F** shows abdominal histoblasts of a stage 17 embryo (*ahb* anterior dorsal histoblast; *phb* posterior dorsal histoblast; *vhb* ventral histoblast), wing disc (*wg*), haltere disc (*hd*) and leg disc (*ld*). Also oenocytes (*oe*) and cap cells of chordotonal organs (*cc*) are labeled. In **G-H**, the tail of a stage 15 embryo with *lacZ* expression in the genital disc (*gd*) is shown. Throughout late embryogenesis, the genital disc is an unpaired, transversely elongated cluster of cells. It is located in the epidermis between the anal pads (*ap*) and the ventral denticle belt of *a8*. *hg* hindgut

Fig. 8.9. Development of the amnio-serosa. **A to B** carry an enhancer trap that highlights the cells of the amnioserosa. **A** is a lateral view of an early stage 11 embryo. The amnio-serosa is folded between dorsal and ventral halves of the segmented germ band. After dorsal closure (B), the cells of the amnioserosa become internalized, and are transiently associated with the heart before they die

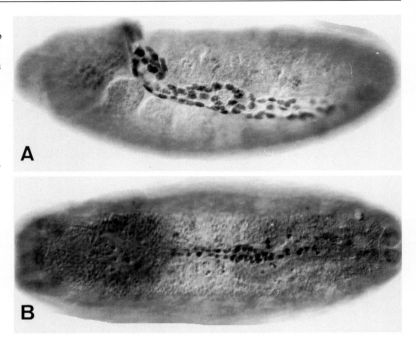

pouch in stage 17 (Fig. 8.8A, C). The genital disc appears as an unpaired cluster of 12-15 cells in the space between the denticle belt of *a8* and the anal pads (Fig. 8.8G-H). Three histoblast nests can be distinguished in each segment by stage 14; two are located dorsally, one anterior to the other at the level of the longitudinal tracheal trunk, and the third is ventral (Fig. 8.8F). The number of *lacZ*-expressing cells within histoblast nests increases during late stages of embryogenesis, from 3-4 cells in stage 14 up to 10-14 cells per nest in stage 17, which is presumably due to recruitment of new *lacZ*-expressing cells rather than to cell divisions.

Chapter 9
Peripheral Nervous System

Sensory Organs and Peripheral Nerves

As in other animals, the peripheral nervous system (PNS) in *Drosophila* has a complex organization. The major part of the embryonic PNS comprises the different types of sensory organs, which are located in or at the epidermis, and their axons. The axons of the sensory receptor cells course through nerves together with the axons of motoneurons which emerge from the developing CNS to innervate the musculature. In addition to sensory organs, and sensory and motor axons, the PNS comprises all the cells of the so-called stomatogastric nervous system. We shall deal with these three parts of the PNS in turn.

9.1
Sensory Organs

The first description of the sensory organs (sensilla) of the *Drosophila* larva was provided by Hertweck (1931). The descriptions by Lohs-Schardin et al. (1979) for the first-instar larva, and Kankel et al. (1980) for the third-instar larva are very similar to that of Hertweck. Indeed, the introduction of novel techniques of investigation, particularly antibodies and enhancer traps, as well as transmission electron microscopy of complete series of sections, has recently allowed the conclusion that the pattern of sensilla of the third-instar larvae is in fact identical to that of the fully developed embryo, and that this pattern basically corresponds to that described by Hertweck (1931). Nonetheless, several new sensilla that had not been seen by previous investigators have been identified recently (Hartenstein and Campos-Ortega 1986; Dambly-Chaudière and Ghysen 1986; Bodmer and Jan 1987; Hartenstein 1988; Jürgens et al. 1986; Schmidt-Ott et al. 1994).

9.1.1.
Types of Sensilla and Sensory Neurons in the Thorax and Abdomen

Two major types of sensillum can be distinguished in the fully developed embryo - external sensory organs, located within the epidermis, and chordotonal organs, located internally in the body wall or close to muscles. The development of the various sensilla is schematically shown in Figs. 9.1 and 9.2, and the complement of sensory organs in the first-instar larva is illustrated in Figs. 9.3-9.7. Sensilla are built of several different cell types (Fig. 9.1). External sensilla contain one or more bipolar neurons, enclosed by three accessory cells. The cell bodies of bipolar neurons lie beneath the epidermis; their dendrites penetrate the epidermis and terminate in the cuticle. Accessory cells are modified epidermal cells which form concentric sheaths around the dendrites and build the stimulus receptor structure. The outer accessory cells form the socket (tormogen cell) and the shaft (trichogen cell) of the sensory organ. The inner cell (thecogen cell) secretes a cuticle-like matrix around the tip of the dendrite(s), called the dendritic cap.

Most external sensilla are mechanoreceptors; these include hairs (trichoid sensilla) and an ill-defined type called a "papilla sensillum" (Ghysen et al. 1986) or "campaniform sensillum" (Campos-Ortega and Hartenstein 1985; Hartenstein 1988). Mechanoreceptors typically possess a single sensory neuron, although one hair and one papilla sensillum of each abdominal segment are innervated by three neurons each. Another type of sensillum which occurs only in the thorax and the telson has been called a basiconical sensillum, characterized by a club-shaped process in an epidermal depression. Basiconical sensilla are innervated by three sensory neurons and, based on ultrastructural observations, are probably hygroreceptors (Hartenstein 1988). Chordotonal organs are stretch receptors attached to the basal (inner) surface of the epidermis. In the embryo, chordotonal organs are formed by one, three or five identical units called scolopidia. Each scolopidium, like the external sensillum, has one bipolar neuron and three accessory cells. Unlike those of external sensilla, accessory cells of chordotonal organs are not organized as sheaths around the sensory neurons, but form ligamentous structures by means of which the chordotonal organs insert at the epidermis.

In addition to the sensilla, which comprise bipolar neurons and accessory cells, several different types of isolated sensory neurons, i.e. devoid of accessory cells, are present internally in the embryo. These cells are also assumed to function as sensory organs (Bodmer and Jan 1987). The neurons possess more than

**External Sensilla
single neuron**

**External Sensilla
multiple neurons**

**Multidendritic
Neurons**

Chordotonal

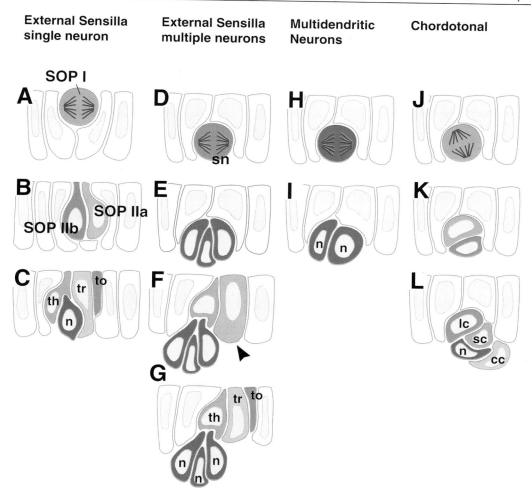

Fig. 9.1. Early morphogenesis of the four types of sensory organs found in the *Drosophila* body segments. The four columns represent different sensory organ types; for each type, several developmental stages are shown from top to bottom. Each panel shows a section of the ectodermal layer (apical surface faces upward). Progenitor cells and their progeny are indicated in different colours. Primary SOPs (SOP I) of singly innervated external sensilla (**A**) are integrated in the epidermal layer. They divide to form secondary SOPs (SOP IIa; SOP IIb). **B** Each of these cells then divides once more, SOP IIa giving rise to tormogen and trichogen cells, SOP IIb to neuron and thecogen cell. Subsequently, the sensory neuron (*n*) delaminates (**C**). The accessory cells (*to* tormogen cell; *tr* trichogen cell; *th* thecogen cell) remain in the epidermis, forming the dendritic sheath and stimulus-receiving structure of the sensillum. SOPs of the other types of sensory organs (**D, H, J**) delaminate prior to division. In the case of the multiply innervated external sensil-la, the SOP (*sn* in panel **D**) most probably gives rise to only the neurons and the thecogen cell (**E, F**), whereas the tormogen and trichogen cell appear to be of a different origin (**G**, arrow in **F**). SOPs of multidendritic neurons (**H, I**) delaminate and divide between one and three times to produce small clusters of neurons. SOPs of chordotonal organs (**J**) divide at least twice with a perpendicularly oriented spindle to form neurons (*n*) and accessory cells (*lc* ligament cell; *sc* scolopale cell; *cc* cap cell)

Fig.9.2 Morphogenesis of the main types of sensillum (**A to C** singly innervated trichoid sensilla; **D to F** multiply innervated basiconical sensilla; **G, H** chordotonal organs; after Hartenstein 1988). The left column (**A, D, G**) shows sensilla at the onset of differentiation (early stage 13); the middle column (**B, E**) represents late embryonic stages (17), and the right column (**C, F, H**) represents the first-instar larva. Sensory neurons and accessory cells are shown in different colours (colour key at bottom). Differentiation of the external sensillum (**A, D**) starts with the delamination of the sensory neuron(s). Apically, sensory neurons form dendrites (dashed magenta lines) which indent the surface of the overlying thecogen cell (violet). Tormogen cells (dark blue) and trichogen cell (light blue) start to form mesaxon-like processes around the thecogen cell (*arrowheads*). At late embryonic stages (**B, E**), the trichogen cell forms the sensillum shaft (*sf*), the tormogen cell forms a socket (*so*; in trichoid sensilla) or the lining of the depression (*de*) in which the shaft is located (in basiconical sensilla). Sensory neurons and cell bodies of the accessory cells sink deep into the embryo (**C, F**). In early chordotonal organs (**G**), both neurons and accessory cells are located beneath the epidermal layer. Sensory dendrites indent the scolopale which is considered homologous to the thecogen cells of external sensilla. The remaining accessory cells, however, do not form concentric sheaths around the dendrites like tormogen and trichogen cells do; instead, they become positioned on either side of the neuron/scolopale cell cluster and form ligaments which connect the neurons/scolopale cells to the epidermis

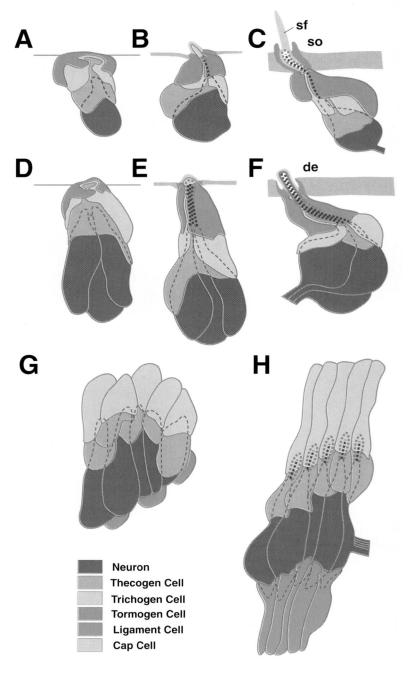

Neuron
Thecogen Cell
Trichogen Cell
Tormogen Cell
Ligament Cell
Cap Cell

one dendrite, and are therefore referred to as multidendritic neurons. Zawarzin (1912) distinguished these neurons as "type II" from the "type I" bipolar neurons. Multidendritic neurons occur as clusters of up to five cells attached to the basal surface of the epidermis or internal organs, such as tracheae, peripheral nerves, and muscles. Three types of multidendritic neurons can be distinguished (Fig. 9.3): those with arborizations underneath the epidermis (*da* neurons), those innervating the tracheal branches (*td* neurons), and those with two opposing dendrites (*bd* neurons) (Bodmer and Ian 1987).

9.1.2.
Morphogenesis and Cytodifferentiation of Sensilla

As a rule, neurons and accessory cells of a given sensillum are derivatives of the same sensory organ progenitor cell (SOP; see reviews on development of sensory organs in insects by Bate 1978; Jan and Jan 1993). In *Drosophila*, formation and proliferation of SOPs has been characterized for the adult microchaetes (Lees and Waddington 1942; Hartenstein and Posakony 1989), as well as for some of the embryonic sensilla (Bodmer et al. 1989; Younossi-Hartenstein et al. 1996).

The progenitor cells of external organs innervated by a single neuron undergo two divisions. The first division yields two so-called secondary SOPs (*SOPIIa* and *SOPIIb* in Fig. 9.1). The *SOPIIa* cell gives rise to two daughter cells which will differentiate as the outer accessory cells (tormogen and trichogen cells). *SOPIIb* always divides slightly later than its sibling; it gives rise to the neuron and thecogen cell. It should be emphasized that the progenitors and their daughter cells are initially an integral part of the epidermal epithelium. However, the prospective neuron will delaminate after mitosis from the epidermal anlage. It differentiates an apical dendrite which extends, forming a depression in the membrane of the overlying cell, which will become the thecogen cell. At the same time, the adjacent trichogen and tormogen cells wrap mesaxon-like processes around the thecogen cell, such that three concentric sheaths are formed around the dendrite. During this process, the cell bodies of the accessory cells shift basally. Both tormogen and trichogen cells retain contact with the apical surface and the apical portions of these cells form the socket and shaft process of the sensory apparatus. The thecogen cell finally loses contact with the epidermal surface, and its former apical membrane comes to face the dendrite and secretes the dendritic cap.

The progenitors of chordotonal organs and multidendritic neurons delaminate from the epidermal anlage into the interior of the embryo (Fig. 9.1). This observation is in line with the fact that, in the mature embryo, both chordotonal organs and multidendritic neurons are subepidermal structures. Proliferation of the SOPs of chordotonal organs and multidendritic neurons begins after they have delaminated and become attached to the basal surface of the ectoderm. Following delamination, the progenitors of multidendritic neurons undergo one to three further divisions to form clusters of up to five neurons. This has been directly followed for the groups of dorsal (*dda A-E*) and ventral (*vda A-D*, see Fig. 9.3) multidendritic neurons (Younossi-Hartenstein et al. 1995b). However, there are exceptions to this, since some of the multidendritic neurons actually derive from SOPs of external sensilla (Brewster and Bodmer 1995). Bodmer et al. (1989) have proposed that the division of the progenitor which gives rise to the four cells of a scolopidium follows a stem cell mode. Thus, according to this view of chordotonal organ development, the progenitor cells of the scolopidia undergo three consecutive mitoses; the ligament cell develops from the first, the cap cell from the second, and the neuron and scolopale cell from the third mitosis.

Progenitors of the tri- and pentascolopidial chordotonal organs have a highly complex proliferation pattern which has not been fully reconstructed. The thoracic *dch3* organs and the posterior three scolopidia of the abdominal *lch5* are serially homologous structures, each of which can be traced back to a single progenitor that delaminates during early stage 10 (see below, 9.1.1.4). This cell, following delamination, divides further into three progenitor cells, each of which will in turn produce the four cells of a scolopidium.

The pattern of proliferation of the progenitors of multiply innervated external sensilla is controversial. In line with older findings in other insects (reviewed in Lawrence 1966; Bate 1978), it was proposed by Bodmer et al. (1989) that the neuronal population arises from additional divisions of the the *SOPIIb* daughter cell. More recent results, however, suggest a different model, according to which the neurons and accessory cells of multiply innervated sensilla have a dual origin (Younossi-Hartenstein and Hartenstein 1996). According to this view, the cells which have been identified as the primary progenitors of multiply innervated sensilla (called *sn* in Fig. 9.1) delaminate during early development and, over the course of two divisions, produce the three neurons and, possibly, the thecogen cell. The outer accessory cells (tormogen and trichogen cell) are apparently derived separately from the overlying epidermal anlage.

9.1.3.
Pattern of Sensory Organs in Thorax and Abdomen

In both thoracic and abdominal segments, a ventral, a lateral and a dorsal group of sensory organs can be readily identified (Figs. 9.3-9.5). Sensory axons deriving from the dorsal and lateral groups of sensilla form an anterior fascicle, called the intersegmental nerve, whereas those from the ventral group course in a posterior fascicle, or segmental nerve (according to Thomas et al. 1984). Each of the three groups is characterized by its relationship to the somatic musculature. Thus, the ventral group lies outside the ventrolateral and ventral muscles, beneath the ventral insertion of the lateral transverse muscles; the lateral group is located in the niche formed between the lateral transverse muscles and the lateral epidermis; finally, the dorsal group lies above the dorsal insertion of the lateral transverse muscles, outside the dorsal intersegmental muscles (Figs. 9.3, 9.18).

9.1.3.1.
Abdominal Segments a1-a7

Ventral group. The ventral group consists of six papilla sensilla (*vp1, vp2, vp3, vp4, vp4a, vp5*), two monoscolopidial chordotonal organs (*vch1* and *vch2*), and several clusters of multidendritic neurons (Figs. 9.3, 9.4; refer to Table 9.1). The ventralmost papilla sensilla, *vp1* and *vp2*, are located in the paraneural space at the posterior border of the corresponding segmental nerve. The neurons of *vp3, vch1* and *vch2,* and *vdap* form a dense cluster located posteriorly in each segment at the level of the ventral oblique musculature. The axons of *vp1-3, vch1* and *vch2,* and *vdap,* together with the motor axons innervating the ventral oblique muscle fibres, form the branch "*c*" of the segmental nerve. A side branch of this axon bundle carries sensory axons of the multidendritic neurons *vdaA-D* and *vbp,* which together form a cluster in the paraneural space anterior to *vp1* and *vp2.* The cells of *vp4, vp4a,* and *vdaa* are located in close proximity to the intersegmental nerve within the anterior half of each segment. Their axons join branch "*b*" of the segmental nerve which also carries motor axons to several ventral longitudinal and oblique muscle fibres. Finally, there is the group of three sensory neurons of *vp5* and the multidendritic *vdap* neuron, which are the lateralmost representatives of the ventral group of sensilla and which are located at the lateral boundary of the ventral musculature. Their axons join branch "*a*" of the segmental nerve.

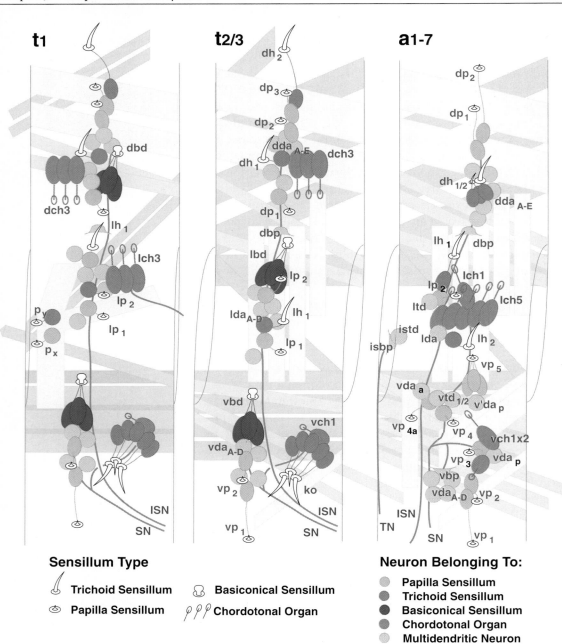

Sensillum Type

⌇ Trichoid Sensillum

⊚ Papilla Sensillum

♘ Basiconical Sensillum

ρρρ Chordotonal Organ

Neuron Belonging To:

○ Papilla Sensillum
● Trichoid Sensillum
● Basiconical Sensillum
● Chordotonal Organ
○ Multidendritic Neuron

Fig. 9.3. The pattern of larval sensilla in *t1, t2-t3* and *a1-a7*. Three segments of a stage 17 embryo are shown. Muscle fibres are shown in grey for reference. Anterior is to the left. Abbreviations. *dbd* dorsal basiconical sensillum; *dbp* dorsal bipolar neuron; *dch* dorsal triscolopidial chordotonal organ; *dda* multidendritic neurons; *dh 1-2* dorsal hair 1-2; *dp 1-3* dorsal papilla 1-3; *isbd* intersegmental bidendritic neuron; *istd* intersegmental trachea-associated neuron; *ISN* intersegmental nerve; *ko* Keilin's organ; *lbd* lateral bipolar neuron; *lch 1* lateral monoscolopidial chordotonal organ; *lch* lateral triscolopidial chordotonal organ; *lch 5* lateral pentascolopidial chordotonal organ; *lda, ltd* lateral multidendritic neurons; *lh 1-2* lateral hair 1-2; *lp 1-2* lateral papilla 1-2; *px, py* papilla sensilla; *SN* segmental nerve; *TN* transverse nerve; *vbp* ventral bipolar neuron; *vch 1-2* ventral monoscolopidial chordotonal organ 1-2; *vda_a vda_p* anterior and posterior ventral multidendritic neurons; *vp 1-5* ventral papilla 1-5.

Table 9.1. A comparison of the terminologies used to designate larval sensilla in *a1-a7*

Present	Hertweck	Kankel et al.	Ghysen et al.[a]	
			Cuticle	Neurons
dc2	b	-	p9	desD
dc1	s	-	p8	desC
dh1	b + st	sh F	h3/4	desA/B
lch1	lch1[x]	lch1[x]		vch1
lch5	lch5[x]	lch5[x]		lch5
lh2[d]	b	sh C	h2	lesC
lh1[d]	H	sh C	h1	lesA
lc1	-	-	p7	lesB
vp5	s	-	p6	ves2
vp4	-	-	p4	vesA
vp4[a]	-	-	p5	vesB
vp3	-	-	p3	vesC
vp2	-	-	p2	vesB
vp1	-	-	p1	vesA
vch1	vch1[x]	vch1[x]		vch

[a] See also Dambly-Chaudière and Ghysen (1987)

Two multidendritic neurons, the intersegmental bidendritic neuron (*isbd*) and trachea-associated neuron (*istd*), are located on the segment boundary (Fig. 9.3). Axons of both neurons form a separate nerve, called the transverse nerve (see below), which extends ventrally along the segment boundary and enters the ventral nerve cord at a dorsomedial position.

Lateral group. The most prominent sensillum of the lateral group of abdominal sensilla is *lch5*, the pentascolopidial chordotonal organ. Its five scolopidia are located posteriorly in the niche formed between the epidermis and the lateral transverse muscle

Fig. 9.4. The pattern of sensory neurons in *a1-a7*. The micrographs of two segments of a stage 17 embryo stained with the 22C10 antibody show from top to bottom, dorsal, lateral and ventral levels; the plane of focus for the panels on the left is interior, that of the panels on the right exterior. Anterior is to the left. Abbreviations. *dbp* dorsal bipolar neuron; *dda* multidendritic neurons; *dh 1-2* dorsal hair 1-2; *dp 1-2* dorsal papilla 1-2; *isn* intersegmental nerve; *lch 1* lateral monoscolopidial chordotonal organ; *lch 5* lateral pentascolopidial chordotonal organ; *lh1-2* lateral hair 1-2; *sn* segmental nerve; *tr* trachea; *vch 1-2* ventral monoscolopidial chordotonal organ 1-2; *vbp* ventral bipolar neuron; vda_a vda_p anterior and posterior ventral multidendritic neurons; *vp 1-5* ventral papilla 1-5

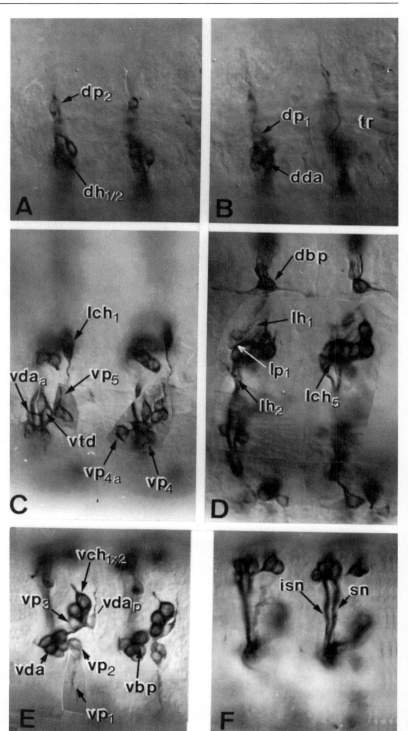

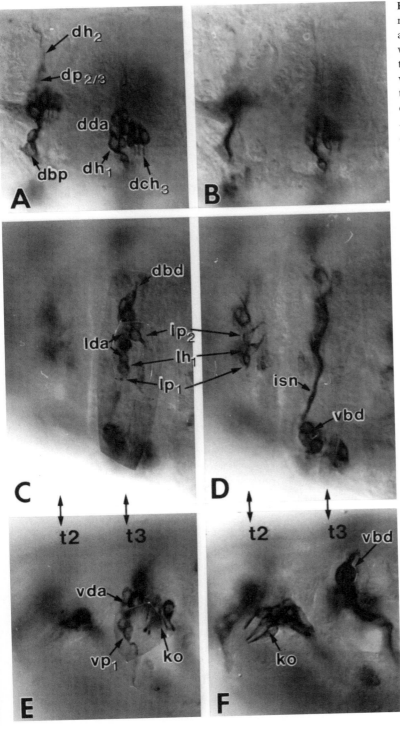

Fig. 9.5. The pattern of sensory neurons in *t1-t3*. The micrographs of *t1* and *t2* of a stage 17 embryo stained with the 22C10 antibody show, from top to bottom, dorsal, lateral and ventral levels; the plane of focus for the panels on the left is interior, that of the panels on the right exterior. Anterior to the left. Abbreviations. *dbd* dorsal basiconical sensillum; *dbp* dorsal bipolar neuron; *dda* multidendritic neurons; *dh 1* dorsal hair 1; *dp 2-3* dorsal papilla 2-3; *isn* intersegmental nerve; *dch 3* dorsal triscolopidial chordotonal organ; *ko* Keilin's organ; *lh 1* lateral hair 1; *lp 1-2* lateral papilla 1-2; *sn* segmental nerve; *vbd* ventral basiconical sensillum; *vda* ventral multidendritic neuron; *vp 1* ventral papilla 1

(Figs. 9.3, 9.4D). These scolopidia are strictly aligned and oriented dorsoposteriorly. The ligament cells anchoring *lch5* to the epidermis are very long (Fig. 8.5B); they insert close to the posterior segment boundary at the level of the dorsal muscle fibres DO2. Sensory neurons of three external sensilla, *lh1*, *lh2* and *lp1*, and two multidendritic neurons, *ltd* and *lda*, flank the pentascolopidial chordotonal organ at its anterior border. The monoscolopidial chordotonal organ *lch1* is located more superficially and dorsally than *lch5*, and is oriented dorso-anteriorly. The axons of all lateral sensilla, except for *lch1*, join the intersegmental nerve; the *lch1* axon projects to join branch "*a*" of the segmental nerve.

Dorsal group. The dorsal group consists of four external sensilla, a small and a large hair sensillum (*dh1* and *dh2*) and two papillar sensilla (*dp1* and *dp2*), arranged along a vertical line in the middle of each abdominal segment, between the epidermis and the dorsal acute muscles. The small dorsal hair *dh1* is innervated by three neurons; the large *dh2*, as well as *dp1* and *dp2*, are each in contact with only one neuron. Sensory neurons of all four sensilla form a dense cluster together with five multidendritic neurons (*ddaA-E*). A sixth multidendritic, bipolar neuron located ventral to this cluster (*dbp*) forms conspicuous longitudinal processes which span the entire segment. Axons of all dorsal sensilla course through the intersegmental nerve.

9.1.3.2.
Thoracic Segments

Ventral group. The ventral group consists of five sensilla in each thoracic segment: two papilla sensilla (*vp1* and *vp2*), a monoscolopidial chordotonal organ (*vch1*), a basiconical sensillum (*vbd*) and Keilin's organ (*ko*; Figs. 9.3, 9.5; refer to Table 9.2). The cell bodies of the neurons of *vbd* and *vp2*, together with four multidendritic neurons (*vdaA-D*), are grouped in a somewhat irregularly shaped cluster in the anterior half of the segment, at the level of the ventral oblique muscles (VO1-3). The five neurons associated with Keilin's organ and the monoscolopidial organ *vch1* can be found more posteriorly. These neurons form part of the leg imaginal disc and persist into the adult period (Jan et al. 1985; Tix et al. 1989a).

Lateral group. The patterns of sensilla of the lateral and dorsal group differ in the three thoracic segments (Fig. 9.3). In the prothorax, the lateral group comprises a tight cluster of cells formed by papilla sensilla *lp1* and *lp2*, hair sensillum *lh1*, a triscolopidial chordotonal organ *lch3*, and four multidendritic neurons (*ldaA-D*). Strikingly, the axons of *lch3* join the segmental nerve of

Table 9.2. A comparison of the terminologies used to designate larval sensilla in *t1-t3*

Present	Hertweck	Kankel et al.	Ghysen et al.[a]	
			Cuticle	Neurons
dh2	b	sh E	h3	desE
dc3	s	-	p7(8)	desD
dc2	s	-	p6(7)	desC
dh1	b	sh D	h2	desB(A)
dc1	b	-	p5	desA(B)
dch3	dch3^x	dch3^x		dch3
dbd	k	sh B	dk	des2
lbd	k	sh B	lk	les3
lch3	lch3^x	lch3^x		lch3
lh1	H	sh C	h1	lesB (C)
lp2	-	-	p4	lesC(B)
lp1	-	-	p3	lesA
vbd	-	sh B	vk	ves3
vp2	-	-	p2	vesB
py	-	-	py	vesB
px	-	-	px	vesA
vp1	-	-	p1	vesA
ko	Fussst.Organ	sh A	KO	vesA-E
vch1	vch1^x	vch1^x		vch1

[a] See also Dambly-Chaudière and Ghysen (1987)

the mesothoracic segment to course towards the CNS. In addition, there are two more papillar sensilla (*px, py*) close to the anterior boundary of the prothorax. In the mesothorax and metathorax these two papillae are absent and instead of the lateral triscolopidial chordotonal organ, there is a basiconical sensillum (*lbd*).

Dorsal group. In the prothorax, the dorsal group of sensilla comprises hair sensilla *dh1* and *dh2*, with one neuron each; three papillae, *dp1-dp3*, with one neuron each; a triscolopidial chordotonal organ (*dch3*), and a dorsal basiconical sensillum (*dbd*). Clustered around the neurons of these sensilla, there are six multidendritic neurons (*ddaA-E; dbp*). In mesothorax and metathorax, the dorsal basiconical sensillum is missing.

The innervation of thoracic sensory organs follows the same scheme as that of abdominal sensilla: the ventral group sends its axons through two different branches into the posterior fascicle or the segmental nerve. The axons of the lateral and dorsal group course through the intersegmental nerve.

Table 9.3. A comparison of the terminologies used to designate larval sensilla in *a8-a10*

Present	Hertweck	Kankel et al.	Ghysen et al.[a]	
			Cuticle	Neurons
dh2	b	sh E	h3	desE
sso	-	-		spA-D
DMSO	sk	sc 3	t7	desn
ALSO	sk	sc 1/2	t4/5	des1/3
lch3/1	lch3/1[x]	lch3/1[x]		dch3/1
PLSO	sk	sc 2/3	t2/3	des3/3+1
ALSO	sk	sc 1	t1	des3+1+1
lch3/1	lch3/1[x]	lch3/1[x]		dch3/1
vs	-	-		vesM

[a] See also Dambly-Chaudière and Ghysen (1987)

9.1.3.3.
Pattern of Sensory Organs of the Telson

The telson of the *Drosophila* larva is formed by three modified segments, *a8*, *a9* and *a10*. Whereas *a8* is similar in size to the more anterior abdominal segments, *a9* and *a10* are rudimentary and fused with each other. Discrete groups of sensilla have been assigned to each of these segments (see Table 9.3). Most sensilla occupy lateral and dorsal positions; with the exception of a small group of multidendritic neurons and one papillar sensillum (*vas*) located ventrally in *a8*, the ventral group of sensory organs present in thoracic and abdominal segments is apparently absent in *a8-a10* (Figs. 9.6, 9.7).

The abdominal segment *a8* is innervated by one nerve which, with respect to its point of entry into the CNS, corresponds to the intersegmental nerve found in all other segments (*na8* in Fig. 9.6). Along the course of this nerve, three clusters of sensory neurons can be distinguished (Fig. 9.6). Ventrally, at the level of the ventral acute muscles, there are three to four multidendritic neurons. Laterally, in the niche formed between the lateral transverse muscle LT1 and the epidermis, there are three to four multidendritic neurons, a cluster consisting of the sensory neurons innervating the triscolopidial chordotonal organ *lch3*, and the basiconical sensillum *ALSOp* (terminology of Jürgens and Hartenstein 1993). A dorsal cluster of neurons, situated at the base of the posterior spiracle, comprises the hair sensillum *DMSOh* and the basiconical sensillum *DMSOp*. Furthermore, four pairs of neurons (*sso*) with

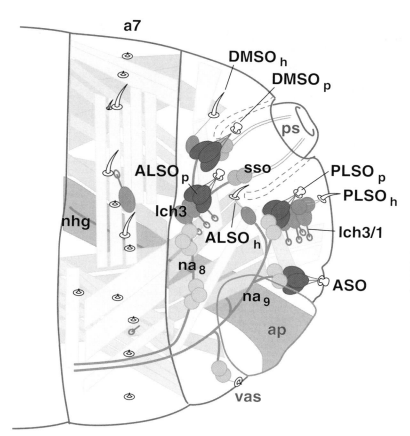

Fig. 9.6. The pattern of larval sensilla and sensory neurons in the telson. The drawing shows a lateral view of *a7* and the telson of a stage 17 embryo. Muscle fibres and hindgut are shown in grey for reference. Anterior to the left. Terminology according to Jürgens and Hartenstein (1993). Abbreviations. *ap* anal pad; $ALSO_h$ hair sensillum; $ALSO_p$ basiconical sensillum; *ASO* basiconical sensillum; $DMSO_h$ hair sensillum; $DMSO_p$ basiconical sensillum; *lch 1* lateral monoscolopidial chordotonal organ; *lch 3* lateral triscolopidial chordotonal organ; *na 8, na 9* segmental nerves; *nhg* nerve of the hindgut; $PLSO_h$ hair sensillum; $PLSO_p$ basiconical sensillum; *ps* posterior spiracle; sso. sensory neurons of the posterior spiracles; *vas* ventral papilla

extremely elongated dendrites, innervate the crown of sensory hairs surrounding the tip of the posterior spiracles. Axons of these neurons also join the *a8* nerve.

The fused segments *a9* and *a10* roughly coincide with the region between the posterior spiracles and anal pads. The terminal oblique muscles (TO1, TO2; Fig. 4.10) form a septum which divides this space into a dorsal and a ventral compartment. Each of these compartments houses a cluster of sensory neurons. The dorsal cluster belongs to *a9* and contains neurons innervating the basiconical sensillum *PLSOp*, the hair sensilla *ALSOh* and *PLSOh*, and the chordotonal organ *lch3/1*. The ventral cluster comprises several multidendritic neurons and three bipolar neurons innervating the basiconical sensillum *ASO* which, based on the findings of Jürgens et al. (1986), can be assigned to abdominal segment *a10*. The sen-

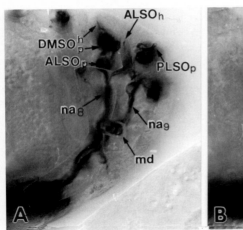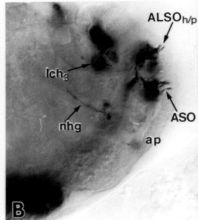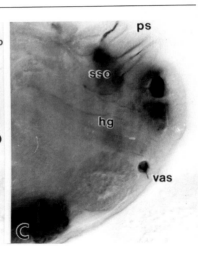

Fig. 9.7 Pattern of sensory neurons in the telson. Lateral view of the telson of a stage 16 embryo in which sensory neurons are labeled with the MAb22C10 antibody. Panels **A** to **C** show three different focal planes (**A.** superficial, **B.** approximately 25 μm deep; **C.** sagittal plane). Abbreviations. *ALSO$_h$*, *ALSO$_p$* hair and peg sensillum of dorsolateral sensory cone; *ap* anal pads; *ASO* caudal (anal) sensory cone; *DMSO$_h$*, *DMSO$_p$* hair and peg sensillum of dorsomedial sensory cone; *hg* hindgut; *lch3* lateral chordotonal organ; *md* multidendritic neurons; *na8* nerve of *a8*; *na9* nerve of *a9*; *nhg* branch of *a9* along hindgut; *PLSO$_h$*, *PLSO$_p$* hair and peg sensillum of dorso-caudal sensory cone; *ps* posterior spiracle; *sso* spiracular sensory neurons (innervate spiracular hairs); *vas* ventral (unpaired) anal sensillum

sory neurons of *a9* and *a10* form a single nerve (*na9* in Fig. 9.6), which enters the ventral ganglion in close proximity to the nerve of *a8*. The *na9* nerve gives off a branch which travels along the hindgut. It probably consists of motor axons innervating the visceral muscles of the hindgut; in addition, this branch may conduct sensory input from the tip cells of the posterior Malpighian cells, located at the tip of the Malpighian tubules, assuming they act as sensory neurons (Hoch et al. 1994). Finally, an unpaired papilla sensillum with two neurons is located in the ventral midline, anteriorly adjacent to the anal pads (*vas* in Fig. 9.6). This location corresponds to the primordium of the genital disc (Hartenstein and Jan 1992). Axons emanating from *vas* join the *na9* nerve of one side.

9.1.4.
Pattern of Appearance of the Sensory Organ Progenitor Cells

The time and position at which SOPs appear in the developing epidermis has been determined for most SOPs in several recent studies (Ghysen and O'Kane 1989; Dambly-Chaudiere et al. 1992;

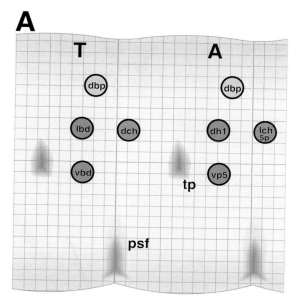

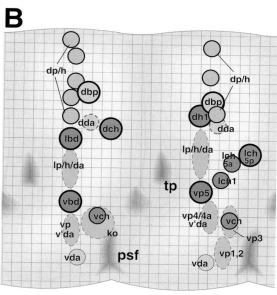

Fig. 9.8. The pattern of sensory organ progenitor cells. **A** shows the early SOPs in a thoracic (**T**) and an abdominal (**A**) segment. Nomenclature of the SOPs follows that of the sensory organs to which the cells give rise. Early SOPs are represented by large circles with thick outlines, SOPs that form later are shown as small circles with thin outlines. *lbd* gives rise to the lateral basiconical sensillum in thoracic segments; *dh1* to the dorsal trichoid (hair) sensillum in abdominal segments. *dch3* gives rise to the three scolopidia of the thoracic triscolopidial chordotonal organ *dch3* and the three posterior scolopidia of the abdominal pentascolopidial chordotonal organ *lch5* (*lch5p*). *vbd* forms the ventral basiconical sensillum in the thorax and the multiply innervated papilla sensillum in the abdomen (*vc5*). *dbp* gives rise to dorsal bipolar neurons. **B** shows the intermediate and late SOPs. *dbp* dorsal bipolar neuron; *dch* dorsal monoscolopidial chordotonal organ; *dda* dorsal multidendritic neurons; *dh 1* dorsal hair 1; *dp/h* neurons for dorsal papilla and hair; *lp/h/da* lateral papilla, hair and multidendritic neurons; *lch5a* anterior two scolopidia of the *lch5*; *vch* ventral monoscolopidial chordotonal organ; *vbd* ventral basiconical sensillum; *vda* ventral multidendritic neurons; *vp 1-5* ventral papilla 1-5. *psf* parasegmental furrow; *tp* tracheal pit

Goriely et al. 1991; Younossi-Hartenstein and Hartenstein 1996). SOPs become evident in waves during stages 10 and 11, such that early, intermediate and late SOPs can be distinguished (ref. to Figs. 9.8, 9.9). Chordotonal organs, multiply innervated external sensilla and multidendritic neurons are derived from early and intermediate SOPs; late SOPs give rise to mechanosensory exter-

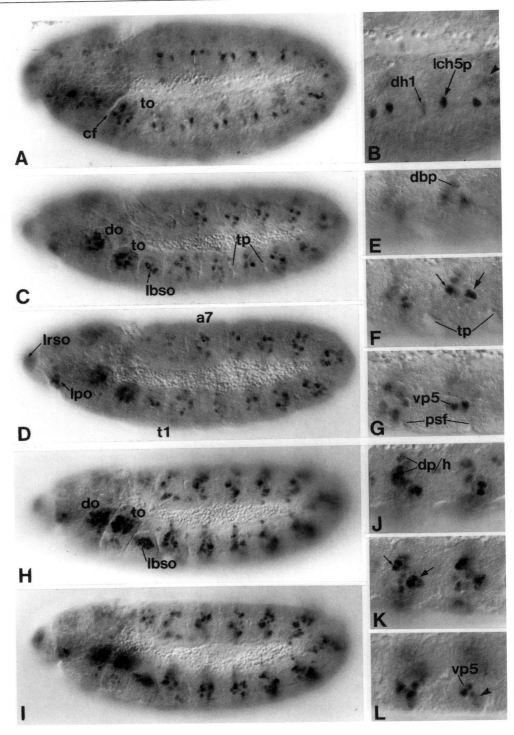

Fig. 9.9 Pattern of sensillum progenitor cells (SOPs), visualized by P-*lacZ* insertion A37 (Ghysen and O'Kane 1989). For complete map of SOPs, see Fig. 9.8. **A, B** Early stage 10 embryo, lateral view. In each segment of the trunk there are two lateral SOPs. The anterior one (*dh1* in **B** which shows abdominal segment at high magnification) gives rise to the multiply innervated dorsal hair sensillum. The posterior SOP (*lch5p*) gives rise to part of the lateral chordotonal organ, most probably its posterior three scolopidia of the lateral chordotonal organ. The first SOPs in the head belong to the terminal organ (*to*), which originates from the maxillary segment. **C-G** Early stage 11 embryo in lateral view. **C** shows a more superficial focal plane than **D**; likewise, **E, F** and **G** show three focal planes (from superficial to deep) of two abdominal segments at high magnification. Besides *dh1* (marked by *small arrow*) and *lch5p* (*large arrow*), there is a dorsal cell (*dbp*) which gives rise to the dorsal bipolar neuron (and possibly other multidendritic neurons) and a ventral cell from which the ventral multiply innervated sensillum *vc5* originates. All SOPs are located within the middle of each segment primordium, as indicated by their spatial relationship to the tracheal pits (*tp*) and parasegment furrows (*psf*). In the head, additional SOPs have appeared in the labium (*lbso* labial sensory complex), intercalary segment (*lpo* lateral hypopharyngeal / hypopharyngeal organ), antennal segment (*do* dorsal organ) and labrum (*lrso* labral sensory organ). **H-L** Late stage 11 embryo in lateral view. **H** shows a more superficial focal plane than **I**; likewise, **J, K** and **L** show three focal planes (from superficial to deep) of two abdominal segments at high magnification. Additional SOPs have appeared at all levels. Shown are the SOPs of the dorsal trichoid and papilla sensilla (*dp/h* in **J**) and the progenitors of the ventral chordotonal organ (*arrowhead* in **L**). Other abbreviations. *a7* abdominal segment 7; *t1* thoracic segment 1

nal sensilla. Our present description is based chiefly on Younossi-Hartenstein and Hartenstein (1996).

Early SOPs. The first recognizable SOPs are two cells which appear in the lateral epidermis during early stage 10 (Ghysen and O'Kane 1989). The anterior one gives rise to the lateral basiconical sensillum (*lbd*) in the thoracic segments and to the multiply innervated dorsal trichoid sensillum (*dh1*) in the abdominal segments. In each metamere, the posterior SOP forms the three scolopidia of the thoracic *dch3* and the three posterior scolopidia of the abdominal *lch5* chordotonal organ. In late stage 10, two more SOPs appear. The ventral SOP forms the ventral basiconical sensillum (*vbd*) in the thorax and the multiply innervated *vc5* papilla sensillum in the abdomen. The dorsal SOP gives rise to a subset of dorsal multidendritic neurons (Fig. 9.10).

Intermediate SOPs. Additional SOPs appear during early stage 11, from which most of the multidendritic neurons and the remaining chordotonal organs will derive. SOPs of *lch1* in the abdomen, the two anterior scolopidia of *lch5* in the abdomen, and *vch1* in thorax and abdomen, have been positively identified in early stage 11 embryos.

Late SOPs. Progenitors of the singly innervated hair and papilla sensilla can be recognized during late stage 11. The progenitors of ventral multidendritic neurons appear at around the same time as, and in close proximity to, the progenitors of the ventral papilla sensilla.

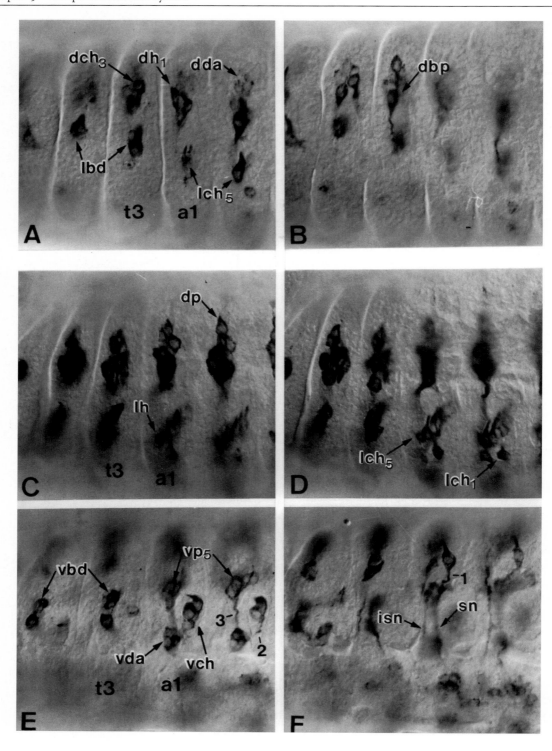

Fig. 9.10. The development of sensory neurons in the trunk. Micrographs are of thoracic and abdominal segments of embryos stained with the 22C10 antibody. The top panels (**A, B**) show a stage 13 embryo, the middle and bottom panels show dorsolateral (**C-D**) and ventrolateral (**E-F**) views, respectively, of a stage 15 embryo. For the panels on the left, the plane of focus is interior, that of the panels on the right exterior. Anterior to the left. Abbreviations. *dbp* dorsal bipolar neuron; *dch₃* dorsal triscolopidial chordotonal organ; *dda* multidendritic neurons; *dh 1* dorsal hair 1; *isn* intersegmental nerve; *lch 1* lateral monoscolopidial chordotonal organ; *lch 5* lateral pentascolopidial chordotonal organ; *lh* lateral hair; *sn* segmental nerve; *vch* ventral monoscolopidial chordotonal organ; *vda* anterior ventral multidendritic neurons; *vp* ventral papilla

The appearance of SOPs correlates with the stage in the cell cycle that the surrounding epidermal cells have reached. Most, if not all, SOPs segregate while they are in late G2 phase, and undergo mitotic division within minutes after having left the ectoderm. This behaviour is in fact similar to that of the neuroblasts, which delaminate when the surrounding cells undergo mitosis. That is to say, with respect to their neighbours, SOPs and neuroblasts delay their entry into mitosis until delamination has been completed. The early SOPs appear during the second postblastoderm mitosis (M15) of the lateral ectoderm, the intermediate and late SOPs during the third postblastoderm division.

9.1.5.
Sensory Organs of the Larval Head

As further detailed in Chapter 16, the head of the *Drosophila* larva is formed from a series of highly modified or rudimentary segments which are folded inside the body. Anterior to the thorax, there are three gnathal segments (labium, maxilla, mandible) which fuse and form the lining of the atrium, and the dorsal pouch, a cleft-like space covering the pharynx. Anterior to the gnathal segments are the procephalic segments, which comprise the intercalary and antennal segments, the labrum and the acron (referred to as the ocular segment in recent studies by Schmidt-Ott and Technau 1992, and Schmidt-Ott et al. 1994). The intercalary segment forms the floor of the pharynx (hypopharynx), the antennal segment part of the atrium, and the labral segment the roof of the pharynx (epipharynx), whereas the acron contributes to the dorsal pouch. Each head segment contains a set of sensory organs (Fig. 9.11). During the course of the drastic rearrangements of cephalic structures that take place during late embryogenesis, head sensilla and the nerves formed by their axons become displaced and, in some cases, fuse with each other, giving rise to complex sensory organs and nerves (see Table 9.4). In the following, the

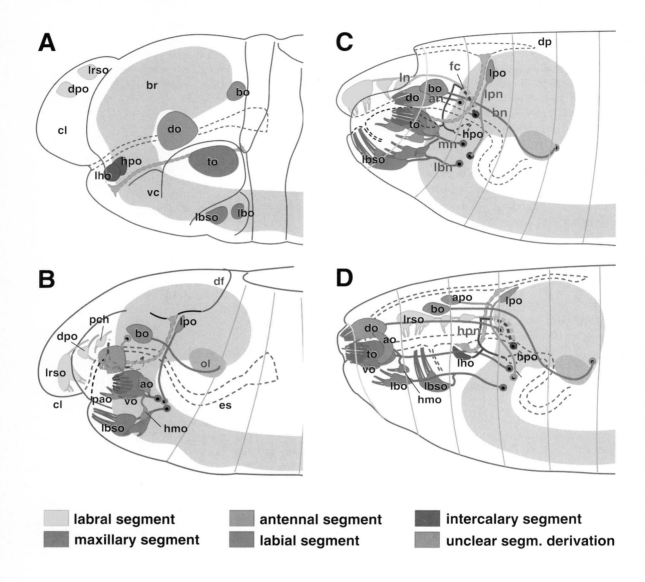

Fig. 9.11. The pattern of head sensilla at stage 13 (**A**), 14 (**B**), 15 (**C**) and 17 (**D**). Each panel shows an embryonic head in lateral view. Segment borders (*solid grey lines*), central nervous system (*grey shading*), foregut (*hatched line*), dorsal pouch (*hatched line*) and tentorium (*dark shading within hatched line*) are given as points of reference. Segmental origin of sensilla and sensory nerves is indicated by different colours (see key at bottom). Head sensilla become distinct during stage 13. By stage 14 (**B**), these sensilla have formed nerves, most of which have already reached their point of entry into the brain or ventral cord (points of entry are marked as black circles in **B-D**). Due to the inward movement of foregut structures, labral sensilla have moved ventrally and posteriorly, hypopharyngeal sensilla have followed the pharynx floor into the stomodeum, and all the other sensilla have moved anteriorly. By stage 15 (**C**), formation of head nerves is complete. Labral and labial sensilla have moved into the atrium. Dorsally, Bolwig's organ and the lateropharyngeal organ have been folded into the dorsal pouch in the wake of beginning head involution. The final pattern of sensilla and nerves is established by stage 17 (**D**). Abbreviations. *an* antennal nerve; *ao* associated organ; *apo* anterior pharyngeal organ; *bn* Bolwig's nerve; *bo* Bolwig's organ; *br* brain; *cl* clypeolabrum; *df* dorsal fold; *do* dorsal organ; *dp* dorsal pouch; *dpo* dorsal pharyngeal organ; *es* oesophagus; *fc* frontal connective; *lbn* labial nerve; *lho* laterohypopharyngeal organ; *ln* labral nerve; *lpn* lateropharyngeal nerve; *hmo* hypomaxillary organ; *hpn* hypopharyngeal nerve; *hpo* hypopharyngeal organ; *lbo* labial organ; *lbso* labial sensory complex; *lrso* labral sensory complex; *mn* maxillary nerve; *ol.* optic lobe; *pao* papilla organ; *pch* dorsal pharyngeal chordotonal organ; *to* terminal organ; *vc* ventral nerve cord; *vo* ventral organ

large sensory complexes of the head (antennomaxillary complex, labial complex, labral complex, Bolwig's organ) will be considered first, followed by a description of the remaining, smaller sensilla.

9.1.5.1.
Antennomaxillary Complex

The antennomaxillary complex (Hertweck 1931) forms a prominent protuberance on either side of the anterior tip of the larva, and comprises three large sensory organs: the dorsal organ, the terminal organ and the ventral organ. Whereas the dorsal (antennal) organ is a derivative of the antennal segment, the other two derive from the fused mandibular-maxillary segments. The organization of the antennomaxillary complex has been described by Hertweck (1931) and by Kankel et al. (1980) in third instar, and by Campos-Ortega (1982) in second-instar *Drosophila* larvae. Electron microscopic studies by Chu-Wang and Axtell (1972) in *Musca* and Singh and Singh (1984) in *Drosophila* allow one to recognize a great deal of similarity in the organization of the antennomaxillary complex in both dipteran insects. In *Drosophila* the analysis of cuticle and wholemount preparations labeled with neuron-specific markers has revealed the following composition of the antennomaxillary complex (Figs. 9.11-9.13).

Table 9.4. A comparison of the terminologies used to designate larval sensilla in the head

Present	Hertweck	Kankel et al.	Singh and Singh
do	antennales Organ	antenno-maxillary complex	dorsal organ
to	Maxillarorgan	antenno-maxillary complex	terminal organ
vo	ventrales Organ	ventral organ	ventral organ
lbo	–	labial organ	ventral organ
lbc^x	–	–	–
lbso	Hypophyse	pharyngeal sense organs	ventral group
bso	Epiphyse	pharyngeal sense organs	dorsal group
dpo	–	–	–
hyo	Organ X	Organ X	–

The dorsal organ consists of a central, dome-shaped cuticular process innervated by seven circularly arranged triplets of dendrites. The wall of the dome shows numerous pore channels, suggesting that this is a chemoreceptor. Six small papilla or campaniform-like sensilla surround the base of the central dome. Five of these papillae are innervated by two dendrites each, the sixth contains a single dendrite (Singh and Singh 1984).

The terminal organ, a derivative of the maxillary segment, consists of a distal group of 11 circularly arranged sensillar units, and a dorsolateral group comprising another three units. On the basis of their cuticular processes, which are virtually identical in *Musca* and *Drosophila*, the sensillar units of the terminal organ can be classified as papilla, knob, spot and pit sensilla. Some of them show ultrastructural specializations indicative of chemoreceptors (e.g. pore channels, innervation by three or five dendrites), others may represent hygro/thermoreceptors (poreless process containing two dendrites) and mechanoreceptors (poreless, single dendrite with tubular body). Additionally, there are two monoscolopidial chordotonal organs associated with the antenno-maxillary complex, the neurons of which are located at the lateral periphery of the terminal organ and project their dendrites medially and rostrally, respectively.

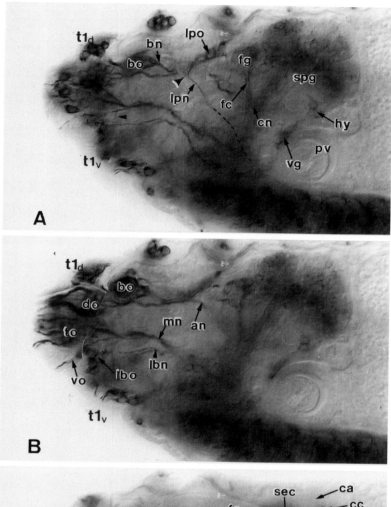

Fig. 9.12. The pattern of cephalic sensory neurons. Optical sections through the head of a stage 17 embryo stained with the 22C10 antibody. Lateral views. Abbreviations. *an* antennal nerve; *bo* Bolwig's organ; *bn* Bolwig's nerve; *ca* corpora allata; *cc* corpora cardiaca; *cn* cervical connective; *do* dorsal organ; *dpo* dorsal pharyngeal organ; *es* oesophagus; *fc* frontal commissure; *fg* frontal ganglion; *hpo* hypopharyngeal organ; *hy* hypocerebral ganglion; *lbn* labial nerve; *lbo* labial organ; *lbso* labial sense organ; *lpn* labral pharyngeal nerve; *lpo* lateropharyngeal organ; *lrsn* labral sense organ and nerve; *mn* maxillary nerve; *ncc* nerve of the corpora cardiaca; *ph* pharynx; *rn* recurrent nerve; *sec* supraoesophageal commissure; *spg* supraoesophageal ganglion; *t1* prothoracic segment; *to* terminal organ; *vc* ventral cord; *vg* ventricular ganglion; *vo* ventral organ

Fig. 9.13. The pattern of cephalic sensory neurons. Optical sections through the head of a stage 17 embryo stained with the 22C10 antibody. Dorsal views. Abbreviations. *bo* Bolwig's organ; *bn* Bolwig's nerve; *DPM* dorsal pharyngeal muscles; *do* dorsal organ; *dpo* dorsal pharyngeal organ; *es* oesophagus; *fc* frontal commissure; *fg* frontal ganglion; *hy* hypocerebral ganglion; *lbn* labial nerve; *lbo* labial organ; *lbso* labial sense organ; *lpo* lateropharyngeal organ; *lrso* labral sense organ; *mn.* maxillary nerve; *ncc* nerve of the corpora cardiaca; *pch* pharyngeal chordotonal organ; *pe* paraoesophageal ganglion; *pao* papilla organ; *rn* recurrent nerve; *sec* supraoesophageal commissure; *spg* supraoesophageal ganglion; *to* terminal organ

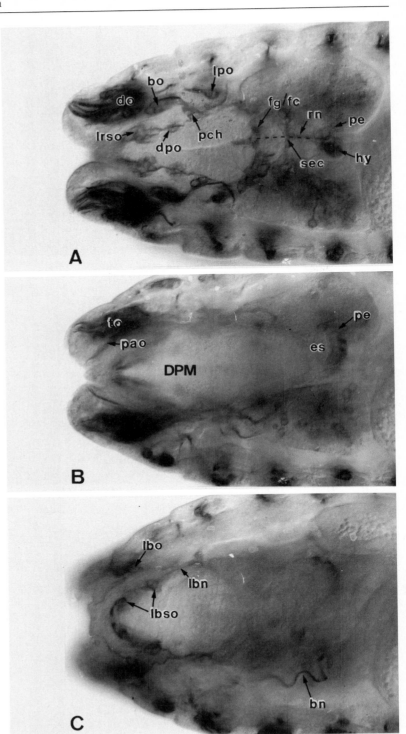

The ventral organ (Hertweck 1931) shows externally three knob-like processes. One of these is innervated by two neurons, the remaining two by one neuron each. The segmental identity of the ventral organ is unclear; it was originally assigned to the mandibular segment (Hertweck 1931), but more recent studies suggest that it is of maxillary origin (Schmitt-Ott et al. 1994). Direct observation does not allow one to make this distinction because the sensilla of the ventral organ can be only identified at a stage when mandibular and maxillary segments have already fused (see Chapter 16).

Sensory neurons innervating the antennomaxillary complex form two large clusters underneath the dorsal and terminal organs, respectively. Neurons of the ventral organ are closely attached to those of the terminal organ. One of the dorsolateral papillar sensilla of the terminal organ is innervated by neurons belonging to the cluster lying underneath the dorsal organ.

9.1.5.2.
Labial Complex

The labial complex is formed by two groups of sensilla, labial organ and labial sensory organ, all of which are derived from the labial segment. During development, the lateral part of the labial segment, i.e. the labial protuberance, shifts ventro-rostrally to become incorporated into the atrium and form its base and the lower lip of the external opening. The labial organ is located on the lower lip. It consists of three sensillar units, one with three neurons and the other two with one neuron each, of unknown sensory modality. The labial sensory organ was previously named the hypophysis (Hertweck 1931). However, since this term, like "epiphysis" (see next section), is misleading, we prefer to call it the labial sensory complex, based on its development from the labial segment. It is located in the floor of the atrium anterior to the opening of the salivary duct (Figs. 9.11-9.16), and is composed of four sensillar units on either side (Singh and Singh 1984). Neurons of the labial organ and labial sensory complex form a large cluster flanking the salivary duct.

9.1.5.3.
Labral Sense Organ

This sensory structure was formerly called the epiphysis (Hertweck 1931). All sensory neurons of the labral sense organ are derived from the labral segment and form a single cluster wedged bet-

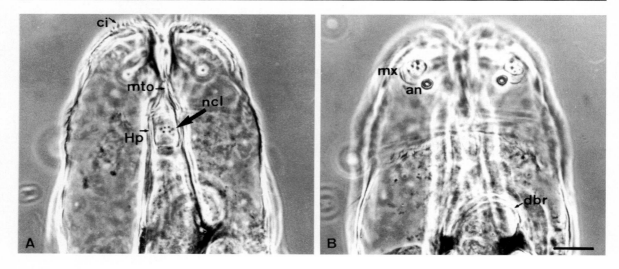

Fig. 9.14. **A** and **B** are two planes of focus of the cephalic region of a first-instar larva, to illustrate the cuticle specializations of the labial (hypophysis) and the labral (epiphysis) sensory complexes, located in the neck clasps (*ncl*), and of maxillary (*mx*) and antennal (*an*) divisions of the antennomaxillary complex. Abbreviations. *ci* maxillary cirri; *dbr* dorsal bridge; *Hp* H-piece; *mto* median tooth. *Bar* 20 μm

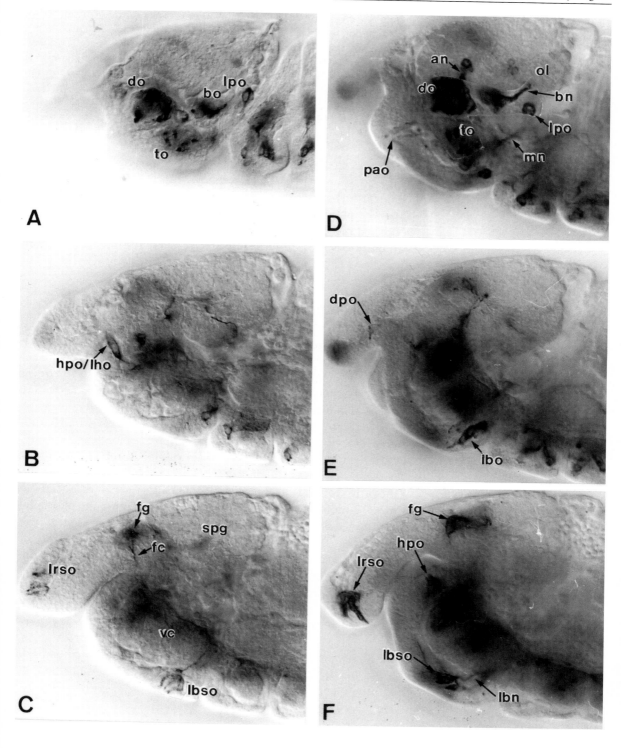

Fig. 9.15. The development of sensory neurons in the head. Micrographs of the cephalic region of embryos stained with the 22C10 antibody. **A to C** and **D to F** are three different planes of focus through a stage 13 and a stage 14 embryo, respectively. Abbreviations. *an* antennal nerve; *bo* Bolwig's organ; *bn* Bolwig's nerve; *do* dorsal organ; *dpo* dorsal pharyngeal organ; *fc* frontal commissure; *fg* frontal ganglion; *hpo* hypopharyngeal organ; *lbn* labial nerve; *lbo* labial organ; *lbso* labial sense organ; *lho* lateral hypopharyngeal organ; *lpo* lateropharyngeal organ; *lrso* labral sense organ; *mn* maxillary nerve; *ol* primordium of the optic lobe; *pao* papilla organ; *spg* supraoesophageal ganglion; *to* terminal organ; *vc* ventral cord; *vo* ventral organ

ween the median tooth and the dorsal pharyngeal musculature (Fig. 9.11-9.16). The electron microscopic study of Singh and Singh (1984) shows that the labral sense organ contains six sensillar units on either side. Two of them are innervated by two dendrites, three by one dendrite, and one by three dendrites. The modality of these units is unclear; it seems probable, however, that they are taste organs.

9.1.5.4.
Bolwig's Organ

The larval photoreceptor organ, called Bolwig's organ after its discoverer (Bolwig 1946), develops from cells of the optic lobe placode (see in Chapter 11). It consists of a dense cluster of 12 neurons which, after head involution, become attached to the outer wall of the dorsal pouch (Figs. 9.11-9.17). Antibodies and P-*lacZ* insertion lines that label the larval photoreceptors do not reveal any additional, closely apposed cells which might be analogous to the support cells found in other types of sensilla or in the ommatidia of the compound eye. The photoreceptors themselves also differ significantly from adult photoreceptors. Thus, instead of a rhabdome, i.e. a highly regular array of microvilli carrying the photopigment molecules, these cells develop at the apical pole several irregularly shaped, highly branched processes which, in view of the fact that they are filled with bundles of microtubules, may be considered as cilium-like structures (Green et al. 1993).

9.1.5.5.
Small Head Sensilla

There exist several small sensilla scattered over the various parts of the head. Some of these have recently been identified by Schmitt-Ott et al. (1994).

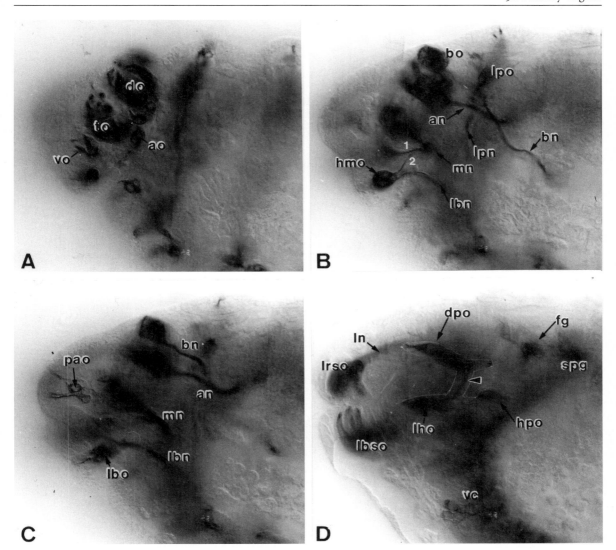

Fig. 9.16. The development of sensory neurons and nerves in the head. *A-D* show four focal planes (*A* superficial; *D* deep) of a stage 15 embryo stained with the 22C10 antibody. Abbreviations. *an* accessory nerve; *ao* antennal organ; *bo* Bolwig's organ; *bn* Bolwig's nerve; *do* dorsal organ; *dpo* dorsal pharyngeal organ; *fg* frontal ganglion; *hmo* hypomaxillary organ; *hpo* hypopharyngeal organ; *lbn* labial nerve; *lbo* labial organ of the labial complex; *lbso* labial sense organ of the labial complex; *lho* lateral hypopharynegeal organ; *lpo* lateropharyngeal organ; *lrso* labral sense organ; *mn.* maxillary nerve; *spg* supraoesophageal ganglion; *to* terminal organ; *vo* ventral organ

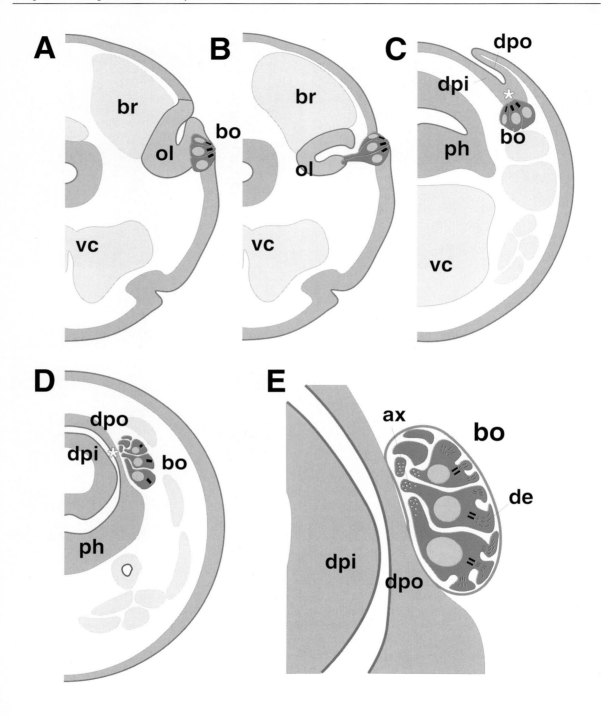

Fig. 9.17. The development of Bolwig's organ. A - D Schematic representations of drawings of sections of embryonic heads at stages 13 (A), 14 (B), 15 (C) and 17 (D). E View of Bolwig's organ at higher magnification. The photoreceptors of Bolwig's organ (bo) are bipolar cells (A, B); black dots in cells indicate basal bodies. The photoreceptors lose contact with the surface during head involution (C). At this stage their apical poles still point towards the apical surface of the epithelium (marked by *asterisk*) lining the dorsal pouch. Between the stages represented in C and D, photoreceptors reorient, so that their former apical pole now points away from the epithelial surface. E illustrates the formation of multiple, microtubule-filled processes (de) at the former apical membrane of the photoreceptors during stage 17. Other abbreviations. ax photoreceptor axons; br brain; dpi inner layer of dorsal pouch; dpo outer layer of dorsal pouch; ol invagination of the optic lobe; ph pharynx; vc ventral cord

The dorsal pharyngeal organ (dpo) and a dorsal pharyngeal monoscolopidial chordotonal organ (pch1) are both derived from the labral segment. The dpo is located posterior to the labral sensory complex and is innervated by three neurons of unknown sensory modality (Figs. 9.12C, 9.13A).

Two sensory structures, the papilla organ (pao) and an associated organ (ao, Figs. 9.11, 9.13B, 9.16) are spatially related to the terminal organ. The pao is probably of mandibular origin. It comprises a multiply innervated sensillum in the lateral atrial wall. The ao is formed by a cluster of four to six sensory neurons attached to the basal aspect of a group of neurons that belong to the terminal organ. These neurons do not look like bipolar sensory neurons in the light microscope; they may be comparable to the multidendritic neurons of thoracic and abdominal segments.

The hypomaxillary organ (hmo) is a multiply innervated sensillum of unknown sensory modality and segmental origin. It is located close to the labial organ, but its axon projects with the maxillary nerve (Fig. 9.11, see below).

The hypopharyngeal organ (Organ X of Hertweck 1931) and the latero-hypopharyngeal organ (lho, Schmitt-Ott et al. 1994) are likely to be derived from the intercalary segment, although Schmitt-Ott et al. (1994) assign these structures to the ocular segment (or acron). Since these sensilla are among the first ones to differentiate in the head, their movement from a position flanking the stomodeum (position of the intercalary segment) to the floor of the pharynx can be easily followed (Figs. 9.15, 9.16). The hypopharyngeal organ is innervated by three or four dendrites. Its axons join an anteriorly directed bundle which runs to the labral nerve (see below); the two or three neurons of the latero-hypopharyngeal organ are located at the junction of the two branches.

The lateropharyngeal organ (lpo) appears at a relatively early stage in the dorsal ridge which is derived by fusion from the dorsal parts of the labium and maxilla (Younossi-Hartenstein et al.

1993). During head involution, the *lpo* becomes incorporated into the lateral fold of the dorsal pouch (Figs. 9.15, 9.16). The *lpo* is located at the lumen of the dorsal pouch, and consists of two or three neurons. Its sensory modality is unknown. Axons of the *lpo* form a separate nerve (see below). This nerve is joined by the axons of the anterior pharyngeal organ (*apo*) which appears in the lateral aspect of the dorsal pouch. Sensory modality and segmental origin of this sensillum are unclear

9.2
Peripheral Nerves

The peripheral nerves of the fully developed *Drosophila* embryo are metamerically organized. Course and distribution of both sensory and motor fibres can be distinguished fairly well on wholemounts stained with neural markers such as MAb 22C10 (Fujita et al. 1982) and anti-HRP antibody (Jan and Jan 1982). Here we present the results of our own analysis (Campos-Ortega and Hartenstein 1985; Hartenstein 1988), supplemented by those of other workers (Schmidt-Ott et al. 1994).

9.2.1.
Peripheral Nerves of the Mature Embryo

9.2.1.1
The Nerves of Abdominal Segments *a1-a7*

We describe abdominal nerves first because segmental nerves *a1-a7* all exhibit the same organization, and may thus be considered as prototypical. Segmental nerves *a1-a7* consist of two fascicles each, one anterior and the other posterior (Figs. 9.3, 9.18, 11.11, 11.13). Although these fascicles do not constitute separate nerves in the strict sense, because they are surrounded by a common glial sheath, the terms intersegmental nerve for the anterior fascicle and segmental nerve for the posterior fascicle (Thomas et al. 1984) have become common usage. Consequently, we will follow this nomenclature. It is interesting to point out that at early stages, when these nerves first form, the segmental and intersegmental nerves are separated from each other by a considerable distance (Figs. 9.18, 11.13).

The fibres of the intersegmental nerve emerge from two roots, an anterior and a posterior one, which derive from different neu-

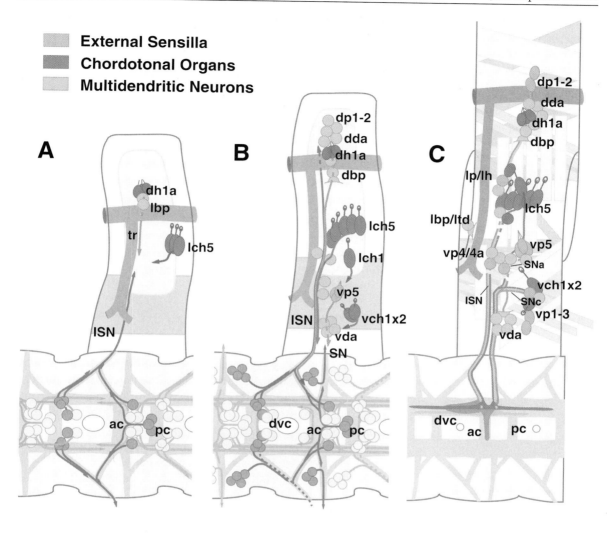

External Sensilla
Chordotonal Organs
Multidendritic Neurons

Fig. 9.18. Development of peripheral nerves. The three panels schematically represent flattened views of abdominal hemisegments (dorsal midline at the top) and associated sections of the ventral nerve cord of stage 13 (**A**), stage 14 (**B**) and stage 17 (**C**) embryos. Sensory neurons of different modalities are shown in different colours (colour key at top left; see also Fig. 9.3). Muscle fibres, tracheal tree, ventral cord neurons and neuropile are shown in grey as

points of reference. During stage 13 (**A**) axons of a small group of dorsal sensory neurons (*dh1a; dbp*) grow along the transverse branch of the tracheal tree (*tr*). At a mid-lateral level, these axons meet and fasciculate with efferent pioneers, sent out by the aCC, U and RP2 neurons (see Fig. 11.13). Axons from the scolopidia of the pentascolopidial chordotonal organ *lch5,* that differentiate first, join this early tract, which pioneers the intersegmental nerve (*ISN*). All

the dorsal and lateral sensory neurons will send their axons with the *ISN*. During stage 14 (**B**), several ventral sensilla which differentiate in sequence (*vp 5, vch 1-2, vda*), as well as the lateral monoscolopidial chordotonal organ *lch1* which originates at a ventral position (see Fig. 9.10D), pioneer the segmental nerve (*SN*). These sensory axons also meet efferent pioneers from the group of LSN motoneurons (see Fig. 11.11) before they enter the ventral nerve cord

Sensory neurons show a modality-specific projection pattern to the ventral cord (**C**). Neurons of chordotonal organs (*purple*) overlap in their axonal terminals over a narrow field which encompasses the homotopic as well as most of the anteriorly adjacent neuromere (neuromere boundaries are marked by the dorso-ventral channel, *dvc*). Multidendritic neurons (*green*) also overlap completely and form a narrow central field medially adjacent to the chordotonal neuron pathway. External sensilla (*blue*) terminate in an area which is broad in the mediolateral axis, but restricted in the longitudinal axis to the anterior part of the homotopic and posterior part of the anteriorly adjacent neuromere. Some evidence for a somatotopically ordered projection has been reported (Merritt and Whitington 1995), in that dorsal sensilla project to more anterior levels, ventral sensilla to more posterior levels. Other abbreviations. *ac* anterior commissure; *dbp* dorsal bipolar neuron; *dda* dorsal multidendritic neurons; *dh₁ₐ* multiply innervated hair sensillum; *dp₁₋₂* dorsal papilla sensilla; *dvc* dorso-ventral channel; *ISN* intersegmental nerve; *lbp/ltd* lateral bipolar neuron and lateral tracheal-associated neuron; *lch5* lateral pentascolopidial chordotonal organ; *lch1* lateral monoscolopidial chordotonal organ; *lp/lh* lateral papilla and hair sensilla; *pc* posterior commissure; *SN* segmental nerve; *vch1x2* ventral chordotonal organ; *vda* ventral multidendritic neurons; *vp₁₋₄* ventral papilla sensilla; *vp₅* ventral papilla sensillum

romeres: whereas the posterior root originates from the anterior commissure of the homotopic neuromere, the anterior root derives from the posterior commissure of the adjacent, immediately anterior neuromere. That is to say, a given intersegmental nerve comprises axons belonging to two contiguous neuromeres. The two roots converge and fuse shortly before leaving the cortex of the ventral cord. The segmental nerve also derives from two roots which leave the neuropile at the level of the anterior and posterior commissures of the homotopic neuromere. The posterior root is small and we had not documented it in our previous description (Campos-Ortega and Hartenstein 1985).

Segmental and intersegmental nerves of each abdominal segment leave the cortex closely attached to each other. A glial sheath formed by two serially arranged cells, called exit glia cells (Klämbt and Goodman 1991), surrounds both nerves. The space, ventral and lateral to the ventral cord, through which peripheral nerves travel, contains only a few slender muscle fibres of the ventral external oblique muscle; we have called this space the paraneural space (Fig. 8.6). The paraneural space is traversed by the proximal portions of the peripheral nerves which, as a consequence of the condensation of the ventral cord, vary in length from segment to segment.

On encountering the fibres of the ventral oblique musculature, the segmental and intersegmental nerves lose their common glial sheath and diverge. The intersegmental nerve (ISN) continues on a dorsally directed course, passing between the epidermis and the ventral oblique muscles, forming several different branches that project to various sensilla and muscles (Fig. 9.18). After passing the ventral musculature, the ISN gives off two motor branches,

ISBb and *ISNd*, to the ventral oblique and ventral muscle fibres. Further laterally, a sensory branch penetrates the lateral transverse muscles to reach the lateral group of sensilla (i.e. lateral pentascolopidial chordotonal organ, lateral hair sensillum). The intersegmental nerve continues dorsally, attached to the inner surface of the lateral transverse muscles. Upon entering the dorsal muscle group, the intersegmental nerve splits up into two terminal branches. The external branch leads the afferent axons derived from the dorsal group of sensory organs, whereas the other, more anterior branch carries efferent fibres for the dorsal musculature.

After detaching from the anterior fascicle the segmental nerve gives off a thick, posteriorly directed branch (*SNc*), which leaves the main segmental nerve trunk at right angles. *SNc* splits up into three fine branches that carry axons originating from the ventral sensilla (*vp1-3, vch 1+2, vda$_{A-D}$, vdap, vbp*) as well as a branch carrying efferents to the ventral acute musculature. The segmental nerve itself continues dorsally; Its terminal branch, *SNa*, supplies the lateral transverse and oblique muscles with efferent axons and carries the axons of the sensilla *vp5* and *lch1*.

The Transverse Nerve

In most insects, a transverse nerve (*TN*) is present in abdominal segments which contains the axons of motoneurons and neuroendocrine neurons (Carr and Taghert 1988). The following description is based on the report by Gorczyca et al. (1994). In *Drosophila*, the transverse nerve, described as the segment boundary nerve by Bodmer and Jan (1987), is present in abdominal segments *a2-a7* and carries the axon of the motoneuron for the ventral transverse muscle; it exits the CNS from the dorsal midline, and is thus clearly distinguishable from the intersegmental and segmental nerves which exit dorsolaterally (Fig. 9.3). In the body wall, the transverse nerve is associated with two neurons, the lateral bipolar dendrite (Bodmer and Jan 1987) on the segmental border muscle (Bate 1993), and the tracheal neuron, which extends further to reach the basis of the alary muscles (Gorczyca et al. 1994). The growth of the transverse nerve is prefigured by the two cells, located on the dorsal surface of the ventral cord in each abdominal segment, which have been called transverse nerve exit glia (Gorczyca et al. 1994). These cells were assigned a mesodermal origin by Beer et al. (1987), on the basis of the results of cell transplantations, and their function appears to be related to transverse nerve pathfinding.

9.2.1.2
The Nerves of *t1-t3*

The organization of thoracic peripheral nerves is very similar to that of the abdominal nerves (Fig. 9.3). Each thoracic segment has a segmental and intersegmental nerve which is spatially related to the neuropile of the ventral ganglion in the manner described for the abdominal nerves.

The prothoracic nerve leaves the ventral cord and extends rostrally, ventral to the salivary gland and salivary duct. After the salivary duct has turned dorsally towards its opening in the pharynx, the prothoracic nerve is covered dorsally by the large ventral longitudinal muscle of prothorax and mesothorax. Upon reaching the border between prothorax and mesothorax, the prothoracic nerve splits into intersegmental and segmental nerves. The latter receives a small anteriorly directed branch (*pf1*) from the ventral monoscolopidial chordotonal organ present in *t1*, from Keilin's organ and *vp1*. The main trunk of the nerve continues anterolaterally, along the ventral surface of the ventral longitudinal muscle, to which it gives off a thick branch (*pf2*). Finally, the terminal branch leads afferents from the prothoracic ventral black dot and *vp2*, and efferents for the pleural transverse musculature. The intersegmental nerve of the prothorax bends anterolaterally and, upon leaving the surface of the ventral longitudinal muscle, receives a branch (*ISN1*) from the lateral group of prothoracic sensory organs (i.e. lateral sensory hair, two papilla sensilla) (Fig. 9.3). Its main trunk continues dorsally, travelling laterally to the internal surface of the pleural transverse muscle to receive axons from the dorsal group of sensory organs (triscolopidial chordotonal organ, two sensory hairs and three papilla sensilla) and send motor axons to the dorsal longitudinal muscle.

The mesothoracic and metathoracic nerves show essentially the same organization, thus the following description will serve for both (Fig. 9.3). The proximal portion of the nerves traverses the paraneural space anterolaterally to reach the ventral rim of the ventrolateral musculature. After coursing for a distance of 10-15 µm along the ventral surface of the ventrolateral musculature, the nerves split up into intersegmental and segmental nerves. The segmental nerve extends rostrally in both segments, lateral to the ventral external oblique muscles, and has a branching pattern identical to that described for the prothorax. The intersegmental nerve extends dorsally at the interior surface of the pleural external transverse and pleural internal oblique muscles. It soon collects a branch (*ISN1*) carrying axons from the lateral group of sensory organs (i.e. lateral basiconical sensillum, two hair sensilla, lateral papillar sensilla 1-2). Dorsally, as reported for the abdominal segments, the intersegmental nerve in mesothorax and metathorax extends between the epidermis and the dorsal external oblique muscles, which it supplies with efferent axons (*ISN3*). Moreover it contains afferents from the dorsal group of sensory organs (*ISN2*) i.e. triscolopidial chordotonal organ, two hair sensilla, dorsal papillar sensilla 1-3.

9.2.1.3.
Nerves of the Telson

The nerve of *a8* comprises a single fascicle, although it derives from three roots, with a pattern corresponding to that of both the intersegmental and segmental nerves of other abdominal segments (Fig. 9.6). The nerve of *a9* only has two roots, reflecting the rudimentary nature of the corresponding neuromere. Before leaving the cortex of the ventral cord, some fibres from the roots of the abdominal nerve *a8* seem to bend posteriorly to join the *a9* nerve. The proximal portions of both nerves remain closely attached to each other as they cross the paraneural space. The *a8* nerve is enclosed between the ventral longitudinal muscles and the lateral transverse muscle of *a8*. After contributing motor branches to these muscles, the nerve of *a8* reaches the groups of the sensory organs located in the lateral and dorsal regions of this segment (*lch3, ALSOp, DMSOh, DMSOp, sso*) whose afferents it carries.

The nerve of *a9* extends further caudally. After passing beneath the gut suspensor muscles (*GS1-3*) the nerve gives off a thick recurring motor branch to the hindgut musculature. This is a conspicuous branch that makes a sharp turn of about 150° to extend along the lateral hindgut wall. The main trunk of the *a9* nerve continues dorsocaudally, and gives off a motor branch to the *a9* terminal oblique (*TO*) muscles. The nerve of *a9* finally splits up into a caudal branch, receiving axons from the *ASO* (presumably of *a10* origin), and a dorsal branch which, besides motor fibres, also carries afferents from the remaining sensory organs of *a9* (*PLSOp, PLSOh, ALSOh, lch3/1*).

9.2.1.4.
Peripheral Nerves of the Head

Suboesophageal (Gnathal) Nerves

Three nerves are related to the suboesophageal ganglion (Figs. 9.11-9.13). The posteriormost of these nerves includes fibres to and from derivatives of the labial segment (labial nerve), the middle nerve (maxillary nerve) results from the fusion of the mandibular and maxillary segmental nerves. The third nerve, called the lateropharyngeal nerve by Schmitt-Ott et al. (1994), exits the suboesophageal ganglion at a dorsal level, between labial and maxillary neuromeres, and carries efferent fibres to the dorsal prothoraco-pharyngeal muscle and afferent axons from the lateropharyngeal and anteropharyngeal sensory organs.

Labial nerve. After leaving the cortex of the suboesophageal ganglion, the labial nerve extends between the lateral wall of the pharynx and the ventral prothoraco-pharyngeal musculature to which it contributes efferent axons. It first receives a branch carrying the axons from the labial sensory complex and continues rostrally to collect the axons from the labial organ (Figs. 9.11, 9.12).

Maxillary nerve. The maxillary nerve is a thick nerve that extends rostralwards parallel to the labial nerve, approximately 20 μm dorsal to it. It penetrates between the fibres of the ventral prothoraco-pharyngeal muscle, to which it also contributes motor fibres, and then runs freely in the space between the pharynx and the prothoracic ventral longitudinal muscle. The axons of the maxillary nerve derive from sensory neurons of the terminal and ventral organs and the small sensilla associated with these (accessory organ, hypomaxillary organ, papillar organ) (Figs. 9.11 - 9.13).

Lateropharyngeal nerve. After leaving the subesophageal ganglion this thin axon bundle curves anteriorly and dorsally, circumnavigating the surface of the brain hemisphere. It crosses the supraesophageal nerves (antennal, Bolwig's nerve) at right angles and approaches the lateral wing of the dorsal pouch. It collects the afferent axons from the lateropharyngeal and the anteropharyngeal sensilla.

9.2.1.4.2.
Procephalic (Supraoesophageal) Nerves

Two unpaired, median nerves and four lateral pairs of nerves are related to the neuropile of the supraoesophageal ganglion (Figs. 9.11-9.13). They comprise, from ventral to dorsal, the antennal nerve, the labral nerve, the hypopharyngeal nerve, Bolwig's nerve, the recurrent nerve and the frontal nerve. The latter two structures are components of the stomatogastric nervous system and will be described in Chapter 10.

Antennal nerve. The antennal nerve originates from a basal portion of the supraoesophageal neuropile. Its entry point marks the position of the deuterocerebrum (Stortkuhl et al. 1994). Throughout its course, the antennal nerve remains in close proximity to the dorsolateral pharyngeal wall, ventral to the dorsal longitudinal thoracic muscle. It appears to be a purely sensory nerve, carrying the axons from the dorsal organ of the antenno-maxillary complex.

Labral nerve. The labral nerve does not enter the supraoesophageal ganglion directly; instead, it fasciculates with the frontal connective which connects the stomatogastric nervous system with the basalmost portion of the supraesophageal ganglion, the tritocerebrum (Hartenstein et al. 1994a; see also Chapter 10). The same relationship between labral nerve and frontal connective has been reported for all insect species investigated so far (for review see Penzlin 1985). In its course, the labral nerve is enclosed between the lateral surface of the dorsal pharyngeal musculature and the bottom of the dorsal pouch. At the level of the meso-metathoracic border, the labral nerve receives two thin branches from the dorsocaudal pharyngeal sensory organ and the pharyngeal chordotonal organ. The remaining fibres stem from the sensory neurons of the labral sensory complex.

Hypopharyngeal nerve. This thin nerve carries afferent axons from the hypopharyngeal sensillum, located in the floor of the pharynx at the level of the cervical connective. The axons travel anteriorly and dorsally and, after receiving a branch from the laterohypopharyngeal organ, they join the labral nerve to travel on towards the tritocerebrum.

9.2.2.
Development of the Peripheral Nerves

9.2.2.1.
Nerves of the Thoracic and Abdominal Segments

The segmental and intersegmental nerves of each segment are pioneered by two components each, one afferent and the other efferent, which extend towards each other and meet at a mid-lateral level. Both components use the transverse segmental branch of the tracheal tree as a substrate on which to grow (Younossi-Hartenstein and Hartenstein 1993). The intersegmental nerve is pioneered by a small bundle of sensory axons formed by a cluster of early differentiating dorsal sensory neurons, presumably *dh1* and *dbp*. These neurons appear adjacent to the longitudinal trunk of the tracheal tree during early stage 13. After extending ventrally for about 10 μm, the growth cones meet the transverse branch of the tracheal tree and follow this structure ventrally (Fig. 9.18).

At the same time as the afferent axon bundle starts to grow ventrally, an efferent axon extends dorsally. This is the axon of the anterior corner cell (Canal and Ferrus 1986; Hartenstein 1988; Johansen *et al.* 1989). This axon pioneers the proximal limb of the intersegmental nerve and also grows along the trachea. Efferent and afferent pioneer axons meet at a lateral level (late stage 13) and fasciculate with each other.

In each embryonic segment there is an invariant number of glial cells which accompany the peripheral nerves. At early stage 13, when these cells can be detected for the first time (Fredieu and Mahowald 1989; Klämbt and Goodman 1991; Hartenstein and Jan 1992; Younossi-Hartenstein and Hartenstein 1993), the peripheral glial cells accompanying the intersegmental nerve come to occupy the following positions (Figs. 11.12, 11.13): the dorsal cell (*pg3*) is basally attached to the dorsal group of sensory neurons; the lateral cell (*pg5*) lies next to the lateral pentascolopidial chordotonal organ; the ventrolateral cell (*pg2*) is located close to the lateral edge of the CNS. Glial sheaths around the peripheral nerves, which are visible both with Nomarski optics in wholemounts of embryos and in the electron microscope, appear to develop in stage 16, long after sensory axons have reached the CNS. Therefore, glial cells do not form a continuous layer on which the sensory axons grow.

The segmental nerve is pioneered by small groups of afferent and efferent axons which do not contact the trachea (Hartenstein 1988). The three axons formed by the lateral chordotonal organ in *t1* have a unique trajectory. These axons cross from the first thoracic into the second thoracic segment to join the second thoracic nerve (Campos-Ortega and Hartenstein 1985; Ghysen et al. 1986; see Fig. 9.3). In crossing from *t1* to *t2*, the axons follow the course of the adjacent anterior spiracle of the tracheal tree (Hartenstein 1988).

9.2.2.2.
Nerves of the Head

As described for sensilla of the trunk, head sensilla send out the first axons during stage 13, prior to head involution. At this stage, the distance between most of the developing head sensilla and the point of entry of their axons into the CNS is very short (10-20 µm). Thus, as documented in Figs. 9.15 and 9.16, neurons of the dorsal organ, terminal organ and the labial complex, are located immediately next to their corresponding neuromeres. Consequently, pioneering of the antennal, maxillary and labial nerves involves merely a short inwardly directed extension of growth cones.

However, other sensilla are much more remote from the developing CNS when they send out axons: for example, the sensilla of the labrum (labral sense organ, dorsal pharyngeal organ, pharyngeal chordotonal organ), the hypopharyngeal sense organ, and, in particular, the lateropharyngeal sense organ. The *lrso*, *dpo* and *pch* are arranged in a row. Their axons, forming the labral nerve, grow out centripetally and fasciculate with each other. Axons of the *pch* have to extend only approximately 10-20 µm before reaching the frontal ganglion with whose axons they will travel towards the tritocerebrum (see Chapter 10). Further rostrally, axons of the *dpo* and *lrso* also have to extend only for a short distance to bridge the gap between these sensilla (Fig. 9.11).
The hypopharyngeal nerve is also associated with the stomatogastric nervous system and develops somewhat similarly to the labral nerve. Two sensilla, the hypopharyngeal organ (*hpo*) and laterohypopharyngeal organ (*lho*), contribute axons to the hypopharyngeal nerve. During stage 14, axons of the *hpo* located in the floor of the pharynx grow anteriorly and reach the neighbouring *lho*. Axons of *lho* grow dorsally for a short distance where they reach the labral nerve.

The thin bundle of axons that originates from the papillar organ (*pao*), and is closely associated with the terminal organ, follows the course of the maxillary nerve, but does not initially adhere to this nerve, and also reaches a point of entry into the ventral cord which is slightly more anterior than that of the maxillary nerve. This trajectory suggests that the axons may actually constitute a rudimentary mandibular nerve which later fuses with the maxillary nerve. Schmitt-Ott et al. (1994) have also noted the early trajectory of the *pao* axon bundle and interpreted it as a rudimentary segmental nerve which, however, they assigned to the intercalary segment .

The lateropharyngeal organ (*lpo*) is far away from the point of entry of its axons into the ventral cord. Evidently, the *lpo* axons use the tentorium (a tube which opens next to the dorsal ridge in which the *lpo* is located; see Chapter 16) as a guide path towards the CNS. Halfway to the CNS, the afferents encounter the efferent axons which pioneer the proximal limb of the lateropharyngeal nerve.

The development of Bolwig's nerve shows some unique features seen in no other sensory nerve of the *Drosophila* embryo (Fig. 9.17). When the primordial optic lobe invagination, which also gives rise to Bolwig's organ, loses contact with the head epidermis during late stage 13, the ventral lip of the invaginated placode remains attached to the basal surface of the anlage of Bolwig's organ. This attachment persists; when the primordial optic lobe and the developing Bolwig's organ move away from each other, the basal parts of the photoreceptor cells elongate into processes which become the axons of Bolwig's nerve. Thus, these do not need to bridge any distance between the neurons and the optic lobe (ref. to Steller et al. 1987, for a different view).

The target of Bolwig's nerve in the fully developed embryo is not the optic lobe itself. Instead, the axons curve around the optic lobe, penetrate the cortex of the brain and closely approach three neurons, adjacent but posterior to the optic lobe. These three cells are among the first brain neurons to differentiate and are presumably the same cells as the "optic lobe pioneers" described by Tix et al. (1989b). Bolwig's nerve fasciculates with the processes of the "optic lobe pioneers". In the third-instar larva, Bolwig's nerve passes over the peripodial membrane of the eye-antennal imaginal disc before reaching the optic lobe (Steller et al. 1987). This spatial relationship between the nerve and the presumptive eye disc is already apparent in the late embryo. Travelling posteriorly, Bolwig's nerve remains in contact with the basal surface of the dorsal pouch epithelium, from which the eye-antennal imaginal disc will form by evagination late in embryogenesis (Chen 1929).

Chapter 10
The Peripheral Nervous System

Stomatogastric Nervous System and the Ring Gland

The stomatogastric nervous system and the ring gland consist of several small clusters of neuroendocrine elements situated posterior to and between the two cerebral hemispheres, in association with the pharynx, oesophagus and proventriculus, dorsal pouch and aorta. Due to the small size of the various ganglia of the stomatogastric nervous system, their identification in the *Drosophila* embryo is difficult. However, the available markers have allowed identification of homologs to the ganglia found in other, larger insects (Fig. 10.1). The ring gland comprises the corpora cardiaca, the thoracic glands and the corpus allatum. The following description is largely based on Hartenstein et al. (1994a), González-Gaitán et al. (1994) and González-Gaitán and Jäckle (1995).

10.1
Organization of the Stomatogastric Nervous System in the Fully Developed Embryo

10.1.1
Frontal Ganglion

The frontal ganglion contains approximately 25-30 neurons, which form a U-shape (Figs. 10.1 - 10.4), wedged between the anterior aspect of the brain hemispheres and the dorsal pharynx. Due to its intimate relationship with the pharynx, we previously used the term "parapharyngeal ganglion" to designate the cells of the frontal ganglion. However, to avoid complicating the nomenclature, we will retain the term frontal ganglion, which is widely used for other insects, to describe the entire population of stomatogastric

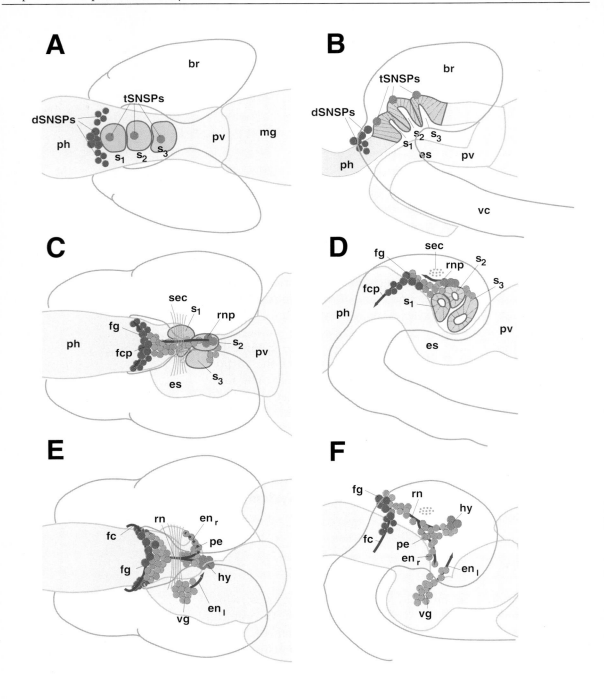

Fig. 10.1 Schematic synopsis of stomatogastric nervous system (SNS) morphogenesis. Panels in the left column represent dorsal views of foregut, brain and associated SNS of embryos at three consecutive stages; panels in the right column illustrate the corresponding stages in lateral view. **A, B** Segregation of SNS progenitor cells during stage 12. Invaginating progenitors form three pouches shaded in different colours (rostral pouch, violet; intermediate pouch, green; caudal pouch, blue, numbered 1-3). Three groups of delaminating progenitors (*dSNSPs*; median group, dark violet; lateral groups, purple) appear in front of the rostral pouch. These cells will contribute to the frontal ganglion. A single progenitor delaminates from the tip of each pouch (*tSNSPs*; in turquoise). The *tSNSPs* most probably become part of the hypocerebral ganglion. **C, D** Dissociation of progenitor cells during stage 14. Cells of the tips of the rostral and intermediate pouches (*1* and *2*, respectively) form an anteriorly directed stream which gives rise rise to the frontal ganglion (*fg*). *dSNSPs* differentiate first and form pioneer tracts for the frontal connective (*fcp*) and recurrent nerve (*np*). A group of early differentiating neurons, most probably deriving from the *tSNSPs*, is located at the position of the hypocerebral ganglion; anteriorly directed axons of these cells pioneer the recurrent nerve (*rnp*). **E, F** Pattern of developing stomatogastric ganglia in a stage 16 embryo. The hypocerebral (*hy*) ganglion is a derivative of the intermediate pouch; the paraoesophageal ganglion (*pe*) comes from the rostral pouch, the ventricular ganglion (*vg*) from the caudal pouch. Other abbreviations. *en_l* left esophageal nerve; *en_r* right esophageal nerve; *es* oesophagus; *fc* frontal connective; *ph* pharynx; *pv* proventriculus; *m* recurrent nerve; *sec* supraesophageal commissure; *spg* supraesophageal ganglion; *vc* ventral nerve cord

neurons located in front of the brain. The somata of the neurons surround a sparse neuropile and are enclosed in a glial sheath. Neurons of the frontal ganglion project laterally and medially. Those axons projecting laterally form the frontal connective which links the frontal ganglion to the brain, and can be considered as part of the tritocerebrum (Hartenstein et al. 1994a). Medially directed axons of the frontal ganglion, upon reaching the dorsal midline, turn either anteriorly or posteriorly. Anteriorly directed fibres, forming the so-called frontal nerve, innervate the pharyngeal muscles. Posteriorly directed axons form the recurrent nerve, which passes beneath the supraoesophageal commissure and connects the frontal ganglion with the hypocerebral and paraoesophageal ganglia. This nerve contains approximately 50 axons, all rather similar in diameter, and is surrounded by flattened cells, probably glial cells and/or scattered stomatogastric neurons.

10.1.2
Hypocerebral and Paraoesophageal Ganglion

Two other stomatogastric ganglia are located between the brain hemispheres, behind the supraoesophageal commissure. Each of these ganglia contains approximately 10-12 neurons, the axons of which enter the recurrent nerve (Figs. 10.1, 10.4) and show a striking asymmetry with respect to their position. The left ganglion, which we term the hypocerebral ganglion, is more compact and usually occupies a slightly more dorsal position. The ganglion of

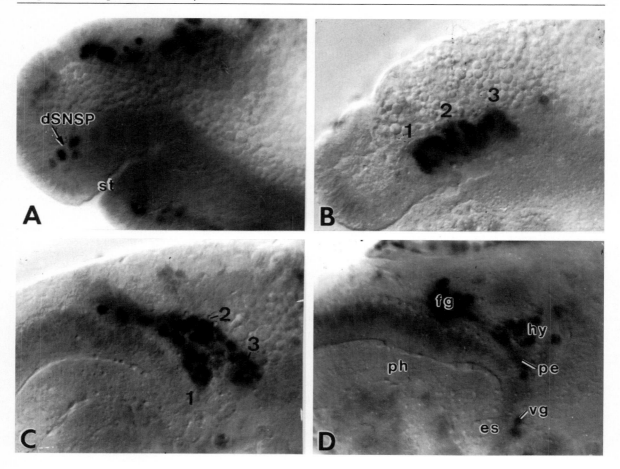

Fig. 10.2 Development of the SNS. The photographs show lateral views of anterior parts of embryos at stages 10 (**A**), 12 (**B**), 14 (**C**) and 16 (**D**). Precursors of the SNS were labeled with an antibody against Asense (**A**) or by the P-*lacZ* insertion 94 (**B-D**). **A** The first group of precursors, called *dSNSPs*, have delaminated from the roof of the stomodeum (*st*). **B** The bulk of precursors forms three pouches (**1-3**) invaginating from the roof of the stomodeum. **C** The epithelial pouches are carried posteriorly by the elongating foregut; the pouches start to dissociate and form an anteriorly directed stream of cells. **D** Descendants of the pouches coalesce into discrete ganglia (*fg* frontal ganglion; *hy* hypocerebral ganglion; *pe* paraoesophageal ganglion; *vg* ventricular ganglion). Other abbreviations. *es* oesophagus; *ph* pharynx

the right side (paraoesophageal ganglion) consists of an elongated group of neurons closely attached to the oesophageal wall.

10.1.3
Ventricular Ganglion

Besides their contribution to the recurrent nerve, other axons of the hypocerebral and paraoesophageal ganglia project posteriorly. The axon bundle coming from the paraoesophageal ganglion (right oesophageal nerve) follows the convex surface of the curved oesophageal wall (Figs. 10.1, 10.2). Several widely spaced neurons (5-10) contribute to this axon bundle, which grows further posteriorly onto the proventriculus during late embryogenesis. Along the oesophagus, axons and neurons are enclosed in a glial sheath and lie in grooves or channels formed by the visceral muscle layer. Shortly before the axons terminate on the proventricular wall, the glial sheath is lost. The posteriorly directed nerve that leaves the hypocerebral ganglion (left oesophageal nerve) is shorter and does not pick up scattered neurons along its trajectory. It extends ventrally to reach a solid cluster of about 10 neurons attached to the proventriculus. This group of neurons probably corresponds to the ventricular ganglion (the proventricular ganglion of Gonzalez-Gaitan and Jäckle 1995), a structure found in other insects at a similar position (Penzlin 1985). Short axons can be seen leaving the ventricular ganglion posteriorly and fanning out over the surface of the proventriculus. Although no axons have been found to grow onto the main body of the midgut, such axons may possibly grow out during larval stages.

10.2
The Ring Gland of the Mature Embryo

10.2.1.
The Corpora Cardiaca

The corpora cardiaca are paired neuroendocrine structures which in dipterans form part of the ring gland (Poulson 1945) and are not part of the stomatogastric nervous system *sensu stricto*. In the late embryo, the corpora cardiaca are formed by 6-8 small MAb 22C10-positive neurons on each side (Fig. 10.4). Short axons connect the corpora cardiaca with the protocerebrum (*n. corporis cardiaci*) and the corpus allatum, another neuroendocrine struc-

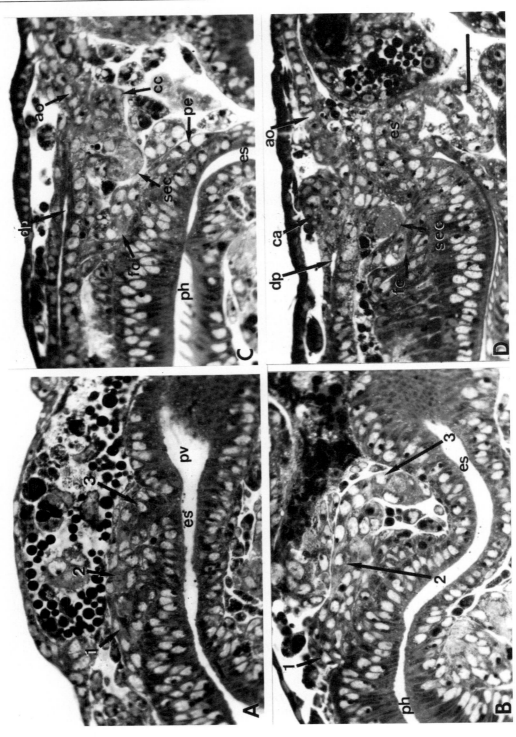

Fig. 10.3 A to D are parasagittal sections to illustrate the development of the stomatogastric nervous system. A shows the primordial cells, derived from three invaginations that appeared at the roof of the pharynx in stage 11. These cells now form three distinct groups, numbered 1, 2 and 3. Mitotic divisions occur within the cells of the three groups which consequently grow, undergoing profound modifications (**B**). In stage 16 (**C**), about halfway through head involution, the definitive ganglia of the stomatogastric nervous system can already be recognized. The frontal commissure (*fc*) is present, as well as the corpus allatum (*ca*) and corpora cardiaca (*cc*). D shows the final organization in the embryo of stage 17. *ao* aorta; *dp* dorsal pouch; *es* oesophagus; *pe* paraoesophageal ganglion; *ph* pharynx; *pv* proventriculus; *sec* supraoesophageal commissure. *Bar* 20 μm

ture located in the dorsal sector of the ring gland. Nerves connecting the corpora cardiaca with the hypocerebral ganglion (*n. cardiostomatogastrici*, Penzlin 1985), which have been described in other insects, are not apparent in the late *Drosophila* embryo. However, we cannot exclude the possibility that these nerve connections are formed postembryonically in *Drosophila*.

10.2.2.
Prothoracic Gland and Corpus Allatum

Detailed descriptions of these structures exist for larval stages (King et al. 1966) but, to our knowledge, nothing has yet been published on their architecture during embryonic stages (Fig. 10.4). Prothoracic glands form a cluster of 10-20 cells on either side of the dorsal vessel. Ventrally, they are continuous with the corpora cardiaca, dorsally with the corpus allatum, an unpaired median cluster of approximately 10 cells on top of the dorsal vessel. Axons of the corpora cardiaca extend along the external surface of the prothoracic glands and terminate in the corpus allatum.

10.3
Development of the Stomatogastric Nervous System and Ring Gland

10.3.1.
Segregation and Morphogenesis of Progenitor Cells

All the ganglia described above derive from groups of progenitor cells which invaginate from the roof of the stomodeum, at the level of the oesophagus, during early stage 12. The origin and subsequent dispersal of these cells was followed by using a P-*lacZ* insertion which is expressed in these cells, as well as the anti-

Fig. 10.4 Development of the ring gland. Diagram shows a dorsolateral view of head of a stage 14 embryo (**A**) and a stage 17 embryo (**B**). Parts of the ring gland are indicated in different colours (see colour key). Several other structures (stomatogastric nervous system, foregut, supraoesophageal ganglion, dorsal vessel, lymph gland) are outlined in grey as points of reference. Precursors of the corpora cardiaca (lilac) and corpus allatum (blue) originate in deep layers of the dorsal part of the gnathal segments. They migrate dorsally during dorsal closure (*arrow* in **A**) and become associated with the aorta (*ao*). The corpora cardiaca are formed by a bilaterally symmetrical cluster of neurons which send axons dorsally into the corpus allatum (**B**). Precursors of the corpora cardiaca can be distinguished first as a group of cells anteriorly adjacent to the precursors of the corpus allatum. It is possible, as shown in this diagram, that they originate in the deep layers of the gnathal segments. However, the issue of their origin has not been resolved. Precursors of the prothoracic gland originate from the dorsal ectoderm of the prothoracic segment. After sinking into the embryo, these cells migrate dorsally and form bilaterally symmetrical vertical plates on either side of the aorta, anterior to the lymph gland (*lg* in **B**). Other abbreviations. *es* oesophagus; *fg* frontal ganglion; *fn* frontal nerve; *hy* hypocerebral ganglion; *mg* midgut; *ph* pharynx; *pv* proventriculus; *rn* recurrent nerve; *sec* supraoesophageal commissure; *sg* salivary gland; *vg* ventricular ganglion. Modified from Hartenstein (1993), with permission of Cold Spring Harbor Laboratory Press

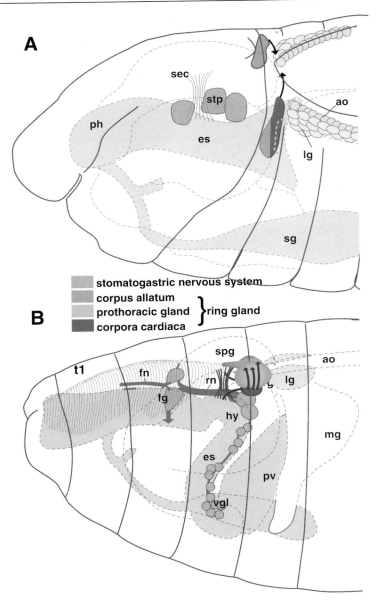

Crumbs antibody (Tepass and Knust 1990; Hartenstein et al. 1994a). During early stage 11, a median group of cells at the dorsal lip of the stomodeum start to express the P-*lacZ* marker (Figs. 10.1 - 10.3). After these cells have moved inside the stomodeum, they invaginate at three contiguous, median positions and synchronously undergo a final round of mitosis, corresponding to the fourth postblastoderm mitosis. This mitosis brings their total number up to about 75, with approximately 25 cells located in each of the three invaginations. The invaginations giving rise to the stomatogastric nervous system remain open to the lumen of the primordial oesophagus until late stage 12 (Figs. 2.19, 2.25). Subsequently, the invaginated cells separate from the oesophagus and form three vesicles located above its roof (Figs. 2.29, 2.33, 10.3). The anterior and intermediate vesicles come very close to each other before the cells lose their epithelial structure (stage 14). The posterior vesicle, which is located at the transition between oesophagus and proventriculus, retains its epithelial structure until stage 15.

From stage 13 onward the three vesicles dissociate and form solid clusters of cells which migrate anteriorly (Figs. 10.1, 10.2, 10.3). During late stage 14, the cells of the former anterior and intermediate vesicles form two elongated clusters which merge anteriorly, at the position where the frontal ganglion will develop. Thus both vesicles contribute cells to the frontal ganglion. The posterior part of the intermediate vesicle appears to give rise to the hypocerebral ganglion. The posterior part of the rostral vesicle remains in close association with the oesophageal wall and will form the paraoesophageal ganglion. The posterior part of the posterior vesicle remains at the junction between oesophagus and proventriculus and develops into the ventricular ganglion. More anterior cells of this vesicle move dorsally and anteriorly; some of these cells seem to contribute to the frontal and/or hypocerebral ganglion.

10.3.2.
Differentiation of Axonal Tracts of the Stomatogastric Nervous System

During stage 15, the cells of the dissociated vesicles assemble into four small ganglia which are closely associated with the brain and foregut. Two small unpaired clusters of neurons become MAb22C10 positive and form axons during late stage 14. One cluster is located anteriorly in the frontal ganglion. These cells send out their axons laterally and thereby pioneer the frontal connectives that link the frontal ganglion to the brain. Interestingly, the further movement of the neurons whose axons pioneer the frontal connective reveals a striking asymmetry, in that, during subsequent development, the cluster of pioneer neurons shifts away from the midline to one side (usually the right side).

The posterior cluster of early differentiating stomatogastric neurons (2-4 cells) forms part of the prospective hypocerebral ganglion (Fig. 10.2). The axons of these neurons pioneer the recurrent nerve, and are referred to as "recurrent nerve pioneers". The developmental origin of the individual components of the stomatogastric nervous system and the formation of pioneer tracts is summarized in Fig. 10.1.

10.3.3.
Development of the Ring Gland

We know very little about the development of the ring gland. The corpus allatum derives from a group of cells that become distinguishable as early as stage 13, on the basis of expression of particular enhancer trap insertions (Fig. 10.5), between the head mesoderm and the epidermis of the gnathal segments. During dorsal closure, the cells move dorsally and finally meet and fuse above the cardioblasts. For other insects, it has been reported that the corpus allatum is a derivative of the ectoderm of the gnathal segments (Snodgrass 1935). We have not been able to in distinguish between a mesodermal and an ectodermal origin for these cells. The thoracic glands originate from the dorsal ectoderm of $t1$. Their cells segregate from the ectoderm during stage 15 and form bilateral clusters near the heart precursors. Later, these cells fuse with the corpus allatum. The origin of the ventral part of the ring gland (corpora cardiaca) could not be analyzed, since no specific marker which labels this structure from early stages award is known at the present time.

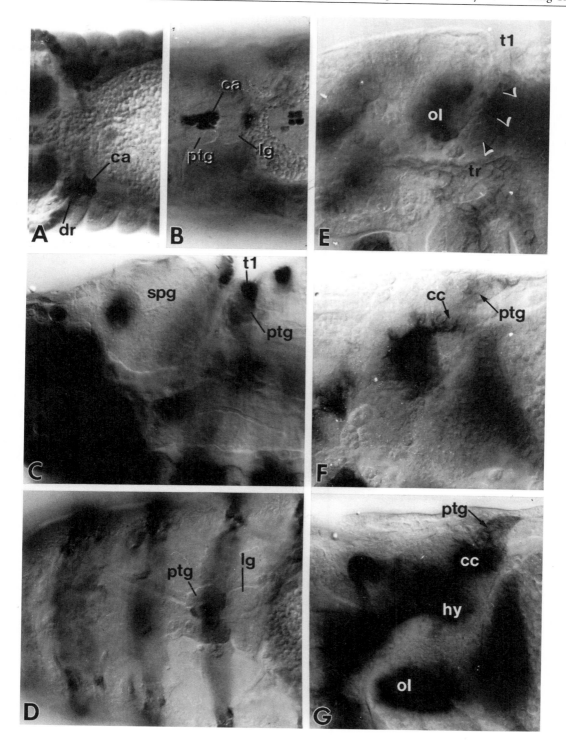

Fig. 10.5 Development of the ring gland. **A, B** Dorsal view of anterior portion of stage 13 (**A**) and stage 16 (**B**) embryos labeled with the P-*lacZ* insertion E2-3-9 (Bier et al. 1989) which is expressed in the corpus allatum (*ca*). Precursors of this structure originate as a bilateral cluster of cells located within the mesoderm of the gnathal segments (*dr* dorsal ridge, formed by dorsal part of the gnathal segments). In the wake of dorsal closure, these cells migrate dorsal and fuse in the dorsal midline above the aorta (see also Fig.10.4). **C, D** Lateral view of stage 14 embryo (**C**) and dorsal view of stage 16 embryo (**D**) labeled with P-*lacZ* insertion *rhx25* (Hama et al. 1990); region-specific expression is directed by portions of the *Drosophila engrailed* promoter. From the dorsal part of the prothoracic *en* stripe (*t1* in **C**) a group of cells sink inward and give rise to the prothoracic gland (*ptg*). At later stages (**D**), the prothoracic gland forms two bilaterally symmetric plates on either side of the aorta. The group of cells behind the prothoracic gland (*rhx25* negative) is the lymph gland (*lg*). **E - G** Lateral views of stage 13 (**E**), stage 14 (**F**) and stage 16 (**G**) embryo labeled with anti-FasII antibody (Grenningloh et al. 1991) which stains part of the developing optic lobe (*ol*), corpora cardiaca (*cc*) and corpus allatum Progenitor cells of the corpora cardiaca can first be detected during stage 13 as a compact cluster of cells adjacent to the precursors of the corpus allatum. During dorsal closure, the corpora cardiaca migrate dorsally and finally fuse in the midline beneath the aorta. The corpora cardiaca are immediately above the hypocerebral ganglion (*hy*) which is also stained with anti-FasII. *spg* supraoesophageal ganglion

Chapter 11
Central Nervous System

The development of the central nervous system is a well ordered process that goes through several different steps. In the first step, ectodermal cells are endowed with neurogenic capabilities, differentiating the neuroectoderm from the remaining regions. The second step consists of the segregation of the neural progenitor cells from the epidermal progenitor cells in the neuroectoderm. Two types of progenitor cells for the CNS can be distinguished, the neuroblasts and the midline progenitor cells. The neuroblasts are large cells (Wheeler 1891, 1893) located on either side of the midline, which originate from a well defined area of the ectoderm called the neurogenic region (Poulson 1950) or neuroectoderm. The midline progenitor cells derive from the mesectoderm cord, adjacent to the neuroectoderm, and give rise to the so-called median cord. Neuroblasts delaminate from the neurogenic region early in embryogenesis and come to lie between the ectoderm and mesoderm; these cells, which are arranged in a regular spatial pattern, form the primordium of the CNS. Segregation of neuroblasts is a continuous process that takes approximately 3 h; it occurs in pulses, each of which gives rise to different populations of neuroblasts. Clear differences exist in the behaviour of segregating neural cell progenitors between the procephalic lobe and the germ band, which force us to give separate descriptions of neurogenesis in these regions. The third step in neurogenesis leads to the production and cytodifferentiation of neurons. After segregation, ventral cord neuroblasts go through several rounds of asymmetrical, stem cell-like divisions giving rise to ganglion mother cells (Bauer 1904; Poulson 1950), which then produce progeny cells through symmetrical mitoses.

The CNS comprises neurons and glial cells. All of these cells, approximately 320 per abdominal hemineuromere (Hartenstein et al. 1987), develop from the cells generally called neuroblasts. Thus, it has been shown that the progeny of some neuroblasts, for example NB1-1, include glial cells as well as neurons (Becker and

Technau 1990; Udolph et al. 1993; Bossing et al. 1996a, b; see also Fredieu and Mahowald 1989, for in vitro studies). Udolph et al. (1993) denominate NB1-1 a neuroglioblast. Moreover, Jacobs et al. (1989; see also Doe et al. 1988) have reported that the longitudinal glia cells derive from a lateral neuroblast which, accordingly, they call a glioblast. The neuroblasts therefore constitute a rather heterogeneous population of cells, which include progenitors of glial cells (glioblasts), progenitors of neurons and glial cells (neuroglioblast), and progenitors of neurons. The pattern of neuroblast divisions varies depending on whether the cell considered is located in the germ band or the procephalic lobe. Generalizations regarding the cell division pattern of neuroblasts are difficult because of these, and other, regional differences. Nevertheless, both symmetrical divisions, giving rise to two neuroblasts, and asymmetrical divisions, from which a so-called ganglion mother cell (Bauer 1904) and a neuroblast will originate, have been observed within the procephalic lobe, whereas only divisions of the asymmetrical type have been observed among the neuroblasts of the germ band.

In the following we shall consider the formation of the neurogenic region from the undifferentiated ectoderm, the process of neuroblast segregation, and the pattern formed by the neural progenitor cells after internalization (Hartenstein and Campos-Ortega 1984; Doe 1992; Campos-Ortega 1993).

11.1
Early Neurogenesis in the Segmented Germ Band

11.1.1
The Formation of the Neurogenic Region

The neurogenic region becomes manifest due to morphological modifications of the ectoderm, in particular a conspicuous enlargement of the cells within certain areas of the germ layer. Chronologically, the first modifications are observed in the procephalic lobe before the onset of mitotic activity in the ectoderm, shortly after formation of the cephalic furrow. A strip of about 80 large cells becomes apparent on either side of the procephalic lobe, which extends dorsoventrally parallel to the head fold (Figs. 2.8, 2.9, 11.8; see section below). The ectoderm of the germ band exhibits similar modifications slightly later, during the first mitotic interphase of mesodermal cells (see Chapters 4 and 14). At this

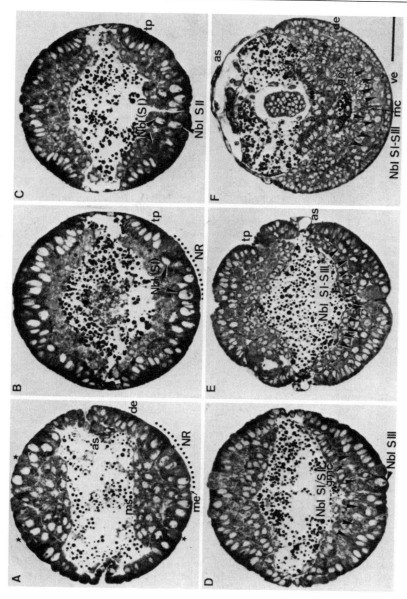

Fig. 11.1. Transverse sections of embryos of increasing age to illustrate segregation of neuroblasts (this is Fig. 4 of Hartenstein and Campos-Ortega 1984). Level of sectioning is about 35% EL. **A** shows the neurogenic region (*nr*) before neuroblast segregation. Notice the size difference between neurogenic ectoderm and primordium of the dorsal epidermis (*de*). Mesodermal cells (*ms*) are in the first interphase. Mitotic cells (*stars*) are present in the centre of the neurogenic region. **B** The first neuroblast subpopulation (**Nbl SI**), *arrows*) segregates during the second mesodermal mitosis. *tp* tracheal placodes. **C** Second neuroblast subpopulation (*Nbl SII*) begins segregation. **D** SII neuroblasts have completed segregation (*arrowheads*), a few ganglion mother cells (*gmc*) have been budded off. SIII neuroblasts (*Nbl SIII*) start segregating. **E** Neuroblast (*arrowheads*) segregation is virtually complete. Tracheal placodes (*tp*) have invaginated. **F** Shortening of the germ band. Ganglion cells (*gc*) form conspicuous neuromeres. Neuroblasts have shrunken considerably by this stage, though some are still visible at the ventral surface of each neuromere. Median cells (*mc*) form a narrow vertical sheet along the ventral midline. Abbreviations. *as* amnioserosa; *de* dorsal epidermis; *ve* ventral epidermis. *Bar* 50 μm

time, the ectoderm in the territory of the germ band becomes organized into two clearly distinct regions that extend on either side along the whole length of the germ band, from the posterior lip of the cephalic furrow down to the developing proctodeum (Figs. 11.1, 11.2). One of these regions is lateral and consists of small cells, and the other is medial and contains large cells. The

Fig. 11.2 A-E show drawings of transverse sections of stage 5 (**A**), stage 8 (**B**), stage 9 (**C**), stage 11 (**D**) and stage 13 (**E**), to illustrate the development of the neuroblasts. The ventral neurogenic region (*VNE*) gives rise to the neuroblasts of the ventral nerve cord. Separating the ventral neurogenic region from the mesodermal anlage (*ms*) is the mesectodermal anlage (*mec*), which consists of a single row of cells on either side of the embryo. In stage 8 (**B**), the cells of the ventral neuroectoderm enlarge. Whereas the primordium of the dorsal epidermis (*DEA*) undergoes its first postblastoderm division, mitosis is delayed in the neurogenic regions. Starting at the end of stage 8, neuroblasts (*nb*) delaminate in pulses, or waves, from the neuroectoderm. The first wave yields two rows of neuroblasts (SI) on either side (**C**). The neuroblasts that delaminate during the second wave (SII) originate from intermediate regions of the neuroectoderm and fill the gap between the two rows of SI neuroblasts. SIII, SIV and SV neuroblasts (Doe 1992) segregate from predominantly medial positions throughout stage 10 and stage 11 (**D**). Shortly after segregation, neuroblasts start dividing with a perpendicularly oriented spindle. Their progeny, the ganglion mother cells (*gmc*), then lie between the neuroblasts and the mesoderm (*ms*). Each ganglion mother cell performs one symmetric division yielding two neurons (**E**). Abbreviations. *mp* midline precursors; *cn* connectives; *mg* midgut; *myo* myoblasts; *pn* peripheral nerve. Modified from Hartenstein (1993), with permission of Cold Spring Harbor Laboratory Press

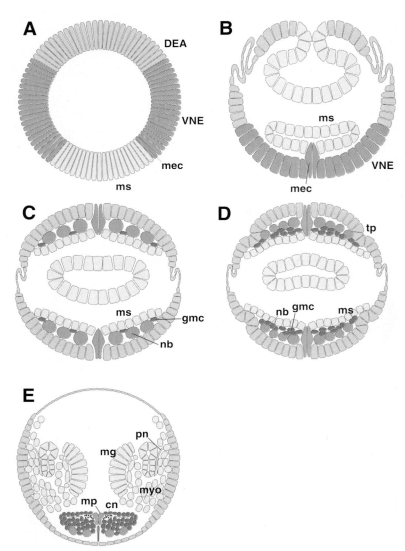

region of small cells comprises the primordia of the dorsal epidermis and of the tracheal tree, while the region of large cells will give rise to the ventral epidermis and the neuroblasts of the ventral cord, and is referred to as the neurogenic region, or neuroectoderm, of the germ band (Hartenstein and Campos-Ortega 1984; Technau and Campos-Ortega 1985).

The cellular organization of the neuroectoderm of the procephalic lobes and germ band exhibits conspicuous regional differences, which are related to both the size of the cells and their mitotic potential (see below). Within the germ band, periodically arranged clusters of similar small mitotic cells can be observed in the middle of the neuroectoderm, forming three distinct subdivisions. We have called these medial, intermediate and lateral regions (Figs. 11.1, 11.2, 14.1, 14.5), and each gives rise to a different number of neuroblasts. The mitotic clusters are segmentally organized, and in fact they represent the first obvious manifestation of metamery during *Drosophila* embryogenesis. Neuroblasts will segregate from the medial and intermediate, zones and, to a far lesser extent, from the lateral subdivision of the neuroectoderm.

11.1.2
The Segregation of Neuroblasts

In the germ band, neuroblast segregation follows a simple scheme. Single cells in the three regions of the neuroectoderm adopt the shape of an inverted bottle with the nucleus displaced basally, while the cytoplasm becomes striking basophilic as the cells leave the ectodermal layer (Poulson 1950; ref. to Fig. 11.1). The apical cytoplasmic process, the neck of the bottle, persists for some time, wedged between the remaining cells of the ectoderm, and is finally withdrawn so that each neuroblast rounds up and starts dividing. Similar cytoplasmic processes are also formed at the basal poles of those cells that remain in the ectodermal layer after the segregation of neuroblasts, and these processes surround the neuroblasts and fill the space between them (see Fig. 6B of Hartenstein and Campos-Ortega 1984; Doe 1992; Fig. 11.1). There is a correlation between segregation of neuroblasts and cell cycle stage, which will be discussed in section 14.2.4.

Within the germ band, neuroblast segregation takes approximately 3 h to complete, and occurs in waves which give rise to different subpopulations of neuroblasts (Figs. 11.3 - 11.6). Hartenstein and Campos-Ortega (1984) distinguished, on the basis of serial sections, three subpopulations: SI, SII and SIII, consisting of different numbers of cells. However, Doe (1992), using a variety of molecular markers, has distinguished two additional pulses of segregation, which consequently produce neuroblasts of the subpopulations SIV and SV. SI neuroblasts segregate within approximately 10 min at the end of stage 8, and are chiefly derived from the medial and lateral neuroectodermal subdivisions. They become arranged in two longitudinal rows on either side of the germ band; in each abdominal segment, two SI cells derived from the intermediate subdivision are found interspersed between the medial and lateral rows, as the first elements of the third, inter-

Fig. 11.3. Pattern of ventral neuroblasts at stages 9 (**A**), 10 (**B**) and 11 (**C**). Broken transverse lines indicate parasegmental metameric boundaries. The regular double row of grey rectangles in the centre represents the mesectoderm. Neuroblasts are shown as coloured circles, with different colours representing different subpopulations (blue, SI; green, SII; yellow, SIII; orange, SIV/V). Nomenclature according to Doe (1992). By stage 9 (**A**), SI neuroblasts form two regular columns with four neuroblasts per metamere on either side of the midline. The pattern of SI neuroblasts is shown for a thoracic metamere; in the abdomen, there are two SI neuroblasts in the space between the medial and lateral columns. Two of the SI neuroblasts do not divide as stem cells, but undergo only one further division (similar to the mesectodermal cells) before their progenies differentiate as neurons. These cells are not shown in **C**. By late stage 10 (**B**), SII and SIII neuroblasts have delaminated. SII neuroblasts mainly form an intermediate column. SIII neuroblasts appear as metameric clusters; three cells appear in the medial column, one in the lateral column. Two additional, smaller progenitor cells delaminate even further laterally than the lateral column; one, perhaps both, of these are glial progenitors and divide differently from the neuroblasts. During stage 11 SIV and SV neuroblasts delaminate as metameric clusters, mainly in between the lateral and intermediate column. One neuroblast, called the median neuroblast, segregates from the mesectoderm; the remaining mesectodermal cells (midline precursors) differentiate into neurons and glial cells without intervening divisions

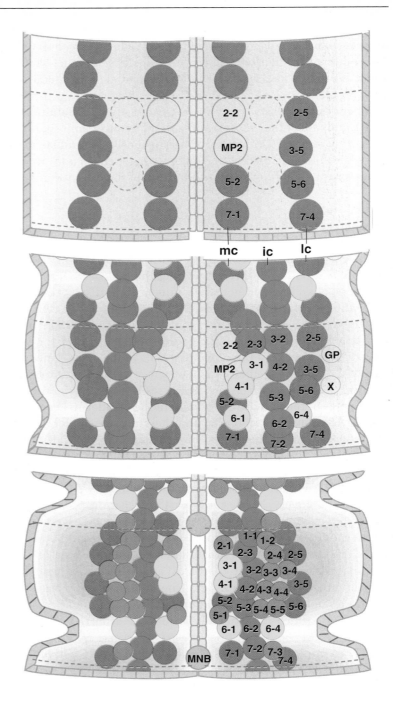

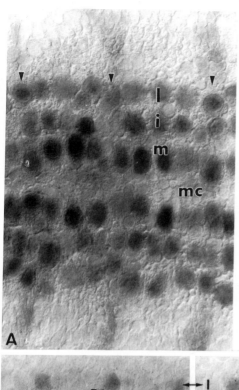

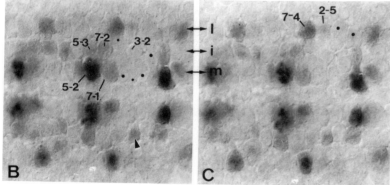

Fig. 11.4 Pattern of neuroblasts during early stage 10. The photographs show ventral views of the mid-section of a stage 10 embryo labeled with 439 (**A**), an enhancer trap insertion in 64A (Campos-Ortega and Haenlin 1992), and H162 which expresses *lacZ* in the *seven-up* pattern (Doe 1992) (**B, C**, represent superficial and deep focal planes of the same preparation). At the stage shown in **A-C**, the S I and S II subpopulations of neuroblasts have delaminated. They form three regular longitudinal rows on either side of the midline (*l, i, m* lateral, intermediate and medial; *mc* mesectoderm). In each metamere, there are four neuroblasts per column, as can be seen directly in **A** where Engrailed stripes (*arrowheads*) demarcate metameric boundaries. The *seven-up* gene is expressed in specific subsets of neuroblasts (compare with Fig.11.3). Large dots in the lateral row (adjacent to neuroblast 2-5 shown in panel C) indicate the Seven-up negative 3-5 and 5-6 neuroblasts; the large dot in the intermediate row (posterior to 3-2 in panel **B**) corresponds to neuroblast 4-2; the small dot in the intermediate row, anterior to 3-2, is neuroblast 1-1; dots in medial row (posterior to 7-1 in C) correspond to 2-2, 2-3, and MP, respectively

mediate row of neuroblasts. The intermediate row will be completed by the segregation of another four SII neuroblasts per hemisegment at the end of stage 9. The arrangement of SI and SII neuroblasts is fairly regular, so that four mediolateral, or transverse, and three longitudinal rows of neuroblasts can be defined per hemisegment. After the segregation of SIII neuroblasts, this regular pattern is disturbed and individual neuroblasts cannot be identified on the basis of their location alone. SIV and SV neuroblasts segregate chiefly from lateral regions of the neuroectoderm,

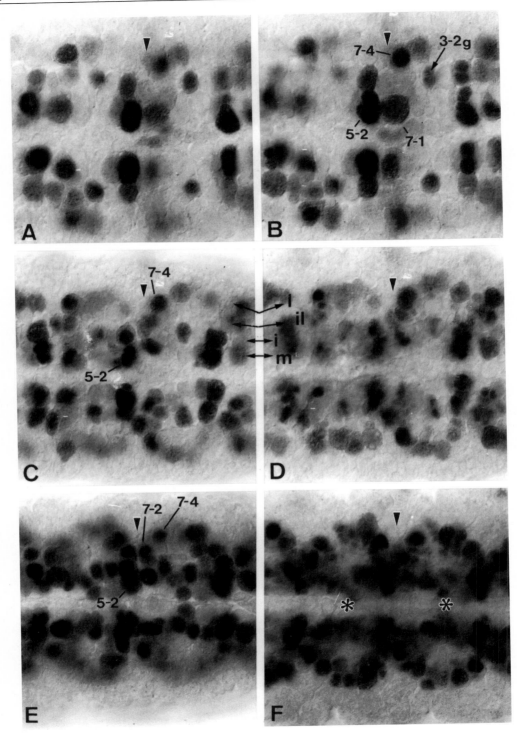

Fig. 11.5 Pattern of neuroblasts during stages 11 (early 1 1. **A, B**; late 11. **C, D**) and 12 (**E, F**). Photographs show ventral views of the midsection of embryos labeled with P-*lacZ* insertion H162 expressing in the *seven-up* pattern. Left panels show a superficial plane of focus, right panels a deep plane of focus. Most SI and SII neuroblasts have undergone two rounds of mitosis, as is evident in the case of 3-2, which has given rise to two ganglion mother cells (3-2g in **B**). During later stages (**C-F**), more neuroblasts delaminate. SIII neuroblasts are added to the medial row. A large fraction of neuroblasts express the Seven-up marker, which makes it difficult to identify individual neuroblasts using this marker alone. The constellation *5-2/7-1* (medial row), *5-3/7-2* (intermediate row) and *7-4* (lateral row) stands out by its somewhat stronger labeling and allows one to follow the development of metameric furrows in the neural primordium (compare also Figs.11.3 and 17.1)

forming two rows in between the SII and lateral SI neuroblasts (Doe 1992; Broadus et al. 1995; see Figs. 11.3, 11.4).

11.1.3
The Pattern of Neuroblasts

The pattern of neuroblasts shown in Fig. 11.3 corresponds to that of Doe (1992) as modified by Broadus et al. (1995). SI and SII neuroblasts are arranged in three regular longitudinal rows (Figs. 11.3, 11.4, 11.5). The pattern formed by these cells seems to be a consequence of the order in which they leave the ectoderm, deriving from the three subdivisions of the neurogenic ectoderm. Up to the segregation of SII neuroblasts, this pattern is highly regular and constant, allowing individual cells to be identified in different embryos on the basis of their position in the lattice. This lattice, at this stage, forms a sheet of cells and, with the exception of the tips of the segmented germ band, i.e. the mandible and *a9*, no differences can be observed between the neuroblast complements of the different prospective segments. Typically the neuroblast lattice (Fig. 11.3) consists of 57 transverse rows on either side of the embryo, of which 53 have three cells each, two rows, the most anterior ones, have two cells, and another two, the most posterior, only one cell; each prospective segment contains at this stage approximately 10 neuroblasts on either side; mandible and *a9* have 4 and 2 neuroblasts, respectively, on either side.

Further differences in the neuroblast complement of the different segments become clearly evident after segregation of SIII neuroblasts; these differences affect maxilla, labium and thoracic segments, which contain more neuroblasts than the abdominal hemisegments. However, these differences appear to be only transitory. Doe (1992) has shown by using specific markers that the final complement of neuroblasts is the same in abdominal, thoracic and gnathal segments, but that many thoracic neuroblasts segregate earlier than their abdominal homologues. It should be

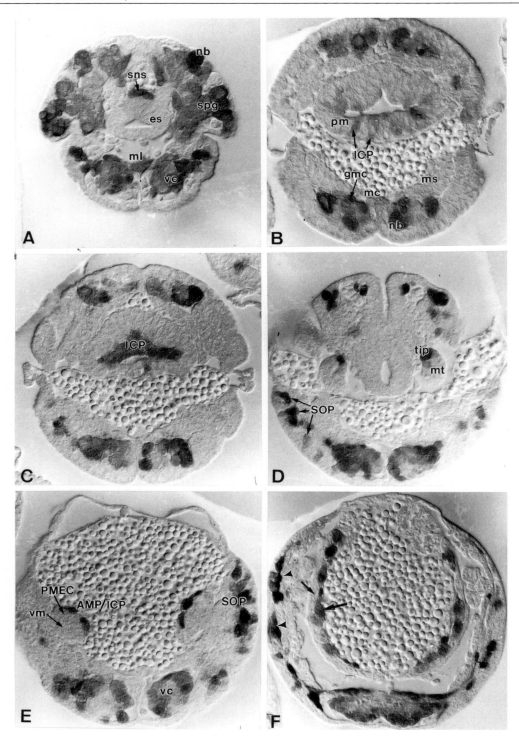

Fig. 11.6. A to F are cross sections of stage 9 (A), 10 (B), 11 (C), 12 (D), 13 (E) and 14 (F) embryos stained with an anti-Asense antibody (kindly provided by A. Jarman and Y.N. Jan). Within the developing central nervous system (*spg* supraoesophageal ganglion; *vc* ventral cord), Asense is expressed in neuroblasts (*nb*) and sensory organ progenitor cells (*SOP*), as well as in ganglion mother cells (*gmc*) and the mesectoderm (*mc*). Within the developing midgut (*pm*), Asense is expressed in the interstitial (*ICP*) and imaginal (*AMP*) cell progenitors, but not in the principal midgut cells (*PMEC*; refer to Fig. 7.3). Finally, it is also expressed in the tip cell (*tip*) in the developing Malpighian tubules (*mt*)

noted that segmental boundaries can be distinguished in the neuroblast lattice as early as the time of the second mitotic interphase in mesoderm cells. Segmental boundaries can be accurately determined because the MP2 cells project about 10 μm above the average neuroblast level. At later stages, when epidermal segmentation becomes apparent, the arrangement of the different neuromeres attains its definitive form (Fig. 11.5). The neuroblasts of the lateral row at this stage exhibit conspicuous indentations, such that the lateral neuroblasts are arranged in a semicircle rather than a straight row (Fig. 11.5; see Chapter 17); these indentations lead to very pronounced segmental clustering of the neuroblasts which completely distorts the regular lattice typical of previous stages.

11.1.4
The Proliferation of the Neuroblasts

Immediately after segregating from the ectodermal germ layer each neuroblast rounds up and begins to divide within 10-20 min. In the prospective ventral cord, neuroblasts divide asymmetrically, i.e. they follow a stem cell pattern of proliferation, giving rise after each mitosis to a ganglion mother cell and a neuroblast that again divides asymmetrically. Ganglion mother cells in various parts of the CNS of insects have been reported to undergo a single symmetrical mitosis that yields two neurons (Bauer 1904; Nordlander and Edwards 1969; White and Kankel 1978; Hartenstein et al. 1987; Hofbauer and Campos-Ortega 1990). In the *Drosophila* embryo, the duration of the cell cycle in ganglion mother cells has been found to be approximately 100 min (Hartenstein et al. 1987). The identification of individual ganglion mother cells is very difficult in normal material, since their topological relationship to the neuroblasts from which they derive is not especially strict. Bueznow and Holmgren (1995) have used FLP recombinase technology to show lineage-specific expression of *gooseberry*. This gene is expressed in particular ganglion mother cells which, accordingly, can be distinguished from other ganglion mother cells that do not express *gooseberry* (refer to Broadus et al. 1995). There is, beyond doubt, great potential for further development of this technique.

Attempts to determine the number of divisions that *Drosophila* neural progenitor cells go through during embryogenesis have been undertaken in two different studies. In one of them fixed material stained with an antibody raised against a nuclear antigen was used (Hartenstein et al. 1987). The observations suggest a fairly stereotypic pattern of neuroblast divisions. The overwhelming majority of neuroblasts were found to carry out up to eight parasynchronous cycles of division, which occur at regular intervals of 40-50 min each. The main period of neuroblast mitoses extends from formation of the stomodeal plate in stage 9, when the first division takes place, to completion of germ band shortening in stage 13.

However, in the neuroblast layer, scattered mitotic figures are still visible at stage 14; in addition, some neuroblasts have been found to incorporate BrdU and divide until stage 16 (Prokop and Technau 1991; see Chapter 14). The other study was based on HRP-labelled neural clones that had developed from cells transplanted to the ventral neuroectoderm of the wildtype (Technau and Campos-Ortega 1986a). Clones of 2-18 neural cells were found, suggesting that neuroblasts perform up to nine mitoses; only four larger clones were found (with 21-26 cells). However, there was significant scatter in clone size, suggesting a larger range of variation with respect to mitoses than that implied by the observations of Hartenstein et al. (1987): some neuroblasts may well perform only one, others up to 10 or 11 mitoses (Technau and Campos-Ortega 1986a; Prokop and Technau 1991). The available studies on the cell lineages derived from neural progenitors, which involved labelling individual neuroectodermal cells with DiI, also indicate some variability in the constitution of these lineages (Udolph et al. 1993; Bossing and Technau 1994; Bossing et al. 1996b; see section 11.1.5).

During their period of proliferation, many, though not all, dividing neuroblasts show a strikingly bilateral symmetric organization, which had already been noticed by Poulson (1950). After each division, the size of the neuroblasts decreases to such an extent that these cells progressively lose their prominent appearance and, at later stages, the majority of neuroblasts are hardly distinguishable from ganglion mother cells and differentiating neurons because the earlier size differences have vanished (Fig. 11.6). Nevertheless, single germ band neuroblasts remain distinguishable for some time, until dorsal embryonic closure, dividing according to the same asymmetrical pattern of previous stages. These late neuroblasts most probably correspond to SIV and SV neuroblasts.

By the time embryonic neuroblasts cease dividing, they have shrunken considerably and cannot be distinguished from the remaining cells of the CNS. However, during the first larval stage, neuroblasts enlarge and again become recognizable, beginning to produce neurons that become incorporated into the imaginal CNS (Truman and Bate 1988). Experimental evidence derived from transplantations of doubly labelled cells indicates that larval and imaginal neurons share common lineages, that is to say, the neuroblasts which proliferate in the larva originate in the embryo and are probably identical with some of the neuroblasts that proliferated in the embryo (Prokop and Technau 1991).

During shortening of the germ band, a reduction in segment length proceeds from rostral to caudal; in addition, by this stage the number of distinguishable neuroblasts has decreased considerably in each segment, while the number of progeny cells has increased. Increase in cell number and decrease in neuroblast size makes the pattern of neuroblasts morphologically indistinct during the later stages of neurogenesis. However, as already mentioned, specific molecular markers allow one to distinguish individual neuroblasts even during these late stages (Doe 1992; Broadus et al. 1995; see Fig. 11.4 - 11.6). As shortening of the germ band occurs, a few lateral neuroblasts start to shift further laterally in each segment, particularly in the labial neuromere, where two neuroblasts show striking lateral displacement. There is no obvious reason for such a lateral displacement of germ band neuroblasts. A further reduction in the extent and volume of the CNS becomes manifest during late embryogenesis, after head involution has been completed. In particular, the ventral cord contracts considerably, such that its posterior tip comes to lie at the level of *a4-a5*.

11.1.5
The Neuroblast Lineages

Studies on neuroblast lineages are scarce. The reason for this is to be found in the technical difficulties associated with unravelling the composition of these lineages in *Drosophila*, due to the small size and the highly complex composition of the ventral cord. Recently, a technique has been developed that permits labelling of neuroblasts by application of DiI to single neuroectodermal cells (Bossing and Technau 1994). This technique is the first available tool for the study of this very difficult problem, and we expect important results to be obtained with it in the future. By using this technique, Bossing and Technau (1994) have analyzed the composition of the mesectodermal cell lineages (see section 11.1.6) and Bossing et al. (1996b) have described first attempts to characterize the lineages associated with neuroblasts derived from the ventral half of the neuroectoderm. These include all the medial and intermediate neuroblasts, i.e. a total of 17 of these cells (NB 1-1, NB 1-2, NB 2-1, MP2, NB 2-2, NB 3-1, NB 3-2, NB 4-1, NB 4-2, NB 5-1, NB 5-2, NB 5-3, NB 6-1, NB 6-2, NB 7-1, NB 7-2 and NB 7-3), which give rise to a total of about 170 interneurons and 20 motoneurons. The authors have studied a great deal of material, consisting of a large number of clones of each of the neuroblasts (up to 59 clones in some cases), and, in spite of some few uncertainties in identifying the neuroblast labelled, inherent to the technique used, the study allows several general conclusions about aspects of the cell lineages. Thus, the size of the cell clones derived from these neuroblasts was found to vary between 2 cells (MP2 and NB 5-1) and 22-24 cells (NB 1-2, NB 5-2 and NB 5-3). Some, albeit not much, variation was observed within the lineage of individual neuroblasts. Thus, the largest variation was found in the lineage of NB 3-2, the size of which was found to vary between 10 and 18 cells in a total of 21 cases in which this particular neuroblast was labelled. However, characteristic features of the lineage of each particular neuroblast could also be distinguished; hence the variability does not affect the typical projection patterns of the lineage, but rather redundant elements.

Bossing et al. (1996b) have succeeded in assigning several previously identified neurons and glial cells to the lineage of particular neuroblasts (see also section 11.1.6). Confirming and extending previous findings, they report that aCC and pCC neurons, as well as the subperineural glia cells A-SPG, B-SPG and LV-SPG (Ito et al. 1995), derive from NB 1-1 (Thomas et al. 1984; Udolph et al. 1993); the thoracic subperineural glia cell A-SPG was found to belong to the NB 2-2 lineage (Bossing et al. 1996b); the RP2 neuron was found to derive from NB 4-2 (Thomas et al. 1984; Doe et al. 1988; Patel et al. 1989; Udolph et al. 1995; Chu-LaGraff et al. 1995); vMP2 and dMP2 derive from MP2 (Thomas et al. 1984; Doe et al. 1988). Bossing et al. (1996b) have also provided evidence for the first time that the RP1, RP3, RP4 and RP5 neurons (Goodman et al. 1984; Sink and Whitington 1991) derive from NB 3-1; that two of the CQ neurons (Patel et al. 1989; Broadus et al. 1995) belong to the lineage of NB 5-3; and that two serotonergic cells (Lundell and Hirsh 1994) are in the lineage derived from NB 7-3.

11.1.6
The Cells of the Midline

Traditionally, the entire population of progenitor cells of the ventral nervous system of insects has been described as being arranged in an unpaired median cord flanked on either side by a lateral cord (see Haget 1977, and Campos-Ortega and Hartenstein 1985, for reviews on the median cord). Although relatively few in number, cells of the median cord have attracted considerable interest, on the one hand due to uncertainties with respect to their origin and developmental fate (Roonwal 1937; Springer and Rutschky 1969; Poulson 1950), and on the other hand due to their experimental accessibility, which allows one to impale them with electrodes or label them by other means during development, both in grasshoppers and *Drosophila* (Goodman et al. 1981; Goodman and Spitzer 1979; Goodman and Bate 1981; Jacobs and Goodman 1989; Klämbt et al. 1991; Bossing and Technau 1994). Cells of the median cord give rise to glial cells (the midline glia) and neuronal progenitor cells (the MP1, UMI and VUM neurons). Intracellular recording and dye injection (Goodman and Bate 1981) have brought detailed insights into the important roles played by the cells of the median cord in neural development of grasshoppers. In different species of grasshoppers the median cells comprise a single large median neuroblast and seven, smaller cells, called midline precursors.

In *Drosophila*, the use of molecular markers, such as enhancer traps and antibodies (Klämbt and Goodman 1991), as well as selective cell labelling with DiI (Bossing and Technau 1994), has permitted precise analysis of midline cell development. Midline precursor cells develop from the mesectodermal ridge (Crews et al. 1988; Thomas et al. 1988) and become evident at stage 11 (Figs. 11.3, 11.7). At this stage, the mesectodermal ridge broadens segmentally between neuromeres, whilst within neuromeres it becomes an extremely narrow vertical sheath (Fig. 11.7). Cells within the broadened mesectodermal regions gradually increase in diameter, developing into median neural precursors. However, since they never attain the large volume of the neuroblasts, it is difficult to distinguish the median progenitor cells from the laterally adjacent progeny of lateral neuroblasts without using specific techniques. Intracellular injections of dyes show that the axons of cells derived from the median progenitors of *Drosophila* follow the same pattern as described for the grasshopper (Thomas et al. 1984). The number of midline progenitors cannot be established unequivocally in fixed preparations without specific markers, because

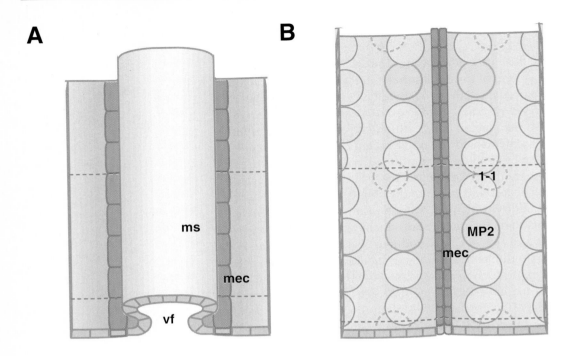

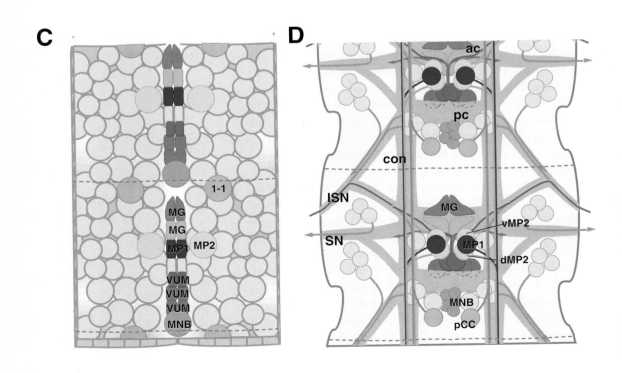

Fig. 11.7. The mesectoderm and derivatives. Part of the neural primordium of embryos at stage 7 (**A**), 10 (**B**), late 11 (**C**) and 13 (**D**) is shown. **A** A single row of cells on either side of the ventral furrow (*vf*) forms the mesectoderm (*mec*). There are 3-4 cells on either side per row per metamere (boundaries are indicated by *hatched lines*). **B** The two cell rows have come together and their cells have undergone their first, and last, mitosis, bringing the number of mesectoderm cells up to about 16 per metamere. A lateral neural progenitor cell which delaminates adjacent to the mesectoderm, MP2, behaves like mesectodermal cells in that it also performs only one mitosis. The MP2 progeny, dMP2 and vMP2, as well as some of the progeny of another lateral neural progenitor, NB1-1, will contribute to the structures formed by the mesectoderm and these progenitors are therefore included in this diagram. By late stage 11 (**C**), mesectodermal cells start to form metamerically reiterated clusters and begin to express their particular fates (symbolized by different colours). By stage 13 (**D**), all mesectodermal derivatives have started to differentiate. Segmental boundaries are represented by hatched lines, connectives (*con*) and commissures (*ac* anterior commissure; *pc* posterior commissure) and peripheral nerve roots (*ISN* intersegmental nerve; *SN* segmental nerve) are shown in a *darker shade of grey*. The median neuroblast (*MNB*) starts dividing in a stem cell mode and, between stages 11 and 14, produces approximately 5-8 progeny cells of known identity. The three pairs of cells anterior to the *MNB* are the *VUM* precursors which form bilaterally projecting motoneurons (see also Fig. 11.13). Anterior to the VUMs is the *MP1* pair which, together with *dMP2*, forms the first, posteriorly projecting axon tract (*MP1/dMP2* pioneer). The first anteriorly projecting axon tract is formed by the axons of *vMP2* and *pCC*. Note that in **D**, midline cell clusters are artificially stretched in the longitudinal axis for clarity; in reality, the *VUM* cluster lies ventrally, beneath the commissures, and the *MP1/dMP2/vMP2* clusters are the only cells between anterior and posterior commissure. Another midline cell pair (left *grey* in **C** but omitted in **D**), forms two previously unidentified, bilaterally projecting interneurons which Bossing and Technau (1994) have called *UMI* (for unpaired median interneurons). The two most anterior pairs of midline cells differentiate as midline glial cells which wrap around the two commissures

these cells are surrounded by the progenies of the older neuroblasts from which they cannot be distinguished, due to their similar shape and size. However, Klämbt et al. (1991), in a careful analysis of the midline cells using a number of enhancer trap lines, report eight mesectodermal cells per segment at the blastoderm stage. Bossing and Technau (1994), based on consideration of a large number of segments, have found a variable number of midline progenitor cells in each segment, ranging from 6 to 10, with an average of 7.5 per segment. By labelling neuroectodermal cells with DiI, Bossing and Technau (1994) have also succeeded in determining the composition of the cell lineages that develop from the midline progenitors. The approximately seven anlage cells per segment comprise six midline progenitor cells, which generally divide only once, and the median neuroblast (MNB), which divides with a stem cell pattern to give rise to 5-8 neurons. The six midline progenitor cells comprise three ventral unpaired median (VUM) progenitors (terminology of Goodman et al. 1981, for grasshopper), which generally give rise to one motoneuron and one interneuron each, (although cases with up to six progeny cells have also been found); two midline glia progenitor cells, each of which gives rise to two glial cells; one midline precursor 1 (MP1), which produces two MP1 neurons; and one progenitor cell that gives rise to two unpaired median interneurons (UMI) (Bossing and Technau 1994).

11.2
Early Neurogenesis in the Procephalon

11.2.1.
The Procephalic Neurectoderm

The brain hemisphere of the *Drosophila* larva corresponds to the supraoesophageal ganglion of other insects. Due to morphogenetic movements subserving head involution, the *Drosophila* larval brain is not overtly subdivided into protocerebrum, deuterocerebrum and tritocerebrum as in other insects. However, the peripheral nerves characteristic of the adult stage are all present already in the late embryo (Fig. 9.11; Campos-Ortega and Hartenstein 1985; Hartenstein 1988; Schmidt-Ott et al. 1994; Hartenstein et al. 1994a).

Neither the extent of the procephalic neuroectoderm nor the pattern of segregation and proliferation of procephalic neuroblasts within it has received the same attention as neurogenesis in the germ band. Neuroblast counts on serial histological sections yielded values of approximately 70-80 procephalic neuroblasts on either side (Hartenstein and Campos-Ortega 1984). However, examination of serial histological sections is certainly not the best method for following neurogenesis in the procephalon, since only a small fraction of the entire procephalic neuroectoderm is morphologically obvious, and identification of the neuroblasts as they segregate becomes very difficult. The particular part of the procephalic neuroectoderm referred to here becomes distinct during stage 7, shortly after formation of the cephalic furrow and before the onset of mitotic activity in the ectoderm, and consists of a bilateral strip of large cells with basally shifted nuclei in the procephalic lobe that extends dorsoventrally parallel to the cephalic furrow (Figs. 2.8, 2.9). This strip contains about 80 cells, and corresponds to mitotic domains B and part of 9 of Foe (1989; see Chapter 14). Some of the first procephalic neuroblasts will indeed segregate from this domain during stage 9 (see below). However, this is essentially all that the study of serial sections has contributed to our knowledge of the extent and composition of the procephalic neuroectoderm.

A much clearer picture of the extent of the procephalic neuroectoderm and its pattern of neuroblasts was obtained when
various molecular markers were used to study head neurogenesis
on embryonic wholemounts. Particularly useful are probes for
expression of genes of the E(SPL)-C and *lethal of scute*; the distribution of transcripts of these genes during stages 9-10 precisely
coincides with the regions from which neuroblasts delaminate
and can, therefore, be used to establish accurately the limits of the
neuroectoderm itself (Knust et al. 1992; Younossi-Hartenstein et
al. 1996; Figs. 11.8 - 11.9).

The procephalon in *Drosophila* is composed of three modified
segments (intercalary, antennal and labral segments) and the
acron (Fig. 11.8; for recent discussions of head segmentation, see
Schmidt-Ott and Technau 1992; Jürgens and Hartenstein 1993).
Although no morphological features permit one to distinguish the
boundaries between these segments in the embryo, the domains
of *engrailed* expression allow visualization of putative posterior
boundaries of the intercalary and antennal segments (ref. to Fig.
16.4). Anterior and dorsal to the antennal *engrailed* stripe there is
a spot-like *engrailed* expression domain; a horizontal line transec-
ting this *engrailed* spot has been tentatively equated with the anterior boundary of the antennal segment (Schmidt-Ott and Technau
1992; Younossi-Hartenstein et al. 1996). The site of the boundary
between the labral segment and the acron is unknown. Accordingly, on the assumption that the *engrailed* stripes delimit segmental boundaries, a tentative assignment of procephalic neuroblasts to the pre-oral segments can be made using an anti-Engrailed
antibody. On this basis, the large majority of procephalic neuroblasts derive from the acron/labrum and can be assigned to the
protocerebrum. The "head spot" of *engrailed* expression (Fig.
16.4), which has been proposed to represent the posterior (ventral) boundary of the acron (Schmidt-Ott and Technau 1992),
gives rise to 2-3 neuroblasts which fall into the ventral one-third
of the population of procephalic neuroblasts (Figs. 11.8, 11.9). The
neuroblasts derived from the antennal region encompass approximately the ventral one-quarter of the population of procephalic
neuroblasts and correspond to the deuterocerebrum. Only few
neuroblasts derive from the putative intercalary segment, corresponding to the tritocerebrum.

The boundaries of the procephalic neuroectoderm have also
been defined experimentally by means of HRP injections in the
late gastrula (Technau and Campos-Ortega 1985; Schmidt-Ott and
Technau 1994). The primordium from which head neuroblasts
segregate occupies in the gastrula a large fraction of the proce-

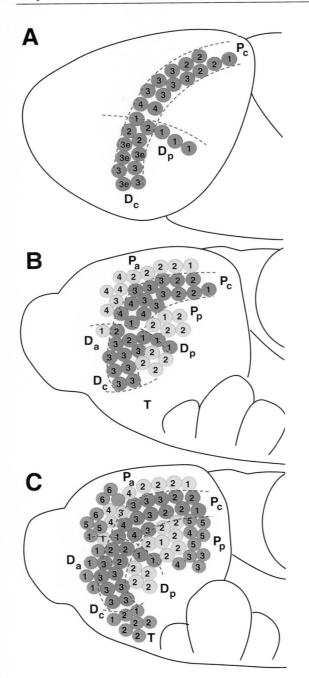

D

Location	Group	9 SI/II	10 SIII	11 SIV	12 SV	svp
Proto-cerebrum *anterior*	Pa1			1		+
	Pa2		3-4			-
	Pa3			2		+
	Pa4			3-5		-
	Pa5				2	+
	Pa6				3-5	-
	Pdm					
central	Pc1		1			+
	Pc2		2-4			-
	Pc3		4-6			+
	Pc4		2-3			-
posterior	Pp1			1-2		+
	Pp2		4-5			-
	Pp3				3	+
	Pp4				2-3	-
	Pp5					4-6
Deutero-cerebrum *anterior*	Da1				2-4	-
central	Dc1		1			+
	Dc2		2-3			-
	Dc3		5-6			+
posterior	Dp1	2-3				+
	Dp2			4-6		-
Trito-cerebrum	T1				2	+
	T2				4-6	-

Fig. 11.8 Map of *Drosophila* brain neuroblasts. This map is derived from the analysis of the expression patterns of *lethal of scute, asense* and *seven-up*. Numbering corresponds to that of Younossi-Hartenstein and Hartenstein (1996). Panels **A to C** represent schematic lateral views of the head of stage 9 (**A**), early stage 11 (**B**) and early stage 12 embryos (**C**), showing groups of neuroblasts delaminating at different stages. The first neuroblasts (SI) are coloured *lilac* in A; at later stages (**B, C**), they are grouped together with SII neuroblasts and shown in *green*. SIII are shown in *yellow*, SIV/SV in *orange*. Numbering and colouring of neuroblasts corresponds to that used in the Table shown in panel **D**. Dashed lines indicate subdivision of head neurectoderm into protocerebrum (*P; Pc* central protocerebral domain; *Pa* anterior protocerebral domain; *Pp* posterior protocerebral domain), deuterocerebrum (*D; Dc* central deuterocerebral domain; *Da* anterior deuterocerebral domain; *Dp* posterior deuterocerebral domain), and tritocerebrum (*T*). **D** Table of brain neuroblasts. Left column lists the different topological subdivisions of head neurectoderm mentioned above. Each gives rise to several groups of neuroblasts listed in the second column. Each neuroblast group is represented by a coloured bar. The left end of the bar indicates the stage (between 9-12; see top) at which corresponding neuroblasts delaminate; this also determines the inclusion of the corresponding neuroblasts within one of the subpopulations SI/II, SIII, SIV/SV (given at the top). Numbers in bars give approximate numbers of neuroblasts belonging to corresponding groups. The last column indicates whether neuroblast groups do (+) or do not (-) express the *seven-up* marker

phalon, from the cephalic furrow to approximately 90% EL, including part of the putative intercalary, antennal and labral segments, as well as the acron. These data, obtained from HRP injections, agree very well with the pattern of expression of the genes of the E(SPL)-C and *lethal of scute* (*l'sc*) during stages 7-11 (Knust et al. 1992; Younossi-Hartenstein et al. 1996). Hence, in the following we use this pattern of expression to characterize the extent of the procephalic neuroectoderm and the dynamics of neuroblast segregation. One problem, however, associated with this approach is that *l'sc* expression cannot be taken as indicative of neurogenic character in all instances, since this gene participates in other developmental processes besides neurogenesis. Therefore, some uncertainty is involved in using this method to gauge the extent and assess the dynamics of modification of the procephalic neuroectoderm.

11.2.2.
Pattern of Procephalic Neuroblasts

Neuroblasts delaminate from the procephalic neurectoderm in a stereotyped spatio-temporal pattern which is tightly correlated with the expression of the proneural gene *l'sc*. The pattern of neuroblasts was reconstructed by using the marker *asense* (*ase*) which is expressed in all brain neuroblasts (Fig. 11.8). Observations using *seven-up* (*svp*), which is expressed in specific subsets of neuroblasts, have contributed to supplement the studies with *ase*. The use of the *ase* and *svp* markers allowed us to identify small

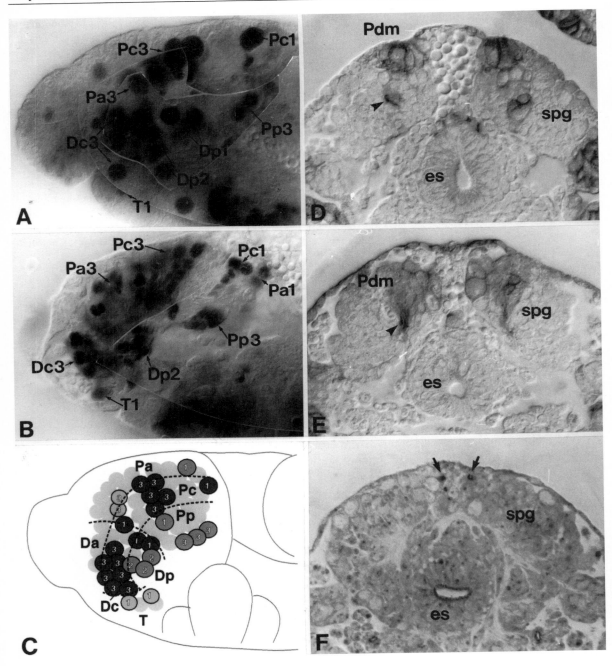

Fig. 11.9 Early neurogenesis in the procephalon. **A, B** Photographs of the procephalon of stage 11 (**A**) and stage 12 (**B**) embryos in lateral view. Subsets of procephalic neuroblasts are labeled by P-*lacZ* H162 (Doe 1992), an insertion in the *seven-up* (*svp*) gene. **C** schematically depicts all *svp*-positive procephalic neuroblasts (compare to neuroblast map shown in Fig.11.8). For identification of the lettered neuroblast groups see Fig.11.8. Note close spatial relationship between neuroblasts and their progeny, exemplified by Pc1 (**A**).

This single neuroblast in the dorso-central procephalon has produced a single ganglion mother cell located ventrally adjacent to it; in **B**, showing an embryo which is about 2 h older than the one in **A**, there are at least two additional ganglion mother cells. **D, E** Segregation of neural progenitors of the dorsomedial protocerebral domain. The two photographs show cross sections of the head of stage 12 (**D**) and stage 13 (**E**) embryos stained with anti-FasII antibody which labels the dorsomedial protocerebral domain (*Pdm*). In **D**, cells of

this domain are still integrated in the ectoderm. In **F**, they have moved interiorly and form the dorsomedial rim of the brain hemispheres (*spg* supraoesophageal ganglion). **F** shows a cross section of a stage 13 embryo stained with the anti-Crumbs antibody which labels apical cell surfaces. Many cells of the dorsomedial protocerebral domain invaginate during stage 12 and form small vesicles (*arrows* point to luminal staining). *es* oesophagus

groups of 1-5 neuroblasts each (Fig. 11.9). We follow the nomenclature used by Younossi-Hartenstein and Hartenstein (1996), which is based on the assumed neuromeric identity (i.e. protocerebral, deuterocerebral and tritocerebral neuroblasts) and the relative positions (i.e. anterior, central, posterior) of the cells.

Most neuroblasts segregate from the procephalic neuroectoderm by delamination. As in the ventral neurectoderm (Poulson 1950; Hartenstein and Campos-Ortega 1984), delamination occurs in waves; in each wave, individual cells constrict apically and move inside the embryo (Fig. 11.6A). The first neuroblasts delaminate from the central protocerebral and deuterocerebral neuroectoderm (*Pc* and *Dc* domains) during early stage 9 (Figs. 11.8, 11.9). They correspond to the *Pc1*, *Pc3* and *Pc4* and some of the *Dc* neuroblasts, and their segregation coincides with that of the SI subpopulation in the ventral neuroectoderm (Fig. 11.3; Hartenstein and Campos-Ortega 1984; Doe 1992). During late stage 9, coincident with the segregation of the SII subpopulation in the ventral neuroectoderm, most or all of the remaining *Pc* and *Dc* neuroblasts, as well as the *Dp1* neuroblasts delaminate.

Later neuroblasts appear concentrically around the *Pc* and *Dc* neuroblasts. During stage 10, corresponding to the time at which the SIII population of ventral neuroblasts segregates, three clusters of 4-5 neuroblasts each (*Pa1/2*, *Pa3/4*, *Ppi1/2*, *Dp2*) segregate from the domains *Pad*, *Pdm*, *eP/Dp* and *Dp*. During stage 11, coincident with the segregation of SIV/SV neuroblasts in the ventral neuroectoderm, neuroblasts segregate from the anterior and posterior protocerebral neurectoderm (*Da, Pdm*). A group of 6-7 tritocerebral neuroblasts appears in the *T* domain (Fig. 11.8). Several groups segregate at later stages from the posterior protocerebral domains .

The position of neuroblasts relative to each other and to other head structures does not change significantly during stages 9-12. Later on, however, there is a general shift of dorsally located protocerebral neuroblasts (e.g., *Pc3*) towards posterior. Dorsoposterior protocerebral neuroblasts (e.g., *Pa1*, *Pc1*), move to more ventral positions. These movements lead to a further bending of the neuraxis towards dorsoposterior (Fig. 11.8).

11.2.3.
Proliferation of Procephalic Neuroblasts

Little is known about the proliferation of procephalic neuroblasts. Most *svp*-positive neuroblasts divide in a typical stem cell mode. Ganglion mother cells are given off at the basal side of the neuroblasts. At later stages, they form chains of increasing length, with the first-born cells farthest away from the neuroblast (Fig. 11.9). The duration of the cell cycle in many *svp*-positive neuroblasts is 45 - 60 min, which is similar to that of ventral neuroblasts (Hartenstein et al. 1987).

11.2.4.
The Dorsomedial Procephalic Domain

The behaviour of the dorsomedial procephalic or protocerebral domain merits particular discussion (Fig. 11.9). During stage 12 and early stage 13, all cells of *Pdm* become internalized by a process which can be best characterized as a combination of mass delamination and invagination: compact clusters of *l'sc*-expressing cells move their nuclei basally and separate from the surface. This cell movement can be easily visualized with the anti-FasII antibody (Grenningloh et al. 1991), which is expressed in cells of the *Pdm* from stage 12 onward. Following segregation, most of the dorsomedial cells become integrated into the brain hemispheres as neurons and glial cells.

This process of mass delamination/invagination of the cells of the dorsomedial domain is accompanied by massive cell death. Thus, from late stage 12 onward, the dorsal aspect of the developing brain is characterized by masses of debris-filled macrophages. In embryos expressing markers that specifically label the dorsomedial domain (e.g. the anti-FasII antibody), most of these macrophages can be seen to contain labeled cellular debris, indicating that the degenerated cells taken up by the macrophages are derived from the dorsomedial domain.

11.2.5.
The Primordium of the Optic Lobe

The imaginal visual system originates from a dorsoposterior strip of the procephalon which largely overlaps the posterior protocerebral division of the procephalic neurectoderm. After undergoing four rounds of mitosis and giving rise to at least one group of neuroblasts (*Pp3*), the cells which will become the primordium of the optic lobe adopt a columnar shape and form a conspicuous placode (Green et al. 1993). The optic lobe placode invaginates during stages 12 and 13 (Fig. 9.17). Its cells constrict apically, leading to the formation of a vertically elongated groove in the center of the optic lobe placode. By the end of germ band retraction, the invagination of the optic lobe primordium forms a deep pouch containing approximately 85 cells, in direct contact with the brain. When the invagination forms, all brain neuroblasts have already separated from the head ectoderm, most of them being located anteriorly and medially to the optic lobe placode. After detaching from the head ectoderm, the optic lobe primordium forms a flattened vesicle attached to the ventrolateral surface of the brain, the cells of which retain their epithelial character.

After the fourth postblastoderm division the cells of the prospective optic lobe enter a mitotically quiescent period which lasts throughout the rest of embryogenesis. Not until the larval stage will they again become mitotically active, forming the inner and outer proliferation zones that will give rise to the visual centers of the adult brain (White and Kankel 1978; Hofbauer and Campos-Ortega 1990).

11.3
Neuronal Differentiation

11.3.1.
Formation of the Cortex and Neuropile

As a result of neuroblast and ganglion mother cell proliferation during stages 9-11, a multilayered aggregate of differentiating neurons develops dorsal to the neuroblasts. Differentiation of neurons begins towards the end of germ band retraction with the outgrowth of axons and expression of neuron- and glia-specific markers which can be recognized by specific antibodies, such as MAb22C10, 44C11, anti-HRP, or anti-Fas II. The majority of neurons produce bundles of short processes that project towards the dorsal surface of the neural primordium, i.e. away from the neuroblast layer. A small subset of neurons extend long axons which form a scaffold of longitudinal and transverse axon tracts that serve as "pioneer tracts" (reviewed in Goodman and Doe 1993) for later differentiating neurons. Short and long axonal processes form an early neuropile core located on the dorsal surface of the layer of neuronal and glial cell bodies. Subsequently, as a consequence of continued neuroblast proliferation, as well as the mass movements of neurons that take place in the wake of nervous system condensation, the cell body layer grows around the neuropile laterally and dorsally. A continuous cortex is thus formed around a central neuropile.

Very little is known about the differentiation of individual neurons in the embryonic CNS of *Drosophila*. Some data have become available from intracellular injections of dyes into differentiating neurons of *Drosophila* (Thomas et al. 1984), showing that such studies are in principle possible in this very small embryo; on the other hand, antibodies and enhancer trap lines have been used to identify a small number of cells which can be easily recognized from embryo to embryo. DiI labelling of individual neuroectodermal cells (Bossing and Technau 1994, see above) will permit characterization of additional neural cells and thus lengthen the list of uniquely identifiable neurons in the *Drosophila* CNS.

Goodman and Doe (1993) have provided a catalogue of the currently identified neurons in the *Drosophila* embryo (Figs. 11.7, 11.10, 11.11). These cells, all of which are members of the groups of early differentiating pioneer neurons described in more detail below, include aCC, MP1, dMP2, vMP2, SP1 and VUM cells, which can be stained with the 22C10 antibody (Grenningloh et al. 1991); RP1 and RP3, stainable with an anti-fasciclin III antibody (Patel et al. 1987); aCC and pCC, and some of the U cells, can be stained with an anti-fasciclin II antibody (Grenningloh et al. 1991); and RP2 is labelled by an anti-Even-skipped antibody (Chu-LaGraff and Doe 1993).

11.3.2.
Development of the Commissures

Two commissures develop in each neuromere, except for the last (*a9*), which has only one commissure (Figs. 11.7, 11.10, 11.11, 11.12).

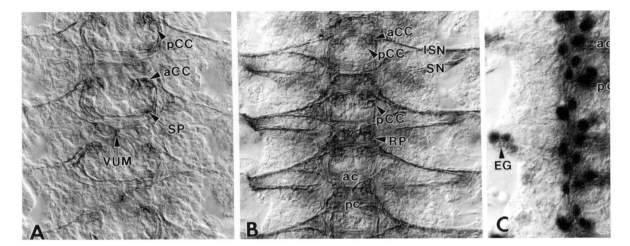

Fig. 11.10 A and **B** show neuromeres of stage 13 and 16 wild-type embryos stained with Mab22C10, to show the development of various identified neurons (*ac* and *pc* anterior and posterior commissures; *aCC* and *pCC* anterior and posterior corner cell; *ISN* intersegmental nerve; *RP* motoneurons; *SN* segmental nerve; *VUM* ventral unpaired neurons). **C.** Dorsal view of the ventral cord of a stage 16 embryo, carrier of an enhancer trap insertion to highlight the longitudinal glial cells, the subperineurial A and B glial cells (Klämbt et al. 1991) and the exit glia cells (*EG*). This embryo was stained with MabA102 to visualize the neuropile and an antibody against β galactosidase. (Slide kindly provided by Thomas Menne and Christian Klämbt; see Menne and Klämbt 1994)

Both commissures lie within the posterior half of each neuromere, with the posterior commissure located approximately at the level of the posterior neuromere boundary. The anterior commissure contains more fibres than the posterior, and occasionally it is split into two fascicles. In fact, other authors (e.g. Thomas et al. 1984) describe two different, fused commissures, A/B, corresponding to what we describe as the anterior commissure. Formation of the supraoesophageal and frontal commissures is delayed with respect to the commissures of the ventral cord, and they do not become evident until about 1 h after germ band shortening (see Chapter 9).

The pioneering of the commissures has been studied in great detail by electron microscopy, dye filling of identified neurons, and the use of specific markers for pioneer neurons (Jacobs and Goodman 1989; Jacobs et al. 1989; Klämbt et al. 1991). The first axons pioneering the anterior commissure are formed by neurons called "Q neurons" by Myers and Bastiani (1993a, b) in the grasshopper embryo; pioneer axons of the posterior commissure are extended by the "RP neurons". The ancestral neuroblasts of "Q"

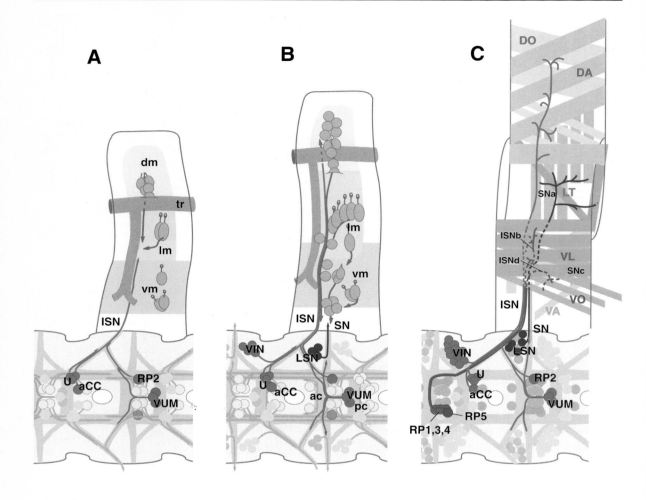

Fig. 11.11. Development of motoneurons and their axonal pathways. The three panels schematically represent flattened views of abdominal hemisegments (dorsal midline at the top) and their associated sections of the ventral nerve cord of stage 13 (**A**), stage 14 (**B**) and stage 16 (**C**) embryos. Sensilla, tracheal tree, peripheral nerves and neuropile are shown in *grey* as points of reference. The ventral nerve cord is viewed from dorsal. Different colours identify groups of motoneurons and muscles innervated by them (*blue* dorsal muscles and the corresponding motoneurons; *green* ventral muscles and the corresponding motoneurons; *purple* lateral muscles and the corresponding motoneurons). It is noteworthy that in the majority of cases, the motoneurons are located in the neuromere anterior to the segment whose muscles they innervate. During stage 13 (**A**) the first motoneurons, which pioneer the intersegmental nerve (*ISN*), differentiate. The axon of the *aCC* neuron grows posteriorly and pioneers the anterior root of the *ISN*, followed by the axons of the *U* neurons and, at a later stage (**B, C**), by the axons of the 14 ventral intersegmental neurons (*VIN*; Sink and Whitington 1991). After leaving the ventral cord, the axons of the *aCC* and *U* neurons use the segmental transverse branch of the tracheal tree (*tr*) as substrate; they then fasciculate with the first ingrowing sensory axons (see also Fig. 9.18) and reach the dorsal acute and oblique muscles (*DA, DO*). The axons of two *VUM* neurons and of the *RP2* neurons pioneer the posterior root of the *ISN*. The segmental nerve (*SN*), which carries the axons to innervate the ventral (*SNc*) and lateral (*SNa*) musculature, is pioneered by two pairs of axons of the VUMs and of the lateral segmental neurons (*LSN*; Sink and Whitington 1991). Shortly after leaving the ventral cord, all these axons meet and fasciculate with the sensory axons of ventral sensilla (see also Fig. 9.18). The axons of the lateral segmental neurons innervate the lateral transverse musculature (*LT; purple*), as well as the ventral acute muscles (*VA; light green*). It is actually not clear which of the two contingents of motoneurons sends out taxons earlier (in the diagram, the lateral group was chosen arbitrarily). The ventral longitudinal (*VL; turquoise*) and ventral oblique (*VO; dark green*) muscles are innervated by the axons of the *RP1, 3, 4, 5* neurons. These neurons are located contralaterally, in the anteriorly adjacent neuromere. After crossing the midline, their axons grow out of the cortex via the anterior root of the *ISN*, leave the *ISN* (*ISNb, d*) and cross over to the *SN* (anastomosis between *ISN* and *SN* called exit junction (*ej*; Goodman and Doe 1993). For several motoneurons, notably the *RPs* and *VUMs*, the exact muscle fibre innervated is known (see Table 11.1). Other abbreviations. *ac* anterior commissure; *dm* primordium of dorsal musculature; *lm* primordium of lateral musculature; *pc* posterior commissure; *vm* primordium of ventral musculature

neurons have not yet been identified. Bossing et al. (1996b) have provided evidence for the first time that the RP neurons (Goodman et al. 1984; Sink and Whitington 1991) derive from NB 3-1 and NB 4-2 (see 11.1.5). The axons that pioneer the anterior commissure grow towards the midline at the level of the anterior pair of midline glia cells (MGA); the pioneers of the posterior commissure grow slightly posteriorly to this, at the level of the anterior VUM precursors. At this stage, the pioneer axons of the anterior and posterior commissure converge onto the MP1 pair of midline cells, adopting the shape of an "X" (Fig. 11.10). Subsequently, the pair of MP1 neurons moves out of the midline to a more ventral and lateral position, thereby creating a gap through which both anterior and posterior commissures migrate. Initially, both commissures remain very close to each other; only the growth cones of the VUM neurons separate them. Subsequently, the cell bodies

Fig. 11.12 Map of glial cells in the central and peripheral nervous system. The right panel represents a flattened view of an abdominal hemisegment (dorsal midline at the top) and its associated section of ventral nerve cord in a mature embryo. Muscles, trachea, sensilla, peripheral nerves and neuropile are shown in grey as reference points. The ventral nerve cord is viewed from dorsal; glial cells attached to it dorsally are shown in solid colour, whereas glial cells at or near the ventral surface are marked by dotted circles. The lower left panel shows a cross section of the ventral cord of a mature embryo. Different classes of glial cells are represented in different colours (see key at upper left; classification according to Ito et al. 1995). Surface glial cells form a continuous covering for the central nervous system and peripheral nerves. They comprise a small number of uniquely identifiable peripheral and exit glial cells (associated with the peripheral nerves), the subperineural glial cells (which enclose the CNS), and the channel glial cells (which line the vertical channels which pierce the ventral nerve cord at segmental boundaries). The perineural sheath, which forms a thin layer beyond the subperineural glial cells, is not depicted. The second main group of glial cells, called interface glial cells, form a sheath around the neuropile, commissures, and nerve roots. Finally, cell body-associated glial cells are located in the cortex of the CNS. Abbreviations. *ac* anterior commissure; *dh1* dorsal hair sensillum; *ISN* intersegmental nerve; *lch 5* lateral chordotonal organ; *np* neuropile; *nr* nerve roots; *pc* posterior commissure; *SN* segmental nerve; *vch 1* ventral chordotonal organ; *vp 5* ventral papilla

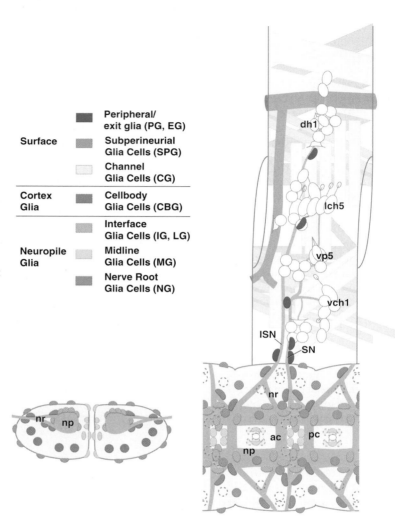

Surface	Peripheral/ exit glia (PG, EG)
	Subperineurial Glia Cells (SPG)
	Channel Glia Cells (CG)
Cortex Glia	Cellbody Glia Cells (CBG)
Neuropile Glia	Interface Glia Cells (IG, LG)
	Midline Glia Cells (MG)
	Nerve Root Glia Cells (NG)

of the MGA midline glia, as well as two other pairs of lateral neurons (RP1, RP3) squeeze in between the two commissures, thereby forcing them apart. As shown in a number of genetic studies, all of the cells that contact the ingrowing commissural pioneer axons are crucially important for normal commissure development (Klämbt et al. 1991; Sonnenfeld and Jacobs 1994).

11.3.3.
Development of the Connectives

The neurons which lay down the pioneer tracts of the connectives are progeny of the midline precursors (MP1) and two neuroblasts that lie adjacent to the midline: dMP2 and vMP2 derive from MP2, and aCC and pCC from NB1-1 (Fig. 11.7). Axons produced by these neurons interact with the longitudinal glial cells (LG). These glial cells (5 per hemisegment) derive from glial progenitor(s), among them glioblast GP, which delaminate from the lateral neurectoderm (Jacobs et al. 1989). Subsequently, they migrate towards medial levels and align themselves on the dorsal surface of the neural primordium, thereby forming a track which is followed by the outgrowing axons. The first axons to grow out sprout from pCC and vMP2. The growth cones are in tight contact with each other, with the vMP2 axon frequently wrapping around the pCC axon. Growing anteriorly, both axons adhere to the longitudinal glial cells, in particular LG5.

A similar pair of fasciculating pioneer axons grows out from MP1 and dMP2 in the posterior direction. These axons also establish contact with the longitudinal glial cells. Eventually, the pCC/vMP2 and the MP1/dMP2 pairs meet and fasciculate with each other over a short distance. Upon reaching their serial homologues in the posteriorly adjacent segment, MP1/dMP2 fasciculate with these axons. Axons of pCC and vMP2 do the same with their homologues, so that two longitudinal fascicles are formed. After the pioneer axons of adjacent segments have linked up with each other, they are displaced from the longitudinal glia layer by later forming axons.

11.3.4
Development of Motoneurons and Their Efferent Tracts

Sink and Whitington (1991) describe the motoneurons of abdominal neuromeres in the stage 16 embryo, labelled by DiI injections in either the nerves or the commissures. Additional studies, using markers specific for subsets of motoneurons and their peripheral

Table 11.1. Identified moneuron-muscle relationships

Motoneurons	Muscles
RP1, RP4	VL2
RP2	DA2
RP3	VL3, VL4
RP5	VL2, VO1, VO2, VO4, VO5
aCC	DA1
U	DA2, DO2
VUMs	VO5, VL2, LT1, 2, DO3, DO4

(after Sink and Whitington 1991)

targets, have also contributed to our understanding of how these cells develop (relevant references are given in Goodman and Doe 1993). Table 11.1 provides a list of motoneurons for which the target muscles are known.

There are approximately 35 motoneurons per segment, organized according to a repeating pattern made up of the following pattern elements (ref. to Figs. 11.10, 11.11).

One ipsilateral neuron is located medially and posteriorly to the intersection of the anterior commissures and the connectives, and corresponds to the aCC cell of Goodman et al. (1984). aCC is the first motoneuron to differentiate. Its axon grows posteriorly, lateral to the connective pioneer tract. After reaching the posteriorly adjacent neuromere, the aCC axon turns laterally and leaves the cortex of the ventral cord, thereby pioneering the anterior root of the intersegmental nerve. aCC innervates the DA1 muscle.

One ipsilateral neuron is located medial to the intersection of the anterior commissures and the connectives, corresponding to the RP2 cell of Goodman et al. (1984). RP2 projects anteriorly, pioneering the posterior root of the intersegmental nerve. Upon reaching the aCC axon, the axon of RP2 fasciculates with, and follows, it towards the dorsal musculature, where it innervates the DA2 muscle.

Four motoneurons projecting contralaterally are grouped at or lie dorsal to the connectives, in the region bounded by the anterior and posterior commissures. This group of cells corresponds to the RP1, RP3, RP4, and RP5 neurons identified by Goodman et al. (1984) in the grasshopper embryo. After crossing the midline in the anterior commissure, their axons project posteriorly, following the ISN path established earlier by the aCC axon. At the level of the ventral musculature the RP axons branch off the ISN (bran-

ches ISNb and ISNd, respectively) and innervate the ventral oblique and ventral longitudinal muscles in a stereotyped pattern.
The RP axon fascicles were previously thought to form branches
of the segmental nerve (SN).

A group of 14 ipsilateral neurons lie in the posterior half of the
segment, most of which are ventral to the level of the connectives;
this has been called the ventral intersegmental group (VIN) by
Sink and Whitington (1991). Two or three of these cells, consistently found more medially than the others, probably correspond
to the U neurons of Goodman et al. (1984). VIN axons follow the
anterior root into the intersegmental nerve and innervate the dorsal musculature.

About nine ipsilateral cells, called lateral segmental neurons
(LSN), are located ventral to the connectives and send their axons
to the segmental nerve which innervates subsets of ventral muscle
fibres.

Three neurons are located between the connectives. Their
axons bifurcate at the midline and exit from the ventral cord
through each of the segmental nerves. Thus, they correspond to
the ventral unpaired median neurons (VUM) of Goodman et al.
(1984). Each VUM neuron innervates a large number of muscles.

Finally, Sink and Whitington (1991) describe three scattered
contralateral neurons, Z (located near the anterior commissure,
dorsal to the level of the connectives), V and M (located at the
level of the posterior commissure in the anteriorly adjacent neuromere). Their exact projection patterns are unknown (Fig. 11.11).

11.4
The Glial Cells

Besides neurons and tracheal cells, the CNS contains a large number of glial cells of a variety of different types (Figs. 11.10 - 11.13).
Using several markers, Ito et al. (1994) have recently analysed the
pattern of glial cells in the abdominal neuromeres of the stage 16
Drosophila embryo and established a comprehensive classification of these cells, which we follow in our description (see Table
11.2).

A total of about 60 glial cells are found in each abdominal segment; these are classified in three main categories, i.e. surface
associated, neuropile associated and cortex associated, and six
different subtypes of glial cells can be distinguished, based on
their location and morphology. There are two subtypes of surface-

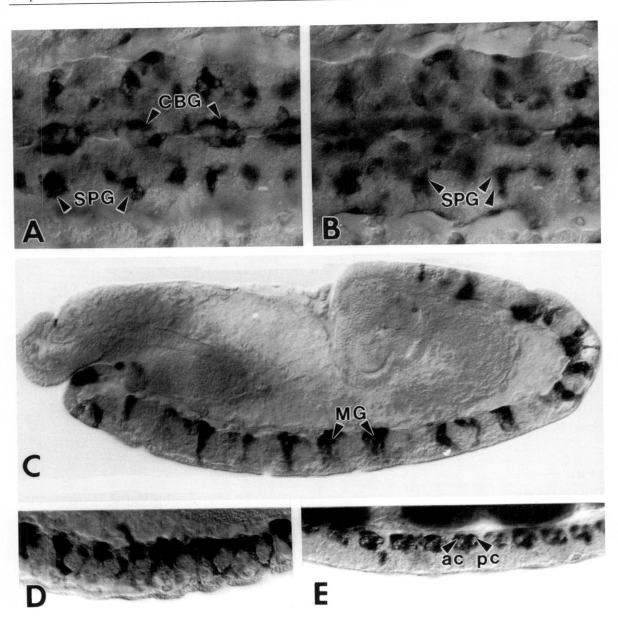

Fig. 11.13 A to B are two planes of focus on the ventral cord of a stage 16 embryo carrying an enhancer trap insertion that stains subperineural (*SPG*) and VUM (M-CBG) glial cells (*CBG*). **C-E** show stage 12, 14 and 16 embryos carrying a *slit* promoter-reporter gene construct that highlights the midline glial cells (*MG*). Notice that the shape of the cells changes to progressively surround the developing commissures (*ac, pc*; slide kindly provided by Evelin Sadlowski and Christian Klämbt)

Table 11.2. Types of glia cells identified by Ito et al. (1994) in Drosophila

	Category	Subtype	Identified glia cells
CNS glia	Surface associated	subperineurial glia, SPG	A-SPG B-SPG MD-SPG LD-SP DL-SP VL-SPG MV-SPG
		channel glia CG	D-CG M-CG cluster V-CG
	Cortex associated	cell body glia CBG	MM-CBG M-CBG VL-CBG L-CBG
	Neuropile associated	nerve root glia intersegmental and segmental ISNG, SNG	M-ISNG L-ISNG M-SNG1 M-SNG2 L-SNG
		interface glia IG	D-IG cluster L-IG cluster V-IG
		midline glia MG	MG cluster
PNS glia	Nerve associated	perineurial cells exit glia peripheral glia	
Trachea		Trachea	

associated glial cells, subperineural and "channel" glial cells; only one type of cortical glial cells; and three subtypes of neuropile-associated glial cells - the intersegmental and segmental nerve root cells, interface glial cells, and midline glial cells.

The subperineural glial cells are the most abundant surface-associated cells and are located immediately subjacent to the perineurium. There are about 16-18 subperineural glial cells in each abdominal hemisegment. Two of these cells are located at the ventral surface, another two at the lateral surface and about four at the dorsal surface of the neuromere. The two most medial of the dorsally located surface glial cells are progeny of neuroblast NB1-1 (Udolph et al. 1993). The channel glial cells comprise 7-8 cells in each segment located along the dorsoventral extent of the channels that form the segmental boundaries between adjacent neuromeres. Two of these channel glial cells lie at dorsal levels, three or four medially and the other two ventrally. There are 6-8 cortical glial cells per segment, distributed in the cortical cell body layer, scattered among the cell bodies of neurons. Particularly striking are the glial cells associated with the VUM neurons (MM-CBG cells of Ito et al. 1994; see Fig. 11.13). Ito et al. (1994) distinguish three types of neuropile-associated glial cells. The nerve root glial cells of abdominal hemineuromeres comprise two cells associated with the intersegmental nerve, one medial and one lateral, and another three associated with the segmental nerve, two of which are medial and the third lateral. The interface glial cells comprise 7-8 cells per hemineuromere located at the interface between neuropile and cortex. Four or five of these interface glial cells are located at the dorsal, two at the lateral, and one or two at the ventral aspect of the hemineuromere. The interface glial cells correspond to the longitudinal glial cells of Jacobs et al. (1989). These glial cells were originally described as being derived from the lateral glioblast.

Halter et al. (1995) have cloned and characterized a homeobox gene, *repo*, specifically expressed in developing glia and closely related cells. Persistent expression of the Repo protein has revealed that not all the longitudinal glial cells are derivatives of the lateral glioblast.

Finally, the midline glial cells, which are mesectodermal derivatives (Crews et al. 1988; Klämbt and Goodman 1991; Bossing and Technau 1994), comprise three or four cells (Bossing and Technau 1994; Ito et al. 1994), rather than four to six, as previously reported by other authors (Fredieu and Mahowald 1989; Klämbt and Goodman 1991). The midline glial cells become intimately associated with the commissures, which they ensheath following complex cell shape modifications during stages 12-15 (Fig. 11.13).

11.5
Origin of the Imaginal Nervous System

The neuroblasts that give rise to the larval and the imaginal central nervous systems segregate during embryogenesis. The proliferative activity of neural cell precursors during embryogenesis results exclusively in the production of larval CNS cells, most of which will be incorporated into the imaginal CNS during pupal metamorphosis. The embryonic origin of some imaginal neurons can be clearly demonstrated by labeling neurons of embryonic origin with [^{3}H]thymidine and performing autoradiography in the adult fly (Campos-Ortega 1982). In addition to these older studies, as already mentioned, Prokop and Technau (1991) demonstrated by transplantation of doubly labelled ectodermal cells that larval and imaginal neuroblasts are identical cells. Most larval neuroblasts shrink at the end of their proliferative period during embryogenesis and become apparent again during the first larval instar (Truman and Bate 1988). Thus, the available data strongly suggest that the same cells share larval and imaginal progenies (Prokop and Technau 1991). Besides neurons of larval origin, the imaginal CNS is build up from several other centres. However, with the exception of the optic lobes, which derive from the optic lobe placode, the precise location of the remaining neural imaginal anlagen in the embryo of *Drosophila* is essentially unknown.

Chapter 12
Tracheal Tree

The tracheal tree develops from ten segmental specializations of the lateral ectoderm which become evident during stage 10: these so-called tracheal placodes (Figs. 2.15B, 12.1; Poulson 1950) will invaginate to form the tracheal pits; the openings of the tracheal tree, the anterior and posterior spiracles, develop from special cells in the prothorax and abdominal segment *a8*, which will fuse to the first and eighth tracheal pits, respectively (see Fig. 12.1-12.5). Before segregation of germ band neuroblasts, the ectoderm is organized into two morphologically distinct regions. The medial region will give rise to the neuroblasts and to the primordium of the ventrolateral epidermis, consisting of large cylindrical cells, whereas the lateral region will give rise to the dorsal epidermis and the tracheal tree, and consists of slender, tightly packed columnar cells. It is within this region that the first mitoses become visible in the ectoderm of the metameric germ band. During stage 10, when ectodermal cells are in their second interphase, mitotic activity ceases transiently in the lateral ectodermal region and eleven slight depressions appear. Shortly afterwards, in stage 11, mitotic figures reappear in the medial cells within these depressed fields and the cells invaginate (Fig. 11.1E) to form ten deep, narrow pits (Fig. 2.17D-E).

At the same time, initial signs of metameric organization have become apparent in the germ band in the form of deep ventromedial parasegmental grooves, thus allowing assignment of the pits to specific metameres. All pits are located in the central one-third of each parasegment, whereby the anteriormost pit opens in parasegment 4 and the posteriormost opens in parasegment 14. The opening of each tracheal pit to the outside only disappears in stage 13, after germ band shortening, when the pits close over. During stages 14-15, the pits in mesothorax and abdominal segment *a8* will fuse with special cells of the prothorax and abdominal segment *a8*, to give rise to the anterior and posterior spiracles, respectively (Fig. 12.5). Specific enhancer trap lines, in particular those with

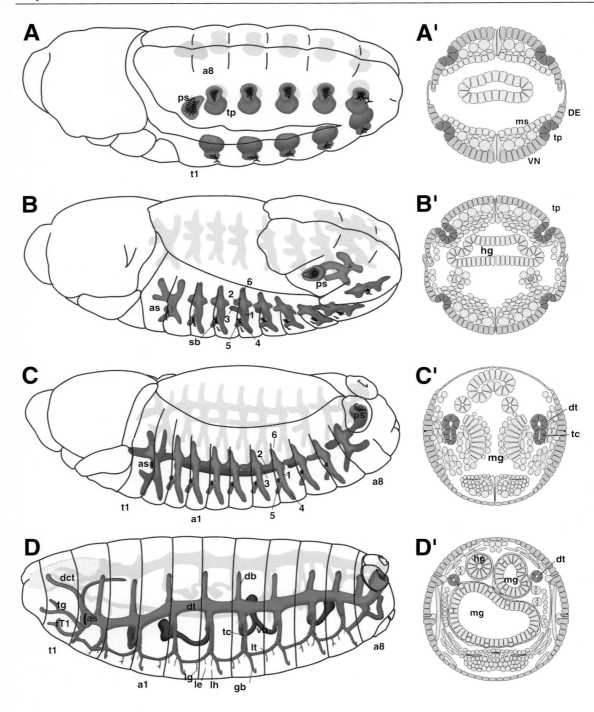

Fig. 12.1. The development of the tracheal tree. **A to D** are stage 11, 12, 13 and 17 embryos, respectively (compare to Fig. 12.3), to illustrate changes in the tracheal pits and progressive formation of tracheal branches. Terminology according to Manning and Krasnow (1993). Abbreviations. *a1* abdominal segment 1; *a8* abdominal segment 8; *as* anterior spiracles; *db* dorsal branch (derived from 6); *dc* dorsal cephalic branch; *dt* dorsal trunk (derived from 2); *gb* ganglionic branch (derived from 4); *hg* hindgut; *lc* lateral cephalic branch; *lt* lateral trunk (derived from 5); *le/g/h* e, g and h branches of the lateral trunk; *mg* midgut; *ps* posterior spiracles; *t1* prothoracic segment; *tc* transverse connective (derived from 1); *tp* tracheal pit; *vb* visceral branch (derived from 3); *vc* ventral cephalic branch. Modified from Hartenstein (1993), with permission of Cold Spring Harbor Laboratory Press

insertions in the FGF receptor homologue, label the tracheal tree from early stages of development on. Manning and Krasnow (1993) describe the tracheal placodes as irregular ovals, elongated dorsoventrally, located one-third of the distance between the amnioserosa and the ventral midline. After completion of mitotic activity, there are about 90 cells in each tracheal metameric unit, with the exception of the first and third, which comprise about 150 and 50 cells, respectively (Manning and Krasnow 1993).

The anterior spiracles of the fully developed embryo open in the anterior (prothoracic) lip of the segmental furrow between mesothorax and prothorax on either side of the embryo (Fig. 2.37D). The spiracles develop from cells of the posterior one-third of the prothorax, which are apparent during stage 14 when the cytoplasm becomes more basophilic (Fig. 2.31F). These prothoracic cells fuse with the invaginated cells of the first tracheal pit (see below) and, therefore, connect to the developing tracheal tree itself. The development of the posterior spiracles is similar, though it occurs much earlier. During stage 12, cells topologically corresponding to the posterior half of abdominal segment *a8* become more basophilic (see Figs. 12.2B, 12.5). These cells form a deep groove which establishes a connection with the posterior arm that develops from the cells of the posteriormost, tenth tracheal pit (refer to Chapter 17 for relationships of tracheal pits to parasegmental furrows).

The tracheal placodes start to invaginate during late stage 10 to form, at stage 11, spherical vesicles that open to the outside right behind the developing segmental furrows (tracheal pits). However, the openings of the tracheal pits disappear around stage 13. Rows of cells remain at the positions of the openings. These cells will form the tracheal histoblasts from which most cells of the adult tracheae will develop. During stage 12, the invaginated tracheal pits elongate to form a dorsal and a ventral stem. The ventral stem bifurcates into a posterior and an anterior branch. The dorsal stem gives off two side branches: one towards anterior, the

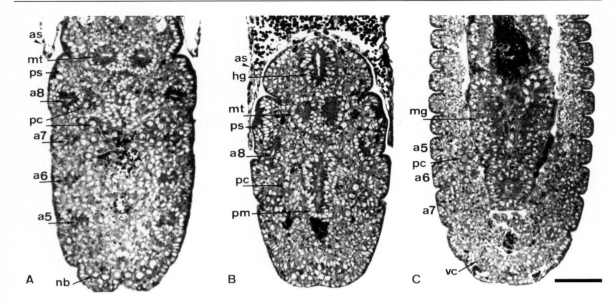

Fig. 12.2 A to C are horizontal sections of embryos of increasing age to illustrate the invagination of tracheal pits and the relationships of pole cells (*pc*) and mesoderm during formation of the gonads. In **A** the tracheal pits of *a8* to *a5* are labelled. Pole cells, clearly distinguishable on the basis of morphological features, can be seen aligned in a row on both sides; neighbouring sections show that pole cells become associated with mesodermal cells of *a8* to *a5*. During shortening of the germ band (**B**) the pole cells become displaced towards anterior. After shortening (**C**) the pole cells can be seen to associate with the mesodermal cells of *a5* and *a6*. Therefore, although the final location of the gonads is at the level of *a5*, the mesodermal cells that form the follicle cells can in principle be recruited from segments *a8* to *a5*. Abbreviations. *as* amnioserosa; *hg* hingut; *mg* midgut; *mt* Malpighian tubules; *nb* neuroblasts; *pm* posterior midgut; *vc* ventral cord. *Bar* 50 μm

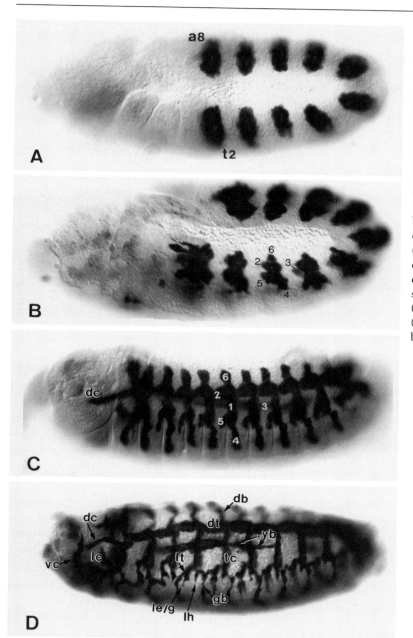

Fig. 12.3 The development of the tracheal tree. **A to D** are stage 11, 12, 14 and 16 embryos, respectively, that express the enhancer trap *sma2* (kindly provided by Norbert Perrimon) to illustrate progressive formation of tracheal branches. Terminology according to Manning and Krasnow (1993). Abbreviations. *a8* abdominal segment 8; *db* dorsal branch (derived from **6**); *dc* dorsal cephalic branch; *dt* dorsal trunk (derived from **2**); *gb* ganglionic branch (derived from **4**); *lc* lateral cephalic branch; *lt* lateral trunk (derived from **5**); *le/g* e and g branches of the lateral trunk; *lh* h branch of the lateral trunk; *t2* mesothoracic segment; *tc* transverse connective (derived from **1**); *vb* visceral branch (derived from **3**); *vc* ventral cephalic branch

Fig. 12.4. The development of the tracheal tree. Lateral (**A**) and ventral (**B,C**) views of stage 17 embryos that express the enhancer trap *sma2* (kindly provided by Norbert Perrimon) to illustrate progressive formation of tracheal branches. Notice that in each segment the *h* branch of the lateral trunk (*lh*) gives off tracheal branches to the epidermis and the ventral surface of the ventral cord, whereas the ganglionic branch (*gb*) provides the dorsal surface of the ventral cord with other tracheal branches. Abbreviations. *dt* dorsal trunk; *lt* lateral trunk; *vb* visceral branch; *va* ventral anastomosis; *vc* ventral cord

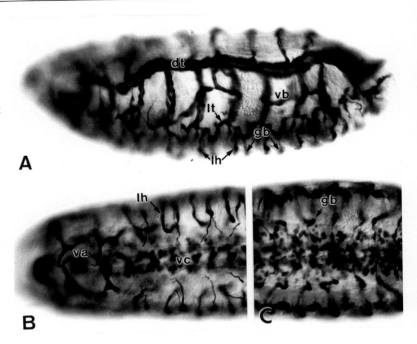

other towards posterior and interior of the embryo. During stage 13, the branched tracheal invaginations of all segments fuse and lay down the basic pattern of the larval tracheal tree (Figs. 12.1, 12.3). The dorsal stems develop into the transverse connectives (nomenclature after Keilin 1944, and Whitten 1980). The anteriorly directed dorsal branches of all segments fuse into the dorsal longitudinal tracheal trunks. The interiorly directed branches form the tracheae of the intestine.

During stage 15, the ventral posterior branches fuse into the lateral longitudinal trunk. The tips of the former dorsal stem and ventral anterior branch continue to elongate towards the median plane. They form the dorsal tracheae and ventral ganglionic tracheae, respectively. The latter grow right over the dorsal surface of the ventral nerve cord. Corresponding dorsal and ventral ganglionic tracheal branches on either side are connected by thin anastomoses.

Further branching of the tracheal tree occurs from stage 15 onwards. Cuticle begins to appear in the lumen of the tracheal branches during stage 15. It is noteworthy that after obliteration of the tracheal pits, as well as during further development, no mitoses seem to occur in the developing tracheal tree. A more detailed description of tracheal branching and development in general is given by Manning and Krasnow (1993), Samakovlis et al. (1996) and Guillemin et al. (1996), to which the interested reader is referred.

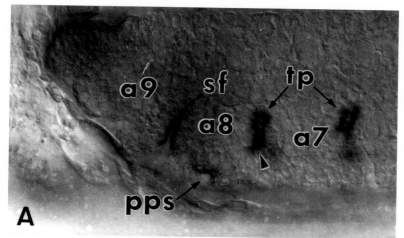

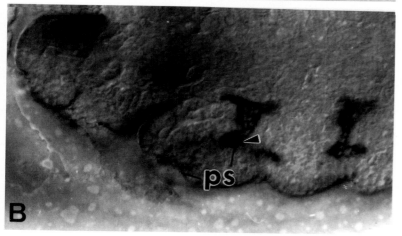

Fig. 12.5. Early (**A**) and late (**B**) stage 12 embryos, stained with an anti-Crumbs antibody, to illustrate the development of the posterior spiracles (*ps*). The caudal opening of the tracheal tree develops by the fusion of the caudalmost tracheal pit (*tp* at *a8*) with the invaginated primordium of the posterior spiracles (*pps*), laterally in *a8*. *sf* furrow between the abdominal segments *a8* and *a9*

Chapter 13
The Gonads

The gonads develop from cells of two different origins; the cells of the germline are recruited from the so-called pole cells, whereas the gonadal sheath and the interstitial cells are derived from the mesodermal layer, probably from that of abdominal segments $a5$, $a6$ and $a7$. Germline precursors and mesodermal elements meet to form the embryonic gonads, two dorsolateral cell clusters visible at the level of $a5$, after completion of germ band shortening (refer to Figs. 13.1, 13.2).

13.1
Formation of the Pole Cells

The pole cells form at the posterior tip of the egg in stage 3, when a group of nuclei reaches this region and enters the polar cytoplasm (Fig. 2.4A). Mahowald and colleagues have thoroughly studied the development of the pole plasm as well as the behaviour of pole cells during embryogenesis, and described their results in a long series of papers (Mahowald 1962, 1971; Illmensee et al. 1976; Mahowald et al. 1976; Turner and Mahowald 1976; Allis et al. 1979; Underwood et al. 1980a). Our observations agree fully with their results. A number of features contribute to the very characteristic morphology of the pole cells, permitting them to be precisely distinguished from all remaining cells of the *Drosophila* embryo. Most important among these features are the shape of the nucleus and cytoplasm, which are regularly spherical, and the staining properties of the cytoplasm, that stains darkly, apparently due to relative lipid deficiency (Allis et al. 1979); the polar granules, which are barely distinguishable with the light microscope; and the nuclear bodies, which are quite conspicuous; both structures are only transiently present in the pole cells, disintegrating in stage 9.

Fig. 13.1. The development of the gonads. **A to E** show dorsolateral views of stage 5 (**A**), stage 8 (**B**), stage 11 (**C**), stage 12 (**D**) and stage 15 (**E**) embryos. In the blastoderm stage (**A**), the pole cells (*pc*) form a cluster of 34-37 cells overlying the anlage of the posterior midgut rudiment (*pmg*). Pole cells will become incorporated into the lumen of the posterior midgut rudiment (in **B**) during gastrulation. When the posterior midgut rudiment loses its epithelial structure, the pole cells move out of the posterior midgut pouch and come to lie between the dorsal surface of the posterior midgut rudiment and the overlying mesoderm (**C**). During germ band shortening (**D**), the pole cells form an elongated cluster on either side. This cluster is in close contact with a set of mesoderm cells which will later give rise to the gonad sheath (follicle progenitor cells; *fp*). After germ band retraction (**E**), pole cells and the surrounding mesodermal sheath cells form the gonads (*go*) at the level of segment *a5*. Only a fraction of the pole cells become incorporated into the gonads; the remainder are lost in the yolk. The genital disc (*gd*) arises as a narrow ridge of cells located in the ventral midline between the denticle belt of *a8* and the anal pads. Modified from Hartenstein (1993), with permission of Cold Spring Harbor Laboratory Press

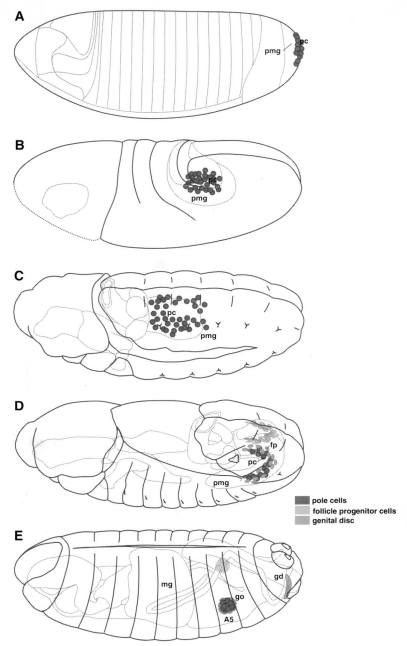

■ pole cells
■ follicle progenitor cells
■ genital disc

However, the movements undergone by the pole cells to form the gonads can be easily followed by means of specific markers, particularly antibodies directed against the product of the gene *vasa* (Fig. 13.2; refer to Hay et al. 1988a, b; Lasko and Ashburner 1988, 1990).

13.2
Migration of Pole Cells

During early development, the pole cells maintain close relationships with the primordium of the posterior midgut. After general cellularization is completed at the blastoderm stage, most of the pole cells lie on the anlage of the posterior midgut (Figs. 13.1A, 13.2A); some of them can be seen among the posteriormost cells of these and a few are found below the blastoderm cells, though clearly separated from the yolk sac. All pole cells, i.e. those in the amnioproctodeal invagination and those in the embryo interior, accompany the posterior midgut primordium when it sinks inwards at gastrulation and later in its progression cephalad as the posterior midgut pocket, at germ band extension. In stage 10, most of the pole cells can be seen at the blind end of the midgut pocket (Figs. 2.13, 13.2B); only a few are located either between the cells of the midgut epithelium or outside, in the immediate neighbourhood of the pocket (Fig. 13.2C). While the second postblastoderm mitosis is in progress, during stage 10, the pole cells will leave the midgut through the dorsal wall of the pocket, and become aligned on both sides of the posterior midgut primordium, at the level of *a6*, *a7* and *a8* (Fig. 12.2A, 13.2D).

Jaglarz and Howard (1994) have shown that exit from the midgut pocket and migration of the pole cells is controlled by factors which are midgut dependent. Callaini et al. (1995) have studied the process of passage of the pole cells through the wall of the posterior midgut with the electron microscope. These authors describe the migration of the pole cells as occurring in several steps. First, the pole cells approach the epithelium of the posterior midgut and establish contact with it by means of irregular processes. At the same time, the epithelial cells of the midgut extend protrusions, which interact with the pole cell projections. Then, the pole cell processes enter the intercellular space of the epithelium, disturbing the apical junctional complexes and finally leaving the pocket. Cell counts carried out at this stage show that practically all pole cells leave the posterior midgut pocket (Hay et al.

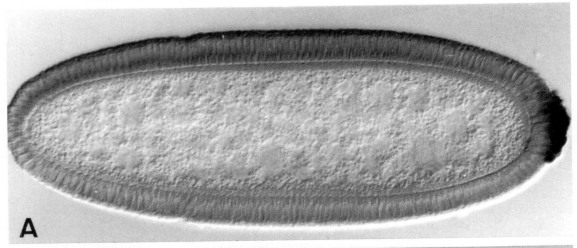

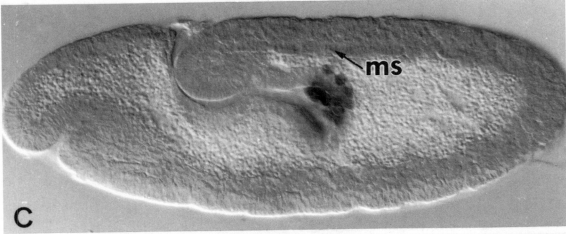

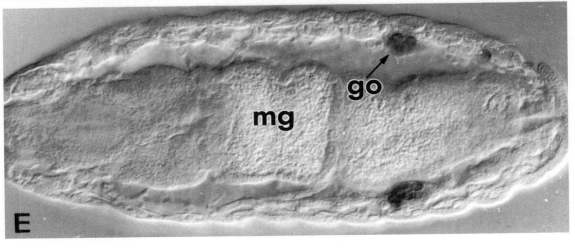

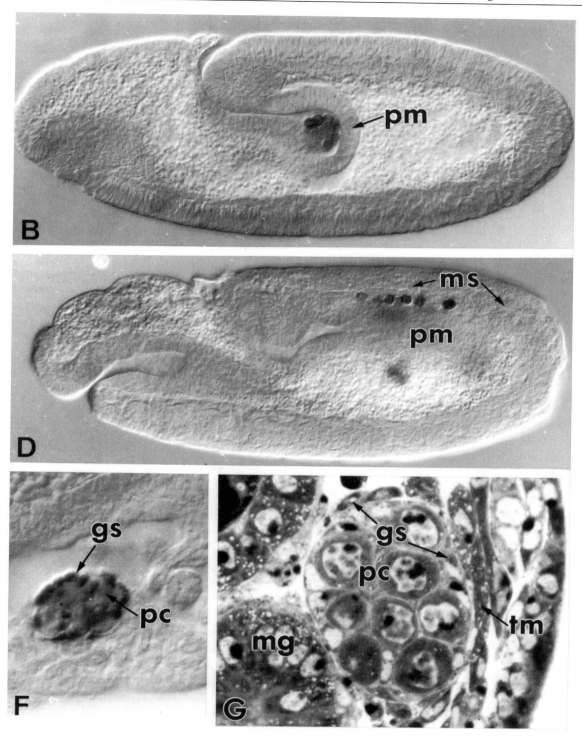

Fig. 13.2 Development of the gonads. **A to E** are embryos in stages 6 (**A**), 9 (**B**), 10 (**C**), early 12 (**D**) and 16 (**E**) stained with an anti-Vasa antibody; F shows the left gonad of an embryo carrying an enhancer trap that highlights the somatic, mesodermal cells of the gonads; G shows a transverse section through a gonad of a stage 16 embryo. At the blastoderm stage (**A**), the pole cells form a cluster of 34-37 cells at the posterior pole of the embryo. During gastrulation, the pole cells will become incorporated into the lumen of the posterior midgut rudiment (*pm* in **B**). When the posterior midgut rudiment loses its epithelial structure, the pole cells migrate out of the midgut rudiment (**C**). In the early stage 12 embryo, pole cells are located between the dorsal surface of the posterior midgut rudiment (*pm* in **D**) and the overlying mesoderm (*ms*). After germ band retraction, pole cells and the surrounding mesodermal sheath cells form the gonads (*go*), located laterally in segment *a5*. Other abbreviations. *gs* gonadal sheath; *mg* midgut; *pc* pole cells

1988a). Due to the fact that pole cells leave the developing posterior midgut dorsally, the pole cells preferentially occupy a position dorsal to the midgut pocket, between the pocket itself and the mesodermal layer (Figs. 13.1C, 13.2D), and only very few of them, most probably those located below the blastodermal cells since cellularization, are found ventral to the posterior midgut primordium.

13.3
Formation of the Gonads

While shortening of the germ band is in progress, pole cells become displaced, apparently passively, by the posterior midgut, particularly by its dorsal division of small cells, and progressively shift their position, eventually to become clustered at the level of *a5* (Figs. 12.2C, 13.1D-E, 13.2E-G). During the period prior to final formation of the gonads, the pole cells actually enter into contact with mesodermal cells of *a5*, *a6*, *a7* and, to some extent, *a8*. Cell counts show several inappropriately located pole cells at this stage, which will eventually be lost (Hay et al. 1988a). It is difficult to figure out the precise origin of the mesodermal component of the gonads, due to the distortion of the general cellular architecture at germ band shortening. In fact the pole cells can be unambiguously seen to be surrounded by, and intermingled with, mesodermal cells for the first time when they are clustered at the level of *a5*; thus, the somatic cells of the gonads most probably derive from *a5*, although some of these cells may well derive from any, or all, of *a5-a8*. Several enhancer trap lines available to us exhibit β-galactosidase activity in the gonadal cells of mesodermal origin (Fig. 13.2F); however, this expression begins too late to allow one to discern the origin of these cells. Brookman et al.

(1992) report that the retrotransposon *412* is expressed in dorsolateral mesodermal cells of parasegments 2-14 during stages 10-13.
Cells expressing *412* in parasegments 10-12 (corresponding to *a5-
a7*) are closely associated with germline cells by mid-stage 11, and
appear to be the only source of gonadal mesodermal cells. Observations on various mutants are consistent with an origin of gonadal mesoderm from *a5-a7* (Brookman et al. 1992).

In contrast to previous descriptions (Poulson 1950), Underwood et al. (1980b) showed experimentally that the gonads are the
only target of the pole cells in the *Drosophila* embryo. Nevertheless, only a small fraction of the total number of pole cells present
at the blastoderm stage reaches the gonads; the remainder are lost
during embryogenesis (Sonnenblick 1950; Technau and Campos-
Ortega 1986; Hay et al. 1988a, b). Most of the cells that do not
reach the gonads are located in the yolk, and either degenerate or
are expelled through the anus; very few may become transiently
trapped between the epithelial cells of the midgut (Illmensee et al.
1976; see, however, Technau and Campos-Ortega 1986, and Hay et
al. 1988a). The total number of pole cells after the two divisions in
the blastoderm stage has been cited by several authors as being
between 30 and 50 (Sonnenblick 1950; Zalokar and Erk 1977;
Underwood et al. 1980a, Technau and Campos-Ortega 1986; Hay et
al. 1988a), and no further divisions can be seen to take place in the
pole cells until formation of the gonads has been completed. Of
these 30-50 cells, only 10 - 15 can be found in the gonads when they
form in stage 14. Cell transplantation experiments (Technau and
Campos-Ortega 1986) provide evidence that the number of pole
cells present in the gonads at stage 16 is regulated, in such a way
that additional pole cells either degenerate or are lost. This regulation seems to occur during and after germ band shortening,
perhaps even after formation of the gonads in stage 14. Sonnenblick (1941) describes a bimodal distribution of pole cells in the
developing gonads, with 5-7 in some embryos and 9-13 in others.
It is not known whether this distribution correlates with the sex
of the embryos. According to the same author the onset of mitotic
activity in the germline cells occurs at 16 h of embryonic development (Sonnenblick 1941).

Chapter 14
The Pattern of Embryonic Cell Divisions

14.1
Overview

Two different periods of mitotic activity can be distinguished during the embryonic development of *Drosophila melanogaster*. The first mitotic period takes place prior to cell formation, and during this period the zygotic nuclei go through thirteen rounds of mitosis (Zalokar and Erk 1977; Foe and Alberts 1983). The first seven zygotic divisions are synchronous, leading to a syncytium of 128 ellipsoidally distributed nuclei in the center of the yolk (Rabinowitz 1941a). During the course of the next three mitotic cycles (8 to 10), most of the nuclei approach the egg surface stepwise to form the syncytial blastoderm, whereas about 200 nuclei (the presumptive vitellophages) remain centrally (Rabinowitz 1941b), and another 17-18 (Zalokar and Erk 1977) are incorporated in the posterior pole plasma to form the pole cells. Three further, parasynchronous (Foe and Alberts 1983) or, rather, metachronous (Foe et al. 1993) mitoses bring the number of syncytial blastoderm nuclei to approximately 5000 (Zalokar and Erk 1977; Turner and Mahowald 1977; Hartenstein and Campos-Ortega 1985), terminating the first period of mitotic divisions. These divisions are referred to as the preblastoderm divisions. The second period of mitotic activity occurs in the embryonic cells after the blastoderm stage. Mitotic activity in the preblastoderm stage has been thoroughly described in papers by Zalokar and Erk (1977) and, particularly, by Foe and Alberts (1983; see also Foe et al. 1993), to which interested readers are referred. We shall not deal with these preblastoderm divisions in any detail, for our present interest is centred on organogenesis. Thus, the major aim of this chapter is to report on the postblastoderm mitotic patterns of embryonic cells.

There are several distinct differences between the pre- and postblastoderm mitoses. The cycles of the preblastoderm nuclear divisions (cell cycles 1 to 13) are very rapid because they lack G1 and G2 phases (Rabinowitz 1941a; Zalokar and Erk 1976; Foe and Alberts 1983); therefore, mitosis ensues immediately after DNA replication (S phase; Edgar and O'Farrell 1989). For each of the three divisions in the syncytial blastoderm stage, a wave of mitotic activity lasting from 0.5 to 2 min sweeps smoothly from the anterior and posterior poles towards the transverse equatorial plane of the embryo (Foe and Alberts 1983). Cell cycles during the postblastoderm period (cycles 14 to 16, and further) by contrast have a G2 phase of variable length that separates DNA replication from mitosis. Due to the differences in length of G2 phases, cells in different regions of the embryo enter mitosis at different stages (Edgar and O'Farrell 1989). Thus, with respect to the time of onset of cell division, the embryo can be subdivided into a number of "mitotic domains" (Foe 1989).

For our initial analysis of the pattern of postblastoderm mitoses (Hartenstein and Campos-Ortega 1985), we used fuchsin-stained wholemounts. Our observations suggested to us that the postblastoderm mitoses are initiated from a number of "mitotic centres", from which mitotic activity appeared to spread into neighbouring regions of the embryo, like an expanding wave. Foe (1989), in a more detailed study of the first postblastoderm mitosis using an anti-tubulin antibody, recognized that the mitotic activity initiated within each mitotic centre does not spread throughout the embryo, as we had initially thought, but is confined within the limits of well defined domains in which mitotic activity is independent of that in other domains. With the material used for our present description (stained with anti-BrdU antibodies following BrdU incorporation), we essentially confirm Foe's results. Therefore, we have adopted her system and, in the following, we will largely use her terminology in naming mitotic domains. One exception, however, will be made for the description of the mitoses in the ventral neuroectoderm, where we will follow a different system for reasons to be explained below. The nomenclature is based on the pattern of the first postblastoderm cell division (cycle 14), which Foe studied. In each of these mitotic domains, "...mitosis appears to start in one or in a small number of cells and to spread, wavelike in all directions from these first cells, until the mitotic wave stops at the domain's boundary" (Foe 1989). Therefore, cells within a given domain complete mitosis at about the same time.

Embryonic cells do not divide very frequently. With the exception of the pole cells, the progenitors of the CNS and the epidermal sensilla, most cells of the *Drosophila* embryo undergo only three or four divisions up to the end of embryogenesis. Following the third postblastoderm division (cell cycle 16), most cells enter a G1 phase for the first time. Some cells undergo additional rounds of division before ceasing mitosis. Eight postblastoderm divisions have been described for the majority of neuroblasts (Hartenstein et al. 1987), but evidence from transplantation experiments, as well as from labelling studies on neuroectodermal cells with DiI, suggests that some neuroblasts perform up to ten or more divisions (Technau and Campos-Ortega 1986a; Prokop and Technau 1991; Bossing and Technau 1994; Bossing et al. 1996). Cells of sensilla undergo at least one more mitosis after cycle 16 (Bodmer et al. 1989). Finally, most cells of the mesoderm (Bate 1993; our unpublished observations), the Malpighian tubules (Janning et al. 1986), and specific ectodermal regions also go through a 17th mitotic cycle. In some cells, e.g. salivary glands and gut, rounds of DNA replication occur that are not accompanied by division of the nucleus or cytoplasm; instead, they lead to the high degree of polyploidy characteristic of dipteran larval tissues.

14.2
The Pattern of the First Postblastoderm Mitosis

Domains were named by Foe according to the sequence of appearance of mitotic activity; thus $\partial 14_1$, for example, designates the first domain in which the 14th mitotic cycle was observed, $\partial 14_2$ the second, etc. Consecutive numbering does not imply contiguity of mitotic domains. As we will describe below, most of the mitotic domains identified during the first postblastoderm division reappear during the two subsequent cell cycles.

14.2.1
Procephalon and Dorsal Ectoderm

Mitotic activity restarts in the postblastoderm embryo at approximately 3 h of embryonic development (stage 8), starting from two regions in the ectoderm located on either side of the procephalon (compare Foe and Alberts 1983; Figs. 14.1, 14.2, 14.4). The anterior domain is located in the middle of the procephalon ($\partial 14_1$), the posterior one is located adjacent to the cephalic furrow ($\partial 14_2$). Other mitotic domains within the procephalon are $\partial 14_3$, $\partial 14_{18}$ and $\partial 14_{20}$, which are dorsal to $\partial 14_1$; parts of $\partial 14_{15}$, and $\partial 14_{23}$ which are anterior to $\partial 14_1$; and the remaining parts of $\partial 14_{15}$ and $\partial 14_8$, which

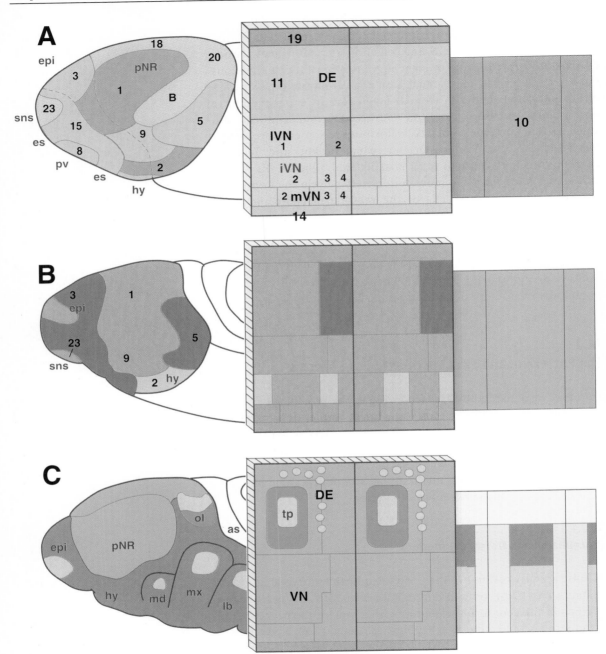

Fig. 14.1. Pattern of mitotic domains in the head (left section), ectoderm (middle section) and mesoderm (right section) during stages 9 (**A**), 10 (**B**), and 11 (**C**). Different colours are used to indicate the cell cycle number (*orange* cycle 14, first postblastoderm mitosis; *purple* cycle 15, second postblastoderm mitosis; *green* cycle 16, third postblastoderm mitosis; *blue* cycle 17, fourth postblastoderm mitosis). Light colouring indicates early interphase, intermediate colouring late interphase, dark colouring mitosis. Numbers identify mitotic domains (Foe 1989; Hartenstein et al. 1994b). Where possible, mitotic domains have been assigned to specific organ primordia. By stage 9 (**A**), the cells of the mesoderm, mesectoderm, dorsal ectoderm and most of the head are in interphase 15; cells of the ventral neuroectoderm and part of the procephalic neuroectoderm, and those in a narrow dorsal strip of the procephalon and germ band, are still in cycle 14. During stage 10 (**B**), mitosis 15 takes place in the primordia of the foregut, posterior procephalon, and dorsal ectoderm. The pattern of mitotic domains in the central procephalon (corresponding to Foe's domains $\partial 141$ and B) during the second postblastoderm mitosis is highly complex and has not yet been worked out. A narrow strip of ventral neurectodermal cells neighbouring the mesectoderm are undergoing mitosis 14. Towards the end of stage 10, the cells of the mesoderm perform mitosis 16. During stage 11 (**C**), cells in most parts of the ectoderm of the germ band and procephalon undergo mitosis 16 and arrest in G1 of cycle 17. Some regions, notably the primordia of the pharynx, stomatogastric nervous system, optic lobe, the mesoderm and parts of the gnathal and thoracic segments go through a fourth postblastoderm mitosis and arrest in G1 of cycle 18. Whereas mesodermal cells divide without any conspicuous pattern during divisions 14-15, the pattern of divisions 17 in the mesoderm is highly complex. Some, if not all, of the mitotic domains correspond to primordia of the various mesodermal derivatives (e.g. visceral mesoderm, fat body). Abbreviations. *DE* dorsal ectoderm; *epi* epipharynx (domain $\partial 143$); *es* oesophagus (domain $\partial 1415$); *hy* hypopharynx (domain $\partial 142$); *iVN, lVN, mVN* intermediate, lateral, medial domain of ventral neurectoderm; *ol* primordium of the optic lobe; *pNR* procephalic neuroectoderm; *pv* proventriculus (part of domain d148); *sns* stomatogastric nervous system; *tp* tracheal pits

are ventral to $\partial 141$. Thus, cells within the entire rostral portion of the procephalon will have carried out the 14th mitosis by around 3:50 h (stage 9). According to our observations, cells in d1423 (which probably corresponds to the anlage of the proventriculus) undergo mitosis 14 at least 30 min earlier than described by Foe (1989).

Mitotic domains $\partial 145$, $\partial 149$ and $\partial 142$ are organized in a ring at the border of the cephalic furrow (Fig. 14.1). Cells within these domains have carried out the 14th mitosis by 3:20 h of embryonic development; however, the cells in a small region (Foe's domain $\partial 1424$), wedged between domains $\partial 145$ and $\partial 142$, appear to divide substantially later than the rest of the posterior procephalon. Strikingly, a strip of cells between $\partial 145$ and $\partial 141$ appears not to undergo mitotic activity during this time (Figs. 2.7 - 2.9). These cells (Foe's domain B, Figs. 14.1, 14.4) form part of the procephalic neurogenic region (Hartenstein and Campos-Ortega 1984) from which neuroblasts of the procephalic lobes will originate.

Two additional mitotic domains appear in the ectoderm of the trunk during stage 8, one immediately posterior to the cephalic furrow ($\partial 146$), and the other ($\partial 144$) bordering the posterior wall of

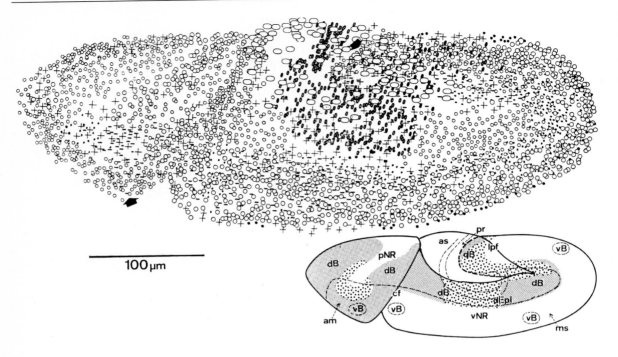

Fig. 14.2. Reconstruction of two *Drosophila* embryos at 3:35 h (**A**) and 1.25 h (**B**) after fertilization, stained with fuchsin as wholemounts; only the left body half is shown (this is Fig. 3 of Hartenstein and Campos-Ortega 1985). Insert shows schematic drawing of the same embryo. Nuclei of ectodermal cells are marked by *large open circles* (before first postblastoderm mitosis), *crosses* (at metaphase of first postblastoderm mitosis), *large black dots* (telophase of first postblastoderm mitosis), and *small open circles* (interphase after first postblastoderm mitosis). *Large open ovals* represent nuclei of amnioserosa cells. Nuclei of meso-dermal/endodermal cells are indica-ted as *small fat ovals* (before first postblastoderm mitosis), *horizontal bars* (metaphase of first postblastod-erm mitosis), and *small black dots* (telophase of first postblastoderm mitosis). **B** Approximately 15 min later a closed dorsal band (*dB*) is for-med by cells that have completed or are still in the first postblastoderm mitosis (*stippled region* and *dots*, respectively). The ventral band of mitotic mesoderm and anterior mid-gut cells (*vB*) is continuous anterior-ly and posteriorly with the dorsal band (*large arrows* in the recon-struction). Amnioserosa cells (*as*) have acquired their typical flat shape and can be differentiated from dor-sal epidermoblasts. Abbreviations. *am* anterior midgut primordium; *as* amnioserosa; *atf* anterior transverse furrow; *cf* cephalic furrow; *lpf* lateral proctodeal furrow; *ms* mesoderm; *pNR* procephalic neurogenic region; *pr* proctodeum; *ptf* posterior trans-verse furrow

the amnioproctodeal invagination. Another mitotic domain can be defined anterior to $\partial 146$, within the posterior lip of the cephalic furrow ($\partial 147$). Posterior to $\partial 146$, mitotic activity is seen throughout of $\partial 1411$, which includes the progenitor cells of most of the dorsal ectoderm of the trunk. Mitotic activity within $\partial 1411$ is initiated in several additional, somewhat irregularly distributed groups of cells, but by 3:20 h mitotic activity covers the entire dorsal ectoderm (see Fig. 14.1B).

14.2.2
Mesoderm and Gut

Cells of the mesoderm and anterior midgut rudiment (Fig. 14.3), which had invaginated during gastrulation, undergo their first postblastoderm mitosis shortly before the appearance of mitotic activity in the cells of the dorsal ectoderm. Division starts along a longitudinal front at the apex of the mesodermal tube (central strip of Foe's domain $\partial 1410$). Mitotic activity then rapidly spreads down the walls of the mesodermal tube to reach the border of the ectoderm at 3:20 h. The double row of cells connecting the mesoderm and ectoderm (mesectoderm; Foe's domain $\partial 1414$) are the last to be involved in this ventral mitotic wave. Therefore, all the cells of the mesoderm, mesectoderm and anterior midgut primordium enter their first postblastoderm mitosis within 10 min, forming a ventral belt of mitotic cells. Mitotic activity stops abruptly after reaching the ventral midline.

The procephalic mesoderm follows a slightly different mitotic schedule from the rest of the mesoderm and the neighbouring anterior midgut rudiment. Cells of the procephalic mesoderm seem to have a dual origin: some appear to arise from the "crossbar" of the ventral furrow; others appear to segregate from the midventral part of the procephalon over a relatively long period of time (stages 8 to mid 10). These latter cells form two vertical plates sandwiched between the anterior midgut primordium (and later stomodeum) medially and the procephalic neuroectoderm laterally. The procephalic mesoderm forms a distinct mitotic domain; cells of this domain undergo mitosis 14 (and later 15) slightly later than the remaining mesoderm.

Posteriorly, the ventral and dorsal mitotic belts are juxtaposed at the position of the amnioproctodeal invagination, which gives rise to the hindgut and posterior midgut. Domain $\partial 144$, which is still located at the outer surface of the embryo when its constituent cells divide, contains the anlage of the distal hindgut; the cells of domain $\partial 1413$ will give rise to the proximal portion of the hind-

Fig. 14.3. Pattern of mitotic domains in the developing gut at stages 8-9 (**A**), 10 (**B**) and 11 (**C**). Different colours are used for the different cell cycles (see Fig. 14.1). By stage 9, all the cells of the anterior midgut rudiment (derived from parts of Foe's mitotic domains $\partial 148$ and $\partial 1410$) are in the interphase of cycle 15, whereas the cells of the posterior midgut rudiment are still in interphase 14. The primordia of the hindgut and Malpighian tubules occupy two mitotic domains, $\partial 144$ (mainly Malpighian tubules) and $\partial 1413$ (mainly hindgut). These domains can be followed throughout cell cycles 14, 15 and 16. During stage 10, the primordium of the proventriculus invaginates along with the stomodeum and, towards the end of this stage, its cells undergo mitosis 15. Likewise, all the cells of both midgut rudiments and of the developing hindgut and Malpighian tubules undergo mitosis 15. During stage 11, the cells of the midgut rudiments go through their third and final postblastoderm division (cycle 16). Interstitial cell progenitors (*ICPs*) and imaginal (adult) midgut progenitors (*AMPs*) will divide slightly later than the bulk of cells forming the larval midgut. Many cells of the primordia of oesophagus, pharynx, Malpighian tubules and part of the hindgut will undergo a fourth division

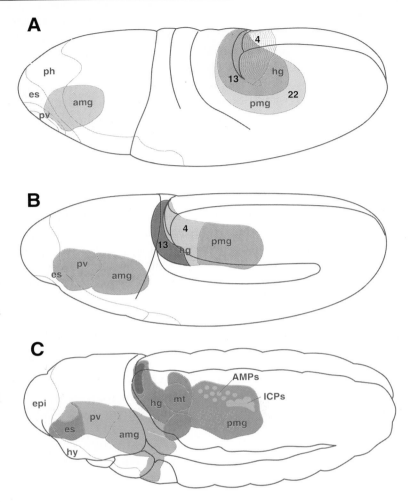

gut, including the Malpighian tubules, and will undergo division as part of the amnioproctodeal invagination. Finally, the primordium of the posterior midgut, which forms the bottom of the amnioproctodeal invagination, divides as mitotic domain $\partial 1422$ during stage 9.

14.2.3
The Pattern of the 14th Division in the Ventral Neuroectoderm

The cells of the ventral neurogenic region slowly enter mitosis 14 according to a very elaborate pattern. Due to its complexity, the pattern of mitosis in the ventral neuroectoderm has only been

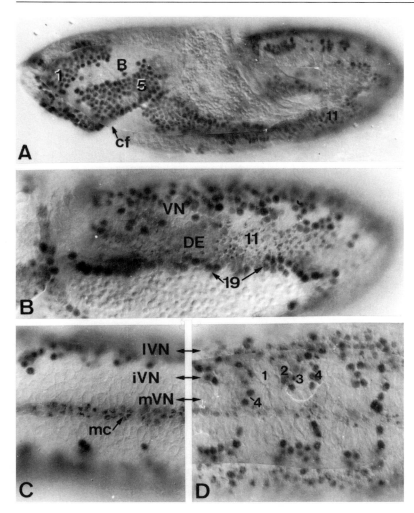

Fig. 14.4. Pattern of first postblastoderm mitosis, as visualized by a 20-min application of BrdU to stage 8-9 embryos, immediately followed by fixation. In this Figure, notice that labelling with BrdU reflects the pattern of the previous division. This is because in the absence of a G1 phase in the early cycles, S phase follows closely on the preceding mitosis; BrdU labeling is spatially correlated with this mitosis. **A** In the ectoderm, the first postblastoderm mitosis begins during stage 8 in the procephalon (Foe's domains $\partial 14_1$, $\partial 14_2$ and $\partial 14_5$) and the dorsal ectoderm ($\partial 14_{11}$). Ventrally, the mesectoderm and metamerically reiterated clusters in the intermediate column of the ventral neurectoderm (*iVN*) are labeled. **B** Stage 9 embryo, showing labeling in the intermediate column of the ventral neurectoderm (*iVN*) and the dorsal edge of the ectoderm ($\partial 14_{19}$). Most of the dorsal ectoderm (*DE*; $\partial 14_{11}$) still shows faint labeling of the nucleoli, the DNA of which replicates at the end of S phase. **C, D** Ventral view of part of ventral neurectoderm of stage 8 (**C**) and early 9 (**D**). In C, DNA replication is restricted to the mesectoderm (*mc*) and the lateral column of the ventral neurectoderm (*lVN*). Later on (**D**), mitotic activity affects cells in more medial parts of the ventral neurectoderm, which replicate DNA in a strictly metameric pattern (*iVN*2, $\partial 14_4$; *mVN*4). Other abbreviations. *B*, mitotic domain B. *cf* cephalic furrow

partially characterized in previous studies (Hartenstein and Campos-Ortega 1985; Foe 1989). A more comprehensive analysis of this pattern is provided by Hartenstein et al. (1994b; ref. to Figs. 14.4, 14.5). In her atlas of mitotic domains in the early embryo, Foe (1989) assigned most cells of the ventral neuroectoderm to two regions which she called M and N, in both of which she distinguished four small, segmentally organized domains ($\partial 14_{16}$, $\partial 14_{17}$, $\partial 14_{21}$, $\partial 14_{25}$). However, BrdU incorporation allows one to define even smaller domains within Foe's domains M and N (Hartenstein et al. 1994). The first two subpopulations of neuroblasts (SI, SII neuroblasts) delaminate from the ventral neuroectoderm shortly after gastrulation (Hartenstein and Campos-Ortega 1984; Hartenstein

Fig. 14.5. The pattern of mitosis and SI neuroblast delamination in the ventral neuroectoderm (*VN*). **A to D** show surface views of part of the *VN* at stage 8 (**A**), early stage 9 (**B**), late stage 9 (**C**) and stage 10 (**D**). The ventral midline is to the left. Thick horizontal lines mark parasegmental boundaries. The upper half of each panel is representative of a thoracic parasegment, the lower half, of an abdominal parasegment. Vertical lines mark boundaries between longitudinal subdivisions of the *VN* (bottom in **A**). Below each panel, a schematic transverse section of the VN at the corresponding stage is drawn. Different colours in the sections and surface views indicate the cell cycle stage the corresponding cells have reached. *Light orange* indicates interphase 14, *dark orange* mitosis 14, *light purple* early interphase 15, *medium purple* late interphase 15, *dark purple* mitosis 15. Large *circles* represent neuroblasts during (*hatched outline*) or after (*closed line*) delamination. During stage 8 (**A**), cells of the mesectoderm (ME) and dorsal ectoderm (DE) divide and enter interphase 15. Only small clusters of cells in the *iVN* (mitotic domains *iVN1* and ∂143; Foe's domains ∂14₁6 and ∂14₁7) of the abdominal parasegments divide at this stage. During early stage 9 (**B**), groups of cells near the parasegmental boundary (domain *lVN2*) divide first. At the same time, SI neuroblasts have started to delaminate, but still maintain contact with the apical surface. During late stage 9 (**C**), metamerically organized cell groups can be seen to divide within the intermediate and medial *VN*; numbers in this panel identify individual mitotic domain in the *VN*. Domain *mVN4* corresponds to Foe's (1989) domain ∂14₂5. SI neuroblasts have fully delaminated and most of them have ente-

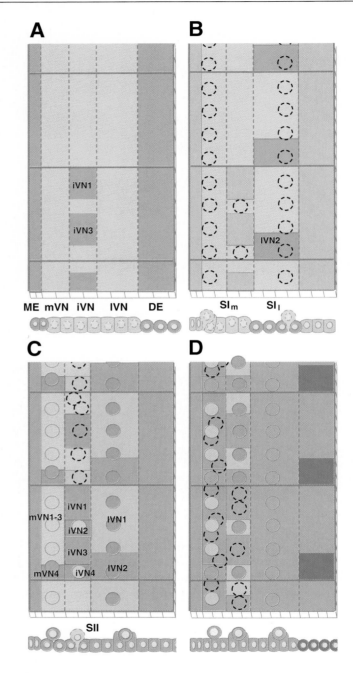

red mitosis. SII neuroblasts start to delaminate in the *iVN*. During stage 10 (**D**), the last cells in the *mVN* undergo mitosis 14; mitosis 15 starts in the *DE*. SII neuroblasts have delaminated and enter mitosis

et al. 1987; Doe 1992). During the same stage, cells of the neuroectoderm perform their first postblastoderm division (mitosis 14); prospective neuroblasts will be discussed in some detail below. To designate the ventral neuroectodermal mitotic domains, we have developed a terminology different from Foe's, based on topological criteria. As mentioned at the beginning of this chapter, proliferating cells in the *Drosophila* embryo do not have a G1 phase, but enter DNA replication immediately after mitosis. Thus, the pattern of DNA replication in cycle 15 (visualized with BrdU) faithfully reflects the pattern of the foregoing mitosis 14 (Fig. 14.4).

The ventral neuroectoderm can be subdivided into three longitudinal zones, each 2-3 cell diameters in width (Fig. 14.5). The lateral zone (lVN) and the medial zone (mVN) give rise to the lateral and medial row of SI neuroblasts, respectively; the intermediate zone (iVN) produces most of the SII neuroblasts, which come to be located in the intermediate row. Cells of lVN divide first (late stage 8/early stage 9), followed by the cells of iVN (mid stage 9) and mVN (late stage 9/early stage 10). In addition to these mediolateral subdivisions, mitotic activity within the ventral neuroectoderm is organized into metamerically repeated domains along the anterior-posterior axis. In the lVN, two of these metamerically organized domains can be identified: one consists of a transverse strip 2-3 cell diameters wide in the posterior part of each parasegment (mitotic domain lVN2); the other consists of all the other, more anteriorly located cells in the parasegment (domain lVN1). This latter domain corresponds to Foe's domain d1421, which she described only for the thoracic segments; the same mitotic domain can, however, be observed in the abdominal tagmata as well. The cells of lVN2 divide before those of lVN1. With respect to iVN and mVN, at least four mitotic domains, iVN1-4 and mVN1-4, respectively, can be recognized (Figs. 14.4, 14.5). Domains iVN1 and iVN3 were identified in the abdominal tagmata by Hartenstein and Campos-Ortega (1984) and designated by Foe (1989) as domains d1416 and d1417, respectively, which she described only for the abdominal segments. Domain lVN2 is in register with the laterally adjacent iVN4 and mVN4. These three domains define the posterior boundary of the parasegment and lie immediately anterior to the *engrailed* expression bands (Foe 1989).

14.2.4
The Pattern of Neuroblast Delamination Correlates with the Mitotic Pattern

Within thoracic and gnathal segments, SI neuroblasts start to delaminate from lVN at the time when cells of this zone enter mitosis (ref. to Fig. 14.5). During this time, the nuclei of the neuroepithelial cells within lVN and iVN exhibit conspicuous movements that accompany the segregation of the neuroblasts and have been described in detail (Hartenstein et al. 1994b). Before the onset of the 14th mitosis, the nuclei of all cells in these two regions of the neuroectoderm are located basally. During stage 8, the nuclei of the overwhelming majority of neuroepithelial cells shift apically in preparation for mitosis; nuclei in a few cells, however, remain basal and postpone division. On the basis of their relative numbers, we believe that the late dividing cells are the prospective neuroblasts, while the early dividing cells correspond to the cells that remain in the neuroectoderm after the segregation of the SI neuroblasts. During late stage 8, the nuclei of most cells of lVN and iVN are either in metaphase or in telophase; presumptive SI neuroblasts have basally located nuclei, but still retain a slender process connected to the apical surface, thus exhibiting the characteristic shape of "bottle cells". Within a few minutes after the lateral SI neuroblasts segregate, losing their connection to the apical surface, they enter mitosis.

The situation is not as clear in mVN, due to its proximity to the mesectoderm. SI neuroblasts within the medial row delaminate at the same time as neuroblasts of the lateral row. The SI neuroblasts are in direct contact with the mesectodermal cells (Fig. 14.5), which divide immediately prior to SI segregation. Medial SI neuroblasts enter mitosis within a short period of time following their segregation. By contrast, non-neural cells of mVN divide considerably later than the medial SI neuroblasts. There is a striking correlation in the time of onset of mitosis between the medial SI neuroblasts and the overlying ectoderm. As mentioned above, cells in the posterior portion of each parasegment divide earlier than cells in the anterior portion. This difference in time of onset of mitosis is most pronounced in the mVN, the posterior cells of which (domain mVN4) divide during mid-stage 9, while the anterior cells divide 30-60 min later. Consistent with this earlier onset of mitosis in the posterior parasegment cells, the neuroblast emerging from mVN4 (NB 5-2; Doe 1992) divides earlier than all other medial SI neuroblasts.

Within abdominal parasegments, mitotic divisions in the ventral neuroectoderm occur in a pattern slightly different from the one described above for thoracic and gnathal segments. Two small clusters of 2-4 cells each, corresponding to domains iVN1 and 3, respectively, in the iVN of each parasegment, undergo mitosis much earlier than any of the surrounding cells. Interestingly, this pattern of division is reflected in the pattern of SII neuroblasts, for initially no SII neuroblasts delaminate from mitotic domains iVN1 and iVN3, leading to gaps in the row of SII neuroblasts. These gaps are filled in at a later stage, so that the final pattern of neuroblasts is ultimately identical in abdominal and thoracic segments (Doe 1992).

In the ventral neuroectoderm, the last cells to divide are located anteriorly in each parasegment in mVN (domains mVN1-3). At the same time that cells divide in mVN1-3, delamination of SIII neuroblasts takes place from mVN. Smaller subpopulations of neuroblasts called SIV and SV by Doe (1992) delaminate even later (stage 11) from intermediate levels of the ventral neuroectoderm. The delamination of SIV and SV appears to be correlated with the pattern of the second postblastoderm mitosis of the ventral neuroectoderm which takes place during stage 11; the larger number and smaller size of neuroectodermal cells during stage 11 has so far made it impossible to reconstruct the precise relationship between late delamination and mitosis.

14.3
The Pattern of the Second Postblastoderm Mitosis

The pattern of the second postblastoderm mitosis (15th mitosis) has not been established with the same degree of accuracy as that of the first one. However, we have made a few observations on this mitosis and found some similarities to the pattern of the first one (Fig. 14.6). Many of the mitotic domains defined for mitosis 14, in particular those within the segmented germ band and the internal tissues, reappear during mitosis 15. This implies that, by contrast to cycle 14, cycle 15 is of equal length for the cells in these regions. Unfortunately, due to the complicated morphogenetic movements that change the topography of the procephalon, we could not verify whether the numerous mitotic domains described for this part of the embryo also reappear during mitosis 15.

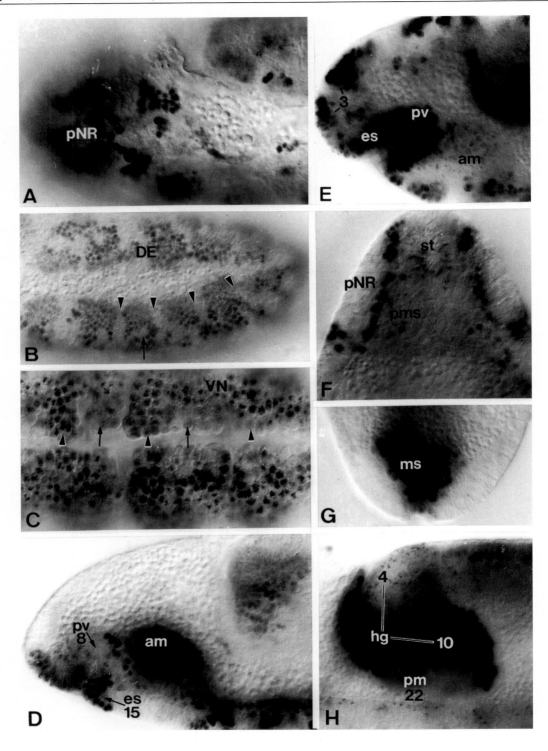

Fig. 14.6. Pattern of the second post-blastoderm division. BrdU was fed to embryos in stages 9-10. A Patchy BrdU incorporation in the procepha-lic neurectoderm (*pNR*), encompas-sing part of former mitotic domains ∂145, B and ∂141 (compare with Fig. 14.4). B The first groups of cells that undergo the second mitosis in the dorsal ectoderm (*DE*), correspond-ing to former domain ∂1411, are metamerically arranged and separat-ed by gaps (*arrowheads*); these lat-ter cells will divide, i.e. replicate DNA, slightly later. Other gaps (*arrow*) may correspond to sensory organ progenitor cells which also divide later than the surrounding epidermal precursors. C In the ven-tral neuroectoderm, the second postblastoderm mitosis is also meta-merically organized. BrdU-labeled cells (*arrowheads*) alternate here with patches of cells in mitosis (*arro-ws*). D In a mid-stage 10 embryo, cells perform the second division in the anterior midgut rudiment (*am*) and in the lip of the invaginating sto-modeum, i.e. the anlage of the oeso-phagus (*es*; Foe's domain ∂1415). The cells in the fundus of the stomode-um, comprising the anlage of the proventriculus (*pv*) and part of the anterior midgut (Foe's domain ∂148), have not not yet replicated DNA. E Mitotic activity will invade the fun-dus of the stomodeum (*pv*) and the anlage of the epipharynx (domain ∂143) slightly later. F Ventral view of the head of a stage 10 embryo, show-ing labeling in the procephalic mesoderm (*pms*). G Dorsal view of labelled mesodermal cells (*ms*) at the posterior pole, which have per-formed their first postblastoderm mitosis. H The cells in the proximal hindgut (domain ∂154) divide earlier, as shown by nucleolar staining in this Figure, whereas those of the posterior midgut rudiment will div-ide slightly later. Other abbreviati-ons. *pNR* procephalic neurectoderm; *st* stomodeum

14.3.1.
Dorsal Ectoderm

The pattern of the second postblastoderm mitosis in the dorsal ectoderm is essentially the same as that of the first mitosis; thus, mitotic domains undergoing mitosis 15 can generally be designa-ted with the same numbers as used for mitosis 14. The mitosis 15 pattern has a much more segmental appearance, however, parti-cularly within the germ band, than does the previous (14th) divi-sion. Mitotic activity of the second postblastoderm mitosis spreads from metamerically organized centres in the trunk to cover two fairly continuous dorsal belts which fuse rostrally and caudally with a midventral belt. At early stage 10, a cluster of cells in the centre of each of segments *t1* to *a9* divides; from each of these centres, mitosis spreads to the segment periphery so that by mid-stage 10, cells organized in a continuous belt covering the dorsal ectoderm have entered the 15th mitosis. In the gnathal and posterior procephalic region of the mid-stage 10 embryo, four mitotic centres can be distinguished (Fig. 14.1). These centres partly overlap with the domains (from posterior to anterior) ∂156, ∂157, ∂152 (dorsal strip of this domain) and ∂155. The posterior-most of these centres (part of ∂156) corresponds to the labium; two other, immediately adjacent domains, partly folded into the posterior and anterior lip of the cephalic furrow, probably corre-spond to the maxilla (part of ∂157) and mandible (part of ∂152), respectively. The fourth domain is located more dorsally, anterior to the cephalic furrow ($\partial 15_5$ and part of $\partial 15_{20}$); its segmental identity, whether part of the antennal segment or the acron, is unclear. As already noted for mitosis 14, proliferation during mitosis 15 stops at the boundary between dorsal and ventral ectoderm. Cells within a narrow ectodermal stripe bordering the amnioserosa (Foe's domain $\partial 15_{19}$) also divide later (late stage 10) than the remaining dorsal ectoderm during mitosis 15.

In the anterior head region, several other mitotic domains can be observed. They coincide approximately with those defined for cycle 14, and undergo mitosis 15 during the second half of stage 10. These are $\partial 15_{15}$ (anlage of oesophagus), $\partial 15_{18}$ (anlage of proventriculus), and $\partial 15_3$ (labrum or epipharynx). In addition, the cells of the ventral part of domain $\partial 15_1$, bordering $\partial 15_{15}$, also divide. The division pattern of the remaining domains of the head is very complex. Domain B and parts of $\partial 15_1$ and $\partial 15_9$ seem to give rise to procephalic neuroblasts (see Chapter 11). Strikingly, numerous small clusters of ectoderm cells within the region in which these three mitotic domains are located initiate their second postblastoderm division during stage 10. In addition to this division, procephalic neuroblasts continue to divide after they have segregated; some neuroblasts even seem to divide while they are still at the surface (see Chapter 11). Due to the complex origin of the head neuroblasts, we could not determine precisely how often the ectodermal cells that will give rise to the dorsal pouch divide, nor how many times the neuroblasts divide.

14.3.2
Mesoderm and Gut

The cells of the mesoderm (included in $\partial 14_{10}$) are the first embryonic cells to perform the second postblastoderm division (Fig. 14.3); they divide in an approximately synchronous fashion at about 4:15 h ($\partial 15_{10}$). During this mitosis, mesodermal cells are all at the same level, arranged in an almost perfect monolayer. This arrangement is lost during interphase of cycle 16, when the cells dissociate. As already mentioned, cells of the procephalic mesoderm divide slightly later than the other mesoderm cells. After mitosis 15, head mesodermal cells also dissociate, to spread out and produce the hemocytes (see Chapter 6).

During early stage 10, mitotic activity is initiated in the anterior midgut primordium and, slightly later, in the amnioproctodeal invagination. The same mitotic domains defined for mitosis 14 (i.e. $\partial 14_4$: posterior hindgut; $\partial 14_{13}$: anterior hindgut, including

Malpighian tubules; $\partial 1422$: posterior midgut rudiment) can be recognized for mitosis 15 (Fig. 14.8).

14.3.3
Ventral Neuroectoderm and Neuroblasts

The tripartite subdivision of the ventral neuroectoderm into lateral, intermediate and medial stripes remains manifest during mitosis 15 (Fig. 14.6C). The cells in the lateral stripe divide during late stage 10; as already mentioned for mitosis 14, division begins in each parasegment in a central domain ($\partial 1521c$), followed by the flanking cells ($\partial 1521i$). The cells of the intermediate stripe seem to divide more synchronously than during mitosis 14, possibly as a consequence of the fact that no neuroblasts delaminate from this region during late stage 10. Finally, the medial stripe, the cells of which divide during early stage 11, exhibits several different, not yet well defined mitotic domains.

SI and SII neuroblasts perform their second postblastoderm division in a parasynchronous manner during early stage 10 (Hartenstein et al. 1987). It is not possible to determine precisely the sequence of divisions of the remaining, later segregating neuroblasts, or the procephalic neuroblasts, since it is not possible to decide how frequently a given neuroblast has divided while in the neuroectoderm prior to segregation.

14.4
The Pattern of the Third and Fourth Postblastoderm Mitoses

Analysis of BrdU incorporation indicates that, with the few exceptions listed below, all embryonic cells perform a third postblastoderm mitosis (the 16th embryonic division), which takes place from late stage 10 through stage 11; some embryonic regions, such as the mesoderm and parts of the foregut, hindgut and head, even undergo a fourth postblastoderm division, a fact which was not appreciated in the first edition of this book. Unfortunately, the data concerning the exact pattern of the third and fourth postblastoderm mitoses are very sparse and mitotic domains cannot be defined for either of these mitoses (Figs. 14.7, 14.8).

A salient feature of the 16th division that can be observed in the germ band, however, is that the 16th division starts from segmentally arranged patches in the dorsal and, later, the ventral ectod-

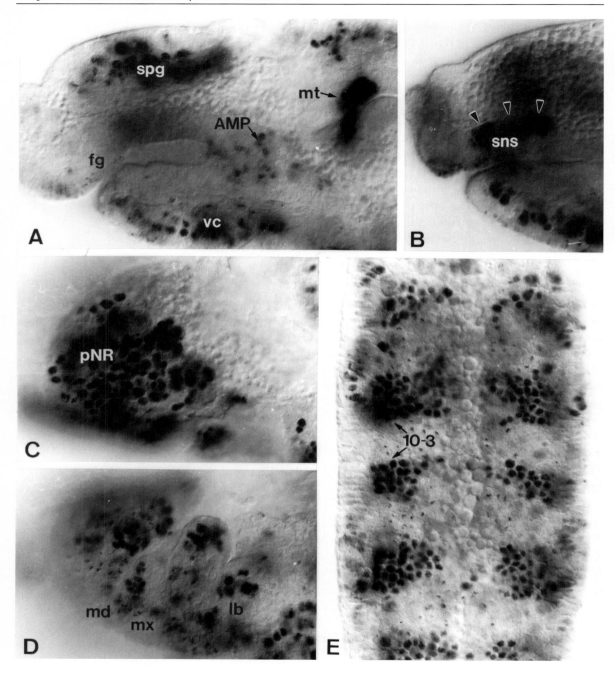

Fig. 14.7. Pattern of the third postblastoderm division. BrdU was applied to stage 11 embryos. Note that all cells labeled in this Figure are those which will undergo a fourth postblastoderm mitosis (M17). Thus, the majority of ectodermal cells, which perform only three divisions and arrest in G1 after the third mitosis, are not labeled. **A** Labeling of adult midgut precursors (*AMP*), Malpighian tubules (*mt*), and nervous system (*vc, spg*). **B** Labelling of invaginating progenitors of stomatogastric nervous system (*sns*). **C** Most cells in the procephalic neurectoderm (*pNR*) of stage 11 embryo are labeled. **D** Labelling of discrete patches of ectodermal cells in gnathocephalon (*md, mx, lb*) and prothorax, which will undergo a fourth mitosis. **E** Labeling in the mesoderm. By contrast with the two foregoing mitoses, in which all cells of the (trunk) mesoderm divided almost synchronously, the pattern of the third division is metamerically organized. Domain 10-3, which corresponds to large groups of muscle progenitors, is also labeled. Other abbreviations. *fg* foregut

erm. This suggests that, like the preceding mitosis, the 16th mitosis also affects embryonic domains of defined fate. Thus, the boundary between dorsal and ventral ectoderm (neuroectoderm) remains well defined by a differential pattern of mitosis. In the neuroectoderm, the pattern of the third division appears less complex than seen in the first and second divisions. This reduced complexity might be related to the fact that neuroblast segregation is complete by the end of stage 11. At least some segregated neuroblasts undergo their third division during late stage 10, their fourth division during mid-stage 11. The first division of the sensillum progenitor cells (see Chapter 9) occurs during the third postblastoderm division of the ectodermal cells; the second division, leading to the groups of sensillum cells, occurs much later, during stages 12 or 13. In the head, cells within dorsal parts of the gnathal buds and posterior procephalic region perform the 16th mitosis during mid-stage 11; the same structures undergo a 17th division during stage 12.

All cells of the gut perform a third postblastoderm division during stage 11 (Figs. 14.3, 14.8). The subdivision of the gut into the domains described above is still apparent. The cells within the primordia of the oesophagus ($\partial 1415$ and $\partial 1515$) and anterior hindgut plus Malpighian tubules ($\partial 1413$ and $\partial 1513$) divide a fourth time during stage 12. Finally, there are two populations of midgut cells which separate during stage 11 from the bulk of the midgut cells: the adult midgut progenitors (Hartenstein and Jan 1992; Tepass und Hartenstein 1994a), which derive from both anterior and posterior midgut rudiments, and the so-called interstitial progenitor cells (Tepass and Hartenstein 1994a, 1995), which originate in the posterior midgut rudiment only. Both groups of cells perform a fourth postblastoderm division during late stage 11 or early stage 12.

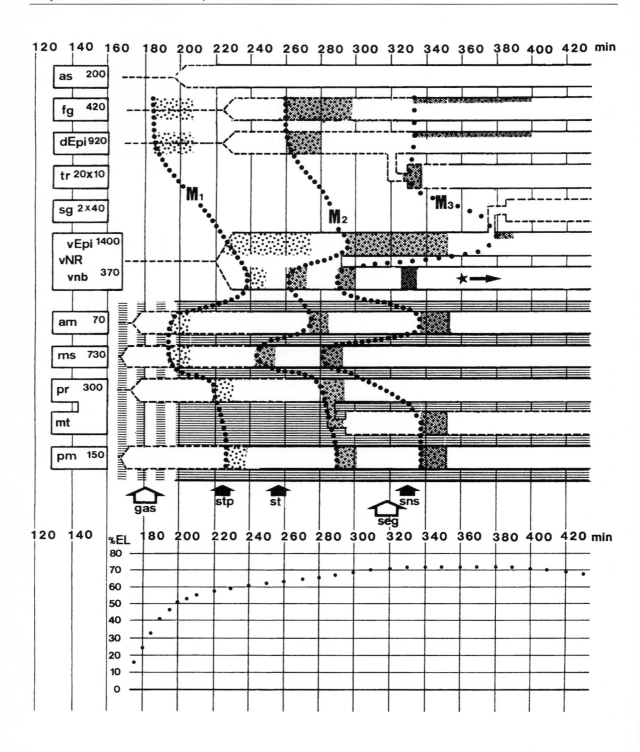

Fig. 14.8. Time table of proliferation during *Drosophila* postblastoderm development (this is Fig. 10 of Hartenstein and Campos-Ortega 1985. Time is given at the top, subdivided in 20 min intervals. Fertilization is time zero. Larval primordia are shown as horizontal bars, starting when the primordia become histologically definable; before this time presumptive primordia are indicated as dashed lines. Hatching applies to primordia invaginating at gastrulation. Onset and duration of mitoses in the primordia are indicated by small open circles and shading. Times of onset of each of the three postblastoderm divisions (*M1-M3*) in the different primordia are connected by dotted lines. Numbers in the boxes refer to the number of blastoderm cells of the corresponding anlagen. The star in the *vnb* bar of the ventral neuroblasts indicates further divisions of the neuroblasts. *Arrows* at the bottom indicate gastrulation (*gas*), formation of the stomodeal plate (*stp*), stomodeal invagination (*st*), segmentation of the epidermis (*seg*); invagination of the primordium of the stomatogastric nervous system (*sns*). In the diagram at the bottom, the extent of germ band elongation (in % EL) is given as a function of time; the dots mark the tip of the germ band. Other abbreviations. *am* anterior midgut; *as* amnioserosa; *fg* foregut (comprising clypeolabrum, pharynx, and oesophagus); *dEpi* dorsal epidermal anlage; *tr* trachea (developing from dorsal epidermal anlage); *ms* mesoderm; *mt* Malpighian tubules; *pr* proctodeum; *sg* salivary gland (developing from the ventral epidermal anlage of the labial segment); *vNR* ventral neurogenic region; *vEpi* ventral epidermal anlage; *vnb* ventral neuroblasts

The third division of the mesodermal cells takes place during late stage 10 or early stage 11, between 4.5 and 5 h (Fig. 14.7E). Unlike the two preceding divisions, during which the entire mesodermal layer behaves essentially like one single mitotic domain (i.e. $\partial1410$ and $\partial1510$), the mesoderm during the third mitosis appears to be organized in four mitotic subdomains per segment; however, it should be noted that all mesodermal cells perform the third mitosis. The first of these mitotic domains to divide ($\partial1610$-1) forms, in each metamere, a transverse stripe that extends from the ventral midline to the lateral edge of the mesoderm; the position of this stripe in the mesoderm corresponds topologically with the position of mitoses within the incipient tracheal pits in the overlying ectoderm. The next domain to divide ($\partial1610$-2) is a continuous longitudinal zone that encompasses the mesodermal crest; $\partial1610$-2 thus includes the precursors of the dorsal vessel and probably part of the prospective visceral mesoderm. Domain $\partial1610$-3 consists of an oval cluster of cells located at the level of the tracheal pits; cells of this domain divide at the time when the tracheal pits are invaginating. The last part of the mesoderm to divide is domain $\partial1610$-4, which consists of metameric clusters of medially located cells (Fig. 14.8).

Immediately after germ band shortening, small clusters of mitotic cells appear in the developing epidermis; these cells most probably correspond to the precursors of the epidermal sensilla. We do not know how many more divisions these ectodermal cells perform. The first clusters are visible on either side of the ventral midline, in the ventral epidermal anlage of the three thoracic seg-

ments, and might be the primordia of Keilin's sensory organs
(Hertweck 1931). Proliferation of the primordia of other sensory
organs (antennomaxillary complex, black dots, chordotonal
organs, sensory organs of the telson) occurs slightly later.

14.5
The Orientation of Mitotic Spindles

The mitotic spindles of cells in the primordium of the dorsal epidermis are preferentially oriented in anteroposterior direction,
during both the first and the second mitosis. In the primordium of
the ventral epidermis, spindles of the first mitosis are oriented
in all directions. No preferential polarity has been observed for
any of the mitoses in the primordia of mesoderm, fore-, mid- or
hindgut.

Chapter 15
Morphogenetic Movements

The cellular architecture of the blastoderm is that of a simple monolayer. Therefore, embryonic development can, in short, be described as the process of transforming this two-dimensional structure into a three-dimensional structure in which the basic body pattern (Sander 1976, 1983) can already be recognized. This topological transformation is achieved by means of morphogenetic movements. Morphogenetic movements may take place either (i) by growth and infolding of certain regions of the monolayer, without alteration of its original arrangement as, for example, occurs in the epidermal anlage; or (ii) by segregation of individual cells or coherent cell arrays from the monolayer into the interior of the embryo, which thereby lose their original contacts with the other regions; examples include the segregation of the neuroblasts and the invagination of the ventral furrow or the midgut. Indeed, most of the elementary morphogenetic movements found in embryos of other animal species, e.g. invagination, involution, ingression, delamination etc., also occur during embryogenesis in *Drosophila*.

Movements associated with cell growth include a variety of processes, e.g. changes in cell shape and size, proliferation, alteration of the position of individual cells relative to each other; these processes may themselves lead to infolding of some areas. Despite these infoldings, the original spatial relationships of neighbouring cells are maintained in this type of movement; that is to say, the cells retain their spatial cohesion throughout the shape changes that occur during later stages. On the other hand, individual cells can separate from the blastodermal monolayer and move into the interior, as is the case with the segregation of neuroblasts. In most instances, however, cells move in groups by means of invaginations; for example, the three germ layers become established by means of invaginations. In any case, unlike the movements associated with growth, this sort of separation implies interruption of the original integrity of the blastoderm, in that cells become displaced from their original position to form a new layer. The spatial relationships of these cells in the blastoderm become modified after these movements in such a way that direct translation from the blastoderm frame onto the new coordinates is not immediately possible.

During *Drosophila* embryonic development, several processes, i.e. gastrulation, germ band elongation and germ band shortening, head involution and dorsal closure, have an eminently morphogenetic character. Features of these very complex morphogenetic movements have already been considered in various chapters. Thus in the following description, in order to avoid extensive repetition, only those aspects of the major morphogenetic movements will be emphasized which contribute to the appearance of the basic body pattern. Due to the abundance of available data related to cephalogenesis and segmentation, these two operations will be considered in separate chapters.

15.1
Gastrulation

Gastrulation leads to the formation of the three germ layers - ectoderm, endoderm and mesoderm. We have seen that the separation of the three germ layers is achieved by means of the invagination of four different anlagen, those of the mesoderm, anterior and posterior midgut primordia, and hindgut. After invagination, these anlagen become organized as a tube and share a common lumen. Three divisions can be distinguished in this tube, each of which gives rise to different organs. The central division, or ventral furrow proper, gives rise to the mesoderm of the trunk; the T-shaped anterior tip of the ventral furrow comprises the endodermal component of the anterior midgut; and the posterior division, or amnioproctodeal invagination, corresponds to the hindgut and posterior midgut anlagen (Figs. 2.7). One important consequence of gastrulation is that large areas of the blastoderm become topologically distinct. The withdrawal from the surface of the blastoderm of 1250 cells, brought about by ventral furrow formation and the amnioproctodeal invagination, leads to the separation of mesodermal and endodermal anlagen from the ectodermal anlage, and from this stage onwards the cellular behaviour of the three germ layers is conspicuously different. For example, the mitotic schedules of ectodermal, mesodermal and endodermal cells are different (refer to Chapter 14 and Fig. 14.8); the endodermal primordia will form groups of mesenchymal cells, whereas the mesoderm initially acquires an epithelial organization; and so on.

Invagination of the ventral furrow is accompanied by conspicuous shape changes of the blastoderm cells, which have been thoroughly analysed by Leptin and Grunewald (1990), Kam et al. (1991), Parks and Wieschaus (1991) and Sweeton et al. (1991). The reader is referred to the review by Costa et al. (1993) for a more detailed description. The most obvious of these changes is a flattening of the apical ends of the cells in a domain called the ventral plate, which comprises the entire mesoderm anlage. As the cells flatten, the blastoderm nuclei become displaced basally, probably being pushed by the cytoplasm as a consequence of a slight shortening of the cells. After flattening, two domains of the ventral plate can be distinguished: the cells in a central domain, between 8 or 10 (Leptin and Grunewald 1990) and 12 cells wide, constrict their apices, probably due to contraction of the apical cytoskeleton (Young et al. 1991), the nuclei are displaced further basally and the bases of the cells expand laterally. The cells within the second, lateral domain follow the medial cells in their movements. In this manner all the cells shorten further and eventually invaginate. Nuclear displacement and apical constriction of the medial cells are asynchronous and affect cells in apparently random positions (Kam et al. 1991). The cells lateral to this central domain move toward the ventral midline as if pushed from their bases and the margins of the furrow meet at the midline to close the ventral furrow and form a tube (Costa et al. 1993).

Invagination of the cells forming the endodermal anterior midgut (Poulson 1950) occurs slightly later than that of the ventral furrow, and the invaginating cells undergo shape changes similar to those described above for the ventral furrow.

The invagination of the amnioproctodeum starts at the beginning of germ band elongation with the dorsal displacement of a cell plate at the posterior pole of the embryo. Its dimensions have been determined by Sweeton et al. (1991): the plate spans about ten cells dorsal and ventral to the pole cells and five cells on each lateral side. The same authors describe the cytological changes accompanying invagination as identical to those in the ventral furrow, with an apical flattening of the cells followed by basal displacement of the nuclei and apical constriction. This apical constriction may be related to an accumulation of cytoplasmic myosin observed at the apical ends (Young et al. 1991), and occurs asynchronously among the cells of the plate, beginning with those immediately dorsal to the pole cells.

With regard to the mesoderm, the original tubular architecture of its primordium in the ventral furrow is lost at the onset of mitotic activity, when the cells dissociate to divide. This disaggregated appearance of mesodermal cells persists during the fast phase of germ band elongation (first and second mesodermal mitoses). During this time, the spatial relationships of individual cells of the mesoderm become difficult to follow. However, there is suggestive evidence from histological observations that indicates that the disturbance of their topological interrelationships is more apparent than real. Most probably the tubular architecture is essentially maintained while the mesodermal cells divide. Although we do not have any evidence other than that derived from the observation of fixed material, extensive mixing of mesodermal cells does not seem to occur during early germ band elongation either in the anteroposterior or in the lateromedial dimension. On the contrary, it appears that the mesodermal cells retain the same topological relationships with the overlying ectodermal cells that their anlage had in the blastoderm. Thus, gastrulation merely displaces the mesodermal cells into the interior, without markedly disrupting their spatial relationships with one another.

After their first mitotic division, the cells of the mesoderm become organized in a regular monolayer. How the spatial coordinates of the tube formed by the mesodermal primordium prior to the onset of mitotic activity translate onto this monolayer is not clear. We assume that after the first mitosis the tube collapses, and that cells of the dorsal walls of the tube intermingle with cells of the lateral walls, like a zipper. Therefore, granted that no gross distortion in the topological order of the mesoderm ensues during the first mitosis, and cells mix during formation of the mesoderm layer as suggested above, the coordinates of the mesodermal tube may be translated onto the mesodermal layer, such that medial and basal portions of the tube correspond to medial regions of the mesodermal layer, lateral regions of the tube to lateral regions of the mesoderm, while the mesectoderm corresponds to the region neighbouring the ectoderm. Other possible transformations can also be conceived.

Beer et al. (1987) report that following transplantation, mesodermal cells may contribute progeny cells to up to three contiguous segments. In other words, after the first mitosis in the mesoderm, cells can indeed mix along the antero-posterior axis. We believe that this may occur as a consequence of passive displacement of daughter cells after each division and that no active cell crawling is required.

15.1.1
A Note on Germ Layers

Gastrulation, that is to say, the process that leads to the formation of the germ layers, is customarily considered to be complete with the invagination of the ventral furrow. However, new data suggest a different endpoint for gastrulation. As discussed in Chapter 7, Technau and Campos-Ortega (1985) found that cells in the fundus of the stomodeum will form part of the anterior midgut cells. Moreover, we have mentioned earlier that cells on the outside of the embryo within the region adjacent to the T-bar of the ventral furrow, join the cells of the posterior lip of the T that ingress to form part of the head mesoderm. According to the conventional definition, the ectoderm comprises those embryonic cells that remain on the outside after invagination of the ventral furrow, that is to say after completion of gastrulation. Therefore, by this criterion both the fundus of the stomodeum and the region adjacent to the T-bar belong to the ectoderm, although the fate of their cells is endodermal in the former case, i.e. the anterior midgut, and mesodermal, i.e. the procephalic mesoderm, in the latter. One is therefore forced to admit that the formation of the germ layers is not actually over when ventral furrow formation is complete, but that it continues for some time, at least as long as the midgut primordium and the head mesoderm continue to recruit cells (see Chapter 3).

15.2
The Cephalic and the Transverse Furrows

These are transient structures that appear when gastrulation is in progress. When the cephalic furrow becomes visible it consists of an almost vertical slit on either side of the embryo at approximately 65% EL, which rapidly deepens and elongates dorsally and ventrally, ultimately spanning the circumference of the embryo. Observations described by Costa et al. (1993) indicate that the infolding of cells to form the cephalic furrow is apparently not preceded by apical flattening. This contrasts with the behaviour of the cells that will invaginate into the ventral furrow cells or into the amnioproctodeal invagination cells. However, both nuclear displacement towards the basal pole and shortening of the cells along the apical-basal axis seem to be involved. Although the cephalic furrow roughly separates the procephalon from the segmented germ band, the furrow is in fact a rather dynamic structure into which different blastodermal regions fold consecutively. Thus, the cephalic furrow does not actually form any definitive border between larval primordia.

The transverse furrows originate through a similar mechanism and persist only during the initial phase of germ band elongation (Figs. 2.6-2.11), extending obliquely over the dorsal half on either side of the embryo. During germ band elongation their dorsal parts are pushed anteriorly and begin to flatten, finally being subsumed into the amnioserosa, which fills the gap that arises between the dorsal and the ventral halves of the expanding germ band. In a similar manner to the cephalic furrow, the transverse furrows sweep over the blastoderm sheet, and thereby sequentially incorporate different embryonic regions.

With our methods of study we cannot pinpoint the cause of cephalic or transverse furrow formation, nor is any evidence available that would support speculations on their morphogenetic consequences. It appears that transverse furrows might be related in some way to the formation of the amnioserosa. However, it is worth mentioning that, despite their transitory character, cephalic and transverse furrows appear in wild-type *Drosophila* invariably at the same relative positions; their appearance is most probably an expression of intrinsic properties of the individual cells rather than being caused by the need for space in which to store cells. Rickoll and Counce (1980) have noticed distortions in the surface of the yolk sac at the time of appearance of the furrows which might be causally related to their formation. This possibility is supported by the fact that cells forming the transverse and cephalic furrows still maintain connections with the yolk sac, and the observed contraction of the yolk sac could indeed pull the cephalic furrow into the interior.

Following the appearance of transverse folds, the amnioserosa forms to cover the embryo dorsally and laterally. In *Drosophila*, the amnioserosa derives from a block 5 cells wide and 40 cells long, which do not divide during embryogenesis (Hartenstein and Campos-Ortega 1985; Technau and Campos-Ortega 1985; Foe 1989). As germ band elongation proceeds, the block is pushed forward, becomes displaced to the sides of the embryo and folds. During this process, the cells undergo considerable modification in shape, finally forming a thin layer.

15.3
Germ Band Elongation

As already mentioned, germ band elongation occurs in two phases, an initial fast phase, that brings the tip of the germ band to 60% EL, and a slow phase, which will continue until the tip has reached 75% EL. During elongation the anlage of the germ band in the blastoderm undergoes gross changes in shape. Its width (dorsoventral extension) will be reduced to about half, whereas its length (anteroposterior extension) increases by about the same degree. These shape changes occur simultaneously over the entire length of the germ band.

15.4
The Behaviour of Blastoderm Cells During Early Morphogenetic Movements

At gastrulation and germ band elongation, the cells of the blastoderm undergo considerable changes in shape, size and position. Various mechanisms may contribute to these topological transformations of the blastoderm cells. For example, movement of neighbouring cells relative to each other, changes in shape and size of individual cells and in the polarity of the mitotic spindle. In an attempt to understand these movements, we analysed gastrulation and germ band elongation quantitatively to distinguish what type of cell behaviour participates in these transformations (Hartenstein and Campos-Ortega 1985). We were forced to use this approach, since individual cells cannot be followed directly in the living embryo.

During gastrulation, 1250 cells leave the surface of the embryo and the remaining 3750 cells redistribute to occupy approximately the area that was formerly occupied by 5000 cells. Since cell proliferation has not yet begun at this time, and gastrulation only effects a slight dorsoventral and longitudinal stretching of the ectoderm, the difference in cell number between 3750 and 5000 must be compensated for by an increase in cell size. In fact the anlage of the germ band increases in mediolateral and anteroposterior extent by about 20%, and concomitantly the area occupied by each of the cells of the germ band anlage increases by about the same degree (from 46 μm^2 to 57 mm^2 per cell). This means that, up to gastrulation, cells do not necessarily have to move relative to each other; growth is sufficient to explain the modifications of the larval anlagen.

Germ band elongation follows, which leads to evident mediolateral narrowing and anteroposterior elongation of the anlage of the metameric germ band, but does not produce major modifications of the procephalon. The germ band elongates from a length of 275 μm at the blastoderm to 605 μm, corresponding to 220 % of the original length. How does such a drastic change occur? We will first consider the dorsal epidermal primordium. In the blastoderm the dorsal epidermal anlage consists of about 460 cells on either side, encompassing the dorsal 35% of the dorsoventral extent (width) of the entire epidermal anlage (75 μm in absolute terms). After elongation the width of the dorsal epidermal primordium is reduced to 50 μm (a reduction of 33%). The diameter of dorsal epidermal cells decreases from 7.5 μm in the blastoderm to 5.7 μm after the first mitosis (a reduction of 24%). In most of the dorsal epidermal cells the mitotic spindle is longitudinally oriented, indicating that the daughter cells will come to lie predominantly on the longitudinal axis, roughly doubling the number of cells on this axis. Thus, the polarity of the first division and reduction in cell size might in principle be related to the longitudinal stretching of 82% and to the narrowing that occurs in the dorsal epidermal anlage during the initial phase of germ band elongation. However, these two events cannot be the cause of germ band elongation, since the rapid phase of germ band elongation takes place normally in *string* mutants, which fail to undergo postblastoderm mitoses (Edgar and O'Farrell 1989; Hartenstein and Posakony 1990). Due to the bending of the germ band at the posterior pole, the length of the dorsal border of the dorsal epidermal anlage is shorter than that of the ventral border. This difference is at least partially compensated for by transient infolding of dorsal epidermal cells in the posterior transverse furrow.

The behaviour of the cells of the ventral (neurogenic) region can best be interpreted if they shift their positions relative to each other in order to accommodate the distortions brought about by the fast phase of germ band elongation. Ventral cells do not divide during this period, and their diameter increases only slightly (from 7.5 μm to 8 μm). In fact, after rapid germ band elongation, the ventral neurogenic region consists of roughly 100 mediolateral rows of about 9 cells each, whereas in the blastoderm anlage the same cells are restricted to a smaller area of about 40 rows of 20-23 cells each. We have not been able to work out the precise pattern of the cell displacements that take place in the ventral neurogenic region during germ band elongation; however, based on circumstantial evidence we assume that these displacements are regular and that individual cells retain the same relationships with their neighbours as in the blastoderm. Proliferation starts among ventral cells when fast elongation is nearly complete. Since mitotic spindles of dividing ventral neuroectodermal cells do not display any preferred orientation (Hartenstein and Campos-Ortega 1985; Foe 1989), proliferation does not seem to be causally involved in the slow phase of germ band elongation. However, neuroblasts of the first subpopulation (Hartenstein and Campos-Ortega 1984) segregate by this stage, comprising approximately 15% of the cells of the ventral neurogenic region. It is conceivable that processes related to their segregation, e.g. irregularities in the position of nuclei and movement of neuroblasts out of the ectodermal layer, may also determine some rearrangement of the remaining cells which contributes to the elongation and narrowing of the ventral neurogenic region. Time-lapse cinematography confirms that, indeed, cells of the ventral neurogenic region intercalate during germ band elongation (Costa et al. 1993; see also Irvine and Wieschaus 1994). These authors propose a model for how cells intercalate, based on Gergen et al. (1986) and Wieschaus et al. (1991). According to Costa et al. (1993), mesodermal cells also intercalate during germ band elongation, and they propose that intercalation is the main mechanism of elongation of the mesodermal tube.

During the slow phase of germ band elongation, the metameric germ band increases in length by an additional 38%, and continues elongating until it has reached 220% of its original length (Costa et al. 1993). During this period, epidermoblasts divide for a second time and consequently decrease in size. Taking into account the orientation of the mitotic spindles, as well as cell dimensions before and after the second mitosis, it can be concluded that the polarity of the second mitosis can compensate by itself for the stretching that affects the germ band during the final phase of elongation.

15.5
Morphogenetic Movements Associated with Organogenesis

During the period of extended germ band, several morphogenetic modifications occur in the embryo; for instance, the invagination of the stomodeum, the appearance of parasegmental and intersegmental furrows, invagination of the tracheal primordia or salivary glands, Malpighian tubules, etc. These movements have in common the fact that no cell displacements relative to each other apparently occur, since the original epithelial architecture persists.

The stomodeal primordium can be defined morphologically at the end of the first postblastodermal cell cycle, as a conspicous cell plate. This stomodeal plate encompasses two mitotic domains, $\partial 1415$ peripherally and $\partial 1423$ centrally in the plate, whose cells divide consecutively. The plate invaginates and becomes transformed into a shallow groove. It is unlikely that mitosis of the cells in $\partial 1415$ and $\partial 14123$ is causally associated with the invagination, since invagination of the stomodeum occurs normally in *string* mutants (Hartenstein and Posakony 1990), which do not perform any of the postblastoderm mitoses. A similar situation can be observed with respect to the tracheal placodes of the wildtype, which appear as small depressions near the border between ventral and dorsal epidermal primordia in the stage 10 embryo (Fig. 2.15). Shortly afterwards all cells of the tracheal placodes divide and, simultaneously, invaginate to form deep pits (Fig. 12.1, 12.2). In this case, mitotic division and invagination are so closely associated as to suggest some interdependency. However, tracheal pits also form in *string* mutants (V. Hartenstein, unpublished). Still another invagination unrelated to cell division is that of the salivary glands, which develop from conspicuous placodes located in the primordium of the ventral epidermis of the third gnathal (labial) segment, on either side of the midline. The salivary gland placodes invaginate at approximately 6:15 h (Fig. 2.17D; Panzer et al. 1992). Although the mechanism of invagination in these three examples is not known, one possibility is that it is the same as that used in formation of the ventral and transverse furrows, discussed above.

15.6
Germ Band Shortening

Between 7:30 and 9:30 h of embryonic development the germ band retracts (Figs. 2.21-2.24). The major consequence of this retraction is the establishment of the normal, definitive anatomical relationships of the germ band organs, in that the proctodeum comes to be located at the posterior tip of the egg. Shortening of the germ band also occurs in two phases, a slow initial and a fast final; there is no obvious reason for this difference in the speed of retraction. During germ band shortening, thoracic and abdominal segments become longer in dorsoventral and shorter in rostrocaudal extent, thus reversing the shape modifications undergone by their primordia during germ band elongation. This remodelling of the shape of the segments is apparently attained through shape changes of the individual cells rather than by reversing the intercalation of the cells that took place during elongation.

15.7
Dorsal Closure

After shortening of the germ band, the embryo is 'open' dorsally, that is, both the midgut walls and the dorsal epidermal primordium have still to grow up to the dorsal midline in order to achieve dorsal embryonic closure. In the period of extended germ band, the amnioserosa exhibits deep infolds over its entire extent. During shortening, though, the extraembryonic membrane will unfold and be stretched considerably to separate the yolk sac from the vitelline envelope and, therefore, constitute the dorsal covering of the embryo at this stage. Laterally the primordium of the dorsal epidermis, evidently segmented by deep furrows, is contiguous and continuous with the amnioserosa.

Based on light microscopical observations, we previously assumed that the dorsal epidermal cells lose continuity with, and slide over, the layer of amnioserosa cells. However, electron microscopy (Ruggendorf et al. 1994) has revealed that the continuity of the epidermis and amnioserosa persists during virtually the entire process of dorsal closure, and is only interrupted as closure is achieved. A possible mechanistic interpretation of the processes accompanying dorsal closure is that, as it moves towards the dorsal midline, the edge of the dorsal epidermis pushes the amnioserosa; as a consequence, the amnioserosa cells change their shape from squamous to columnar. Since the integrity of the entire layer is preserved during this process, the amnioserosa should then fold in as if invaginating. Of course, one could easily invert the causal sequence, and propose that the amnioserosa invaginates actively and that dorsal closure is a consequence rather than a cause of the internalization of the amnioserosa. The cardioblast precursors then squeeze between the amnioserosa and the epidermis, and the amnioserosa cells detach from the epidermis as it closes dorsally, its cells dying subsequently. The modifications of the midgut leading to its closure occur simultaneously. These modifications have been discussed in Chapter 5. Considerable growth of dorsal epidermal cells follows, which is probably involved in head involution (see Chapter 16).

Chapter 16
Cephalogenesis

The larva of *Drosophila*, like that of other cyclorraphous dipteran insects, is acephalic or, more properly, cryptocephalic; that is to say, most structures of the head are located inside the larva and only a few of them, constituting the so-called pseudocephalon, are visible from the outside. Internalization of head structures occurs through head involution, a complex pattern of movements in which both the gnathal segments and the territory anterior to the cephalic furrow, the so-called procephalon, comprising the clypeolabral lobe, the hypopharyngeal lobe and the procephalic lobe, are involved. Movements related to head morphogenesis thus include: the appearance within the gnathocephalon of gnathal protuberances, sternal region and dorsal ridge; the differentiation within the procephalon of procephalic lobe and clypeolabrum; and the displacement orally of derivatives of the gnathal segments and other structures during atrium formation and head involution. In fact, as we shall see below, cephalogenesis is intimately related to the development of the foregut. We shall discuss the as yet unsolved problem of head segmentation in *Drosophila* in Chapter 17.

16.1
Early Events

16.1.1
Cephalic Furrow

The primordia of the head can be localized as early as the time of formation of the cephalic furrow, when the procephalon and the prospective territory of the metameric germ band become distinguishable. Poulson (1950) considered the cephalic furrow to be the posterior border of the presumptive head region, and Turner and Mahowald (1977) proposed that the anlage of the mandibular segment maps just posterior to the territory of the cephalic furrow. However, this view was later corrected by Underwood et al. (1980a), who demonstrated by experimental means that the presumptive mandibular region lies immediately in front of the furrow. Data of Technau and Campos-Ortega (1985), based on HRP injection into the cells of the cephalic furrow, support this latter view. At the time when the cephalic furrow becomes visible, i.e. at the onset of gastrulation in stage 6, the anlagen of the three gnathal segments are located in a region bordering, and including, the walls of the furrow: the mandibular anlage is located anterior to the maxillary, and this in turn is anterior to the labial anlage. Furthermore, each one of the anlagen occupies a similar fraction of the furrow. However, during subsequent development, the cephalic furrow is tilted posteriorly, the foregut initiates development, and the relationships of gnathal anlagen to the cephalic furrow become more difficult to follow (Figs. 16.1, 16.2). In any event, it should be emphasized that the cephalic furrow does not correspond at any stage to any boundary within the fully developed head.

16.1.2
Gnathal Segments

During the stage of extended germ band, the segments of the metameric germ band show a rather uniform organization, with the ectoderm of $t1$-$a9$ being set off by well defined dorsal and ventral borders from the mesectoderm and the amnioserosa, respectively. Moreover, ventral and dorsal subdivisions of $t1$-$a9$ are clearly defined by the regional specializations formed by their cells: for example, the neurogenic region, the tracheal placodes, etc. In principle, the organization of the three gnathal segments is strikingly similar to that of thoracic and abdominal segments, though they exhibit some peculiarities which are associated with their participation in head formation. The gnathal segments can actually be subdivided into dorsal, lateral and ventral regions. The lateral regions correspond to the sites of origin of limbs (segmental appendages), which in *Drosophila* are represented merely by specific sense organs. The ventral regions are serially homologous to the sternal regions of trunk segments and derive from the former neurogenic region. The dorsal regions fuse together to give rise to the dorsal ridge. Observations of Hartenstein and Campos-Ortega (1985) and Technau and Campos-Ortega (1985) support this similarity of organization between trunk and gnathal metameres. These data, which derive from observations on the mitotic behaviour of the cells and results of HRP injections into cells of the early gastrula, suggested that the gnathal segments comprise the anlagen of the sternum, the gnathal buds and the territory dorsal to the buds, delimiting the procephalic lobe. Whereas the gnathal buds will eventually form the lateral and ventral margins of the larval mouth, the sternum will become progressively incorporated into the foregut, and the area dorsal to the buds, in particular the area dorsal to the labial bud, will become part of the dorsal ridge (Technau and Campos-Ortega 1985).

The basic pattern of the gnathal segments described above is clearly discernible only up to early stage 12 (the so-called phylotypic stage; Sander 1983), when dorsal, lateral and ventral subdivisions are clearly evident in all three segments. However, during later stages, the pattern changes dramatically as the cells of the gnathal segments rearrange in the wake of foregut invagination and head involution (see below). Although the morphological boundaries between individual gnathal segments disappear, the further development of these structures can be followed by using molecular markers expressed in metamerically reiterated stripes, such as *engrailed* (DiNardo et al. 1985; Schmidt-Ott et al. 1994). In fact, the dorsal regions of all three gnathal segments fuse into a triangular territory, part of which is hidden within the deep folds surrounding the gnathal protuberances, while the other part forms the dorsal ridge (stage 12, Fig. 16.1A, B). This view is supported by the expression pattern of *engrailed*, which labels the posterior compartment of each segment (Diederich et al. 1991; Younossi-Hartenstein et al. 1993). Whereas each one of the lateral subdivisions of the gnathal segments is marked by its own *engrailed* stripe at the posterior margin (see Fig. 16.4), these stripes converge and fuse dorsally in the region of the dorsal ridge. The ventral subdivisions of the gnathal segments extend from the protuberances to the ventral midline and, as mentioned above, derive from the former neurogenic ectoderm, from which neuroblasts segregated at an earlier stage. By the phylotypic stage, these neuroblasts and their progenies have already formed three clearly demarcated neuromeres which will later fuse to give rise to the suboesophageal ganglion. Most of the ventral region of the labial segment invaginates to form the salivary gland.

While germ band extension is in progress, extensive cell movement occurs within ventral levels of the cephalic furrow until, by the time of formation of the ventral intersegmental furrows, the ventral arm of the cephalic furrow has vanished and all of its cells have become superficial; they eventually give rise to the gnathal protuberances (Fig. 16.1). This implies a great deal of cellular rearrangement. Stomodeal invagination in stage 10 brings about still

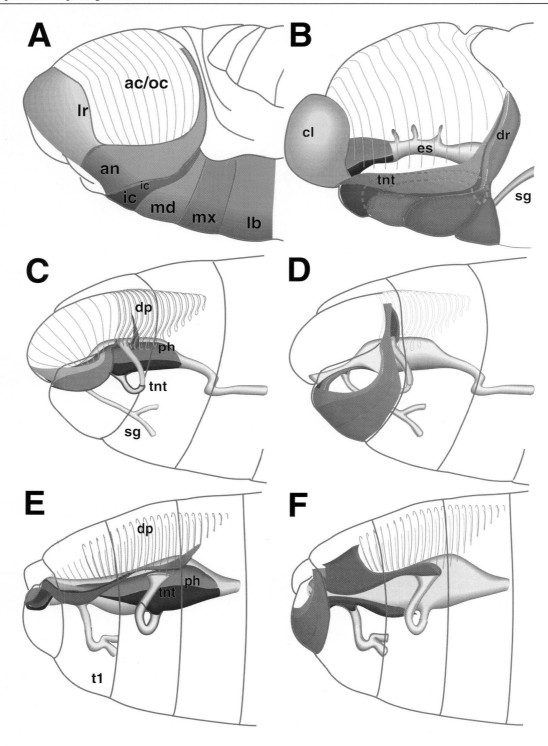

Fig. 16.1 Morphogenesis of the larval head. Panels **A** to **F** show schematic drawings of the embryonic head at three different stages (**A** stage 11; **B** stage 12, prior to head involution; **C**, **D** stage 15, during head involution; **E**, **F** stage 17, after head involution). Six head segments are shown in different colours [labrum (*lr*). *yellow*; antennal segment (*an*). *orange*; intercalary segment (*ic*). red; mandible (*md*). *purple*; maxilla (*mx*). lilac; labium (*lb*). green]. The acron (*ac*) may constitute a seventh head segment called the ocular segment (*oc*) and is shown by *hatching*. At stages 15 (**C, D**) and 17 (**E, F**), preoral segments (*lr, an, ic*) and postoral, or gnathal, segments (*md, mx, lb*) are shown separately in different panels for clarity. Prior to head involution, gnathal segments are drawn anteriorly, preoral segments ventrally (*an, ic*) and posteriorly (*lr*). Parts of these segments are involuted through the stomodeum and form part of the foregut (*lr* epipharynx; *ic* hypopharynx). This trend continues during later stages (**C-F**) when the ventral portions of the gnathal segments also move through the stomodeum and form part of the pharynx and the atrium. In addition, the dorsal portions of the gnathal segments, as well as the dorsal antennal segment, (possibly) the intercalary segment and the acron involute to form the dorsal pouch (*dp*). Other abbreviations. *es* esophagus; *ph* pharynx; *sg* salivary gland; *t1* thoracic segment; *tnt* tentorium

further oral displacement of cell groups, some of which will be withdrawn from the embryonic surface to become incorporated into the atrium and pharynx. Unfortunately, by this time the topological relationships of the different subdivisions of the gnathal anlagen have been modified to such an extent that they are no longer readily recognizable.

16.1.3
Procephalon

The tripartite organization described above for the gnathal segments cannot be recognized in the pre-gnathal head segments. The reason may lie in the topological transformations that the procephalon undergoes. The hypopharyngeal lobe may correspond to the ventral (and, possibly, lateral) region of the intercalary segment, because it bears the primordium of a small sense organ, the hypopharyngeal organ, also called "organ X". By contrast, the antennal segment of *Drosophila* appears to lack a ventral region, which would be consistent with the situation in other insect embryos (see Matsuda 1964; Rempel 1975). The primordium of the antennal (or dorsal) sense organ is located dorsal to the hypopharyngeal lobe; therefore, this could correspond to the lateral region of the antennal segment. By the same criterion, part of the prominent clypeolabral lobe appears to constitute the lateral region of the labral segment. Such a subdivision of the pre-oral head region is in agreement with the expression pattern of *engrailed*, in that individual expression stripes mark the extent of the presumptive intercalary, antennal and labral segments (Diederich et al. 1991; Schmidt-Ott and Technau 1992; Fig. 16.4).

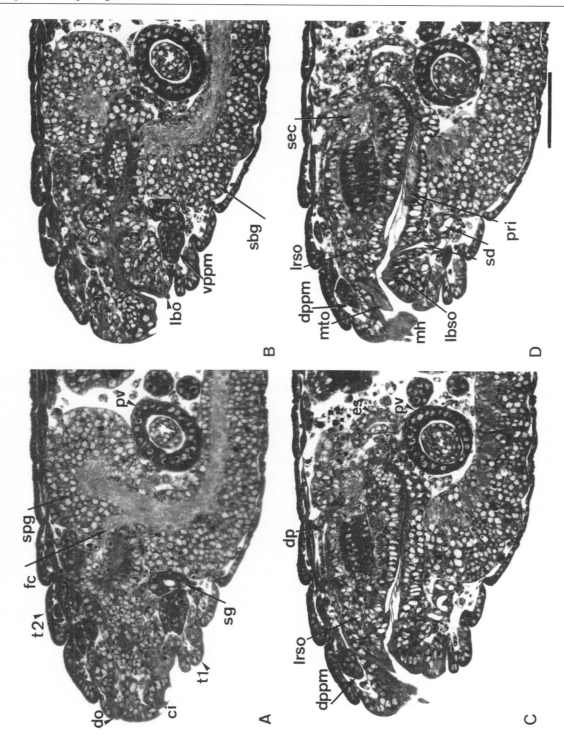

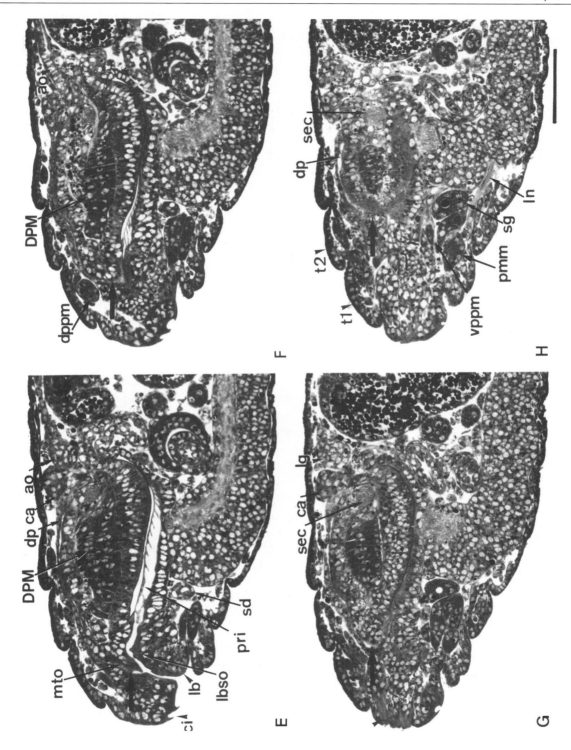

Fig. 16.2. A to H are parasagittal sections of a fully developed embryo to illustrate the organization of the cephalic region. The sections cover the entire pharynx and dorsal pouch, from one side (**A** and **B**), across the midline (**C** to **F**), to the other side (**G** and **H**). In this particular embryo, the cuticle of the cephalopharyngeal skeleton is not yet completely pigmented; thus it appears grey in histological sections.

The large *arrow* in **E** to **H** points to the connections between frontal sac and pharynx. Abbreviations.
an antennal division of the antennomaxillary complex; *aao* aorta; *cc* corpora cardiaca; *ci* maxillary cirri; *dp* dorsal pouch; *dppm* dorsal mouthpart muscle (DM in Fig. 4.6); *ep* labral sense organ (epiphysis); *es* oesophagus; *fc* frontal commissure; *lb* labium; *lbo.* labial sensory organ; *lg* lymph gland; *ln* labial nerve;

mh mouth hooks; *mto* median tooth; *DPM* dorsal pharyngeal musculature; *pmm* prothoracic-metathoracic muscles; *pri* pharyngeal ridges; *pv* proventriculus; *sbg* suboesophageal ganglia; *sd* salivary duct; *sec* supraoesophageal commissure; *sg* salivary gland; *spg* supraoesophageal ganglia; *vppm* ventral mouthpart muscle.
Bar 50 μm

There is no precise landmark for the boundary between the labral segment and the non-segmental acron (Fig. 16.1). The latter has been proposed to cover most of the dorsal part of the embryonic head (see Rempel 1975). Based on the expression pattern of *engrailed* and *wingless*, the acron (or part thereof) has been considered to be a 7th head segment, the ocular segment, interpolated between the antennal and lateral segments (Schmidt-Ott and Technau 1992). On comparative anatomical grounds, the eye and the visual centres of the brain are assigned to the acron. In the *Drosophila* embryo, the primordia of these structures, i.e. the visual organ of the larva (Bolwig´s organ) and the optic lobe, originate at a position dorsal to the antennal sensory complex.

The clypeolabrum is the only morphologically detectable subdivision in the entire procephalon, that is to say, within the pregnathal cephalic territory of the extended germ band embryo. The clypeolabrum, which becomes distinct from the procephalic lobe during stage 10 (Fig. 2.12), forms the tip of the procephalon and is continuous with the roof of the pharynx, being separated from the procephalic lobe proper by a deep transverse furrow. The cells of the clypeolabrum are arranged in a regular cylindrical epithelium, and thus differ very much from the neighbouring procephalic cells, which become considerably flatter after termination of the process of neuroblast segregation. In particular, mid-dorsal cells are extremely stretched, reminiscent of the cells of the amnioserosa. Clypeolabrum and procephalic lobe remain essentially unchanged until relatively late in embryonic development, up until the time of mouth formation and head involution. The anlage of the optic lobes is the only structure distinguishable in the procephalic lobe during this time prior to head involution. The optic lobes derive from a dorsolateral placode of about 75-85 cells (Hartenstein et al. 1985; Green et al. 1993), which invaginates as a whole to join the developing supraoesophageal ganglion during stage 12 (Figs. 2.14-2.19, 9.17).

16.2
Late Events: Larval Structure of the Atrium, Cephalopharyngeal Skeleton and Dorsal Pouch

The atrium of the *Drosophila* larva derives from the gnathal segments, chiefly from their superficial protuberances. The dorsolateral border of the prospective atrium consists of labial, mandibular and maxillary structures, the ventral border derives from the labium. All these structures will constitute the opening of the anterior part of the foregut, the so-called atrium or cibarium, which abuts on the pharynx, both being connected to the dorsal pouch. Atrium, pharynx and dorsal pouch contain the cephalopharyngeal skeleton, a very prominent larval structure, which consists of several different pieces, and has been thoroughly studied and described by Strasburger (1932), Schoeller-Raccaud (1977) and, in particular, by Jürgens et al. (1986) in the larva of *Drosophila* (Figs. 16.1 - 16.4).

In the fully developed embryo, the cephalopharyngeal skeleton comprises an intricately modelled chitinous structure secreted by the cells lining the atrium, pharynx and dorsal pouch (Figs. 16.1 - 16.4). Three main parts can be distinguished in the cephalopharyngeal skeleton: the mouth hooks, the H-piece and the cephalopharyngeal plates or cephalopharyngeal sclerites. Additional small pieces, the median tooth and the dorsal and ventral neck clasps, intimately related to each other and to the H-piece, also contribute to the cephalopharyngeal skeleton (Fig. 16.3). The mouth hooks are movable structures situated at the tip of the skeleton, one on either side of the atrium, each of which bears three processes; one projects to the outside and the other two are directed inward. The middle of the triangle formed by the mouth hooks on either side is thinner than the processes and, based on a description by Wahl (1914), Strasburger (1932) assumed the presence of a sensillum in this thinner region; we have not been able to confirm the existence of a sensillum related to the mouth hooks. During development, the mouth hooks can be seen to derive from the maxillary segment (Schoeller 1964).

The H-piece is situated between the mouth hooks and the cephalopharyngeal plates at the base of the boundary between atrium and pharynx, while the opening of the salivary gland is located immediately caudal to the transverse bar of the H. In fact

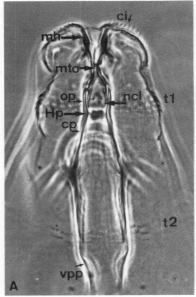 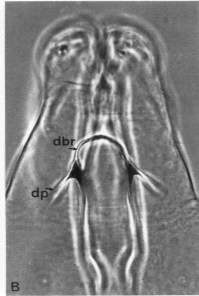 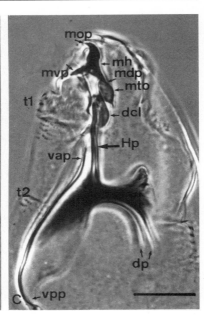

Fig. 16.3. A to C are cuticle preparations of first-instar larvae, to illustrate the organization of the cephalopharyngeal skeleton. **A** and **B** are two different planes of focus of the same larva. The skeleton consists of three main parts, mouth hooks (*mh*), H-piece (*Hp*) and cephalopharyngeal plates, and three smaller parts, the median tooth (*mto*) and the ventral and dorsal neck clasps (*dcl*). The mouth hook has a triangular shape and three processes, oral (*mop*), ventral (*mvp*) and dorsal (*mdp*). The median tooth is situated between the dorsal processes of the mouth hooks and the oral processes of the H-piece (*op*). Immediately behind the median tooth, dorsal (*dcl*) and ventral neck clasps show the cuticle specializations of the labial and labral sense organ complexes. The caudal processes of the H-piece (*cp*) are fused to the anterior processes of the cephalopharyngeal plates (*vap*). The cuticle of the dorsal processes (*dp*) of the cephalopharyngeal plates is secreted by cells at the sides of the dorsal pouch, though it does not extend along the entire width of the space connecting pharynx and frontal sac, but is restricted to its posterior one-third. The dorsal processes are interconnected by means of a thin dorsal bridge (*dbr*). The pharyngeal musculature is located between dorsal and ventral processes of the cephalopharyngeal plates. *ci* maxillary cirri. *Bar* 50 μm

the dorsal processes of the mouth hooks adhere to the oral processes of the H-piece, whereas the caudal processes of the H-piece are directly inserted into the oral section of the cephalopharyngeal plates, from which they cannot be distinguished in the young first-instar larva (Fig. 16.3). The bridge of the H-piece is a derivative of the sternal cells of both labial segments (Jürgens et al. 1986).

The architecture of the cephalopharyngeal plates in the first larval stage is much simpler than in the later two stages, and consists of a ventral section contained in the pharyngeal cavity, and a dorsal section contained in the dorsal pouch, which, are connected laterally (Fig. 16.3). The ventral, pharyngeal section consists of two apposed sheets, which are secreted by the walls of the pharynx. Particularly striking are the pharyngeal ridges, approximately 10 longitudinal folds of the cuticle that extend ventrally along the entire length of the pharynx (Fig. 16.2), which, according to Schoeller (1964) in *Calliphora* and Turner and Mahowald (1979) in *Drosophila,* are derivatives of the hypopharyngeal lobes. The cephalopharyngeal plates lie rostral to the oesophagus, and extend laterally into wing-like, caudally directed processes, which connect the ventral to the dorsal sections of the cephalopharyngeal plates situated in the dorsal pouch. These lateral processes are connected to each other across the dorsal midline by means of a dorsal bridge within the dorsal pouch, derived from the involuted procephalic lobe.

The median tooth is a labral derivative, characteristic of the first larval stage. It is situated at the midline, in the territory of the atrium, at the level of the anterior processes of the H-piece. Dorsal and ventral neck clasps are barely distinguishable as individual entities in the first-instar larva. They are located in the anterior region of the pharynx in intimate association with two large sensory organs, the labial sensory complex (or hypophysis) and the labral sensory complex (or epiphysis), and form the cuticle specializations of these organs. Ventral and dorsal neck clasps are derived from the labium and the pharynx, respectively.

The dorsal pouch has a rather complex organization extending over the pharyngeal musculature and what remains of the clypeolabrum after head involution (Fig. 16.1). The dorsal pouch derives partially from the dorsal ridge, which is mainly derived from the labial segment (Technau and Campos-Ortega 1985), and partially from the epidermis of the procephalic lobe. The dorsal pouch is connected to the pharynx anterior to a broad cellular strip that extends along the caudal border of the lateral processes of the cephalopharyngeal plates, referred to above, up to the border of the cibarium. Thus, in outline the dorsal pouch corresponds fairly

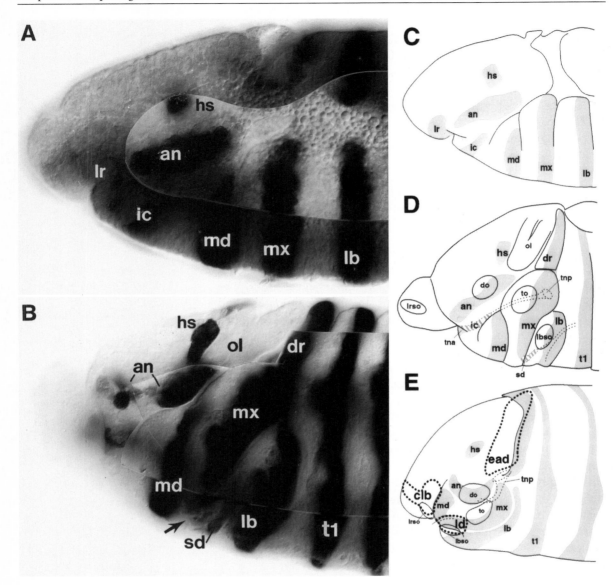

Fig. 16.4 Illustrates the relationships between head segments and head primordia. **A** and **B** show the *engrailed* expression domains of stage 12 and 15 embryos; **C - E** are camera lucida drawings of *engrailed*-stained stage 11 (**C**), late 12 (**D**) and 15 (**E**) embryos; *engrailed* expression domains are lightly shaded. Shown are also the position of the cephalic sensory complexes [*lbso* labial sense organ of the labial sensory complex; *to* terminal (maxillary) organ; *do* dorsal (antennal) organ; *lrso* labral sensory organ] and the position of the presumptive head imaginal discs (*ld* labial disc; *ead* eye-antenna disc; *clb* clypeo-labral bud). Other abbreviations. *an* antennal segment; *dr* dorsal ridge; *hs* head spot; *ic* intercalary segment; *lb* labium; *lr* labrum; *md* mandible; *mx* maxilla; *ol* optic lobe primordium; *sd* salivary duct; *tna* anterior opening of tentorium (intercalary segment); *tnp* posterior opening of tentorium (maxilla); *t1* first thoracic segment

precisely to the profile of the clypeolabrum before head involution. The dorsal pouch is only partially lined by pigmented cuticle; in fact the only cuticle structures contained in the dorsal pouch are the lateral and dorsal processes of the cephalopharyngeal plates.

16.3
Atrium Formation and Head Involution

Atrium formation and head involution are intimately related, and will therefore be described together. During atrium formation, both the sternal region of the gnathal segments and the gnathal buds themselves become displaced anteromedially, eventually coming to lie at the margins of the atrial opening or within the atrium itself; at head involution another gnathal subdivision, the dorsal ridge, fuses in the dorsal embryonic midline with its homologue on the other side to form the dorsal fold, which then becomes displaced, sliding over the procephalon. These two complex morphogenetic movements of gnathal structures have several consequences: the formation of the dorsal pouch, the inclusion in the mouth cavity of the clypeolabrum and part of the labium, and the allocation of maxillary and mandibular cells to the dorsolateral rim of the atrium (ref. to Fig. 16.1).

These morphogenetic movements can be followed easily using any of a number of useful landmarks. The most obvious are the sense organs of the head segments and the acron, which first appear during stage 13 and are easily recognizable throughout head involution by staining with the 22C10 antibody, for example; but the ectodermal invaginations, which produce the salivary gland and duct, and the so-called tentorium, can also be traced. The opening of the salivary duct marks the boundary between the maxillary and labial segments (Younossi-Hartenstein et al. 1993). The tentorium, which has previously been described in some detail for *Calliphora* (Schoeller 1964), is formed by the fusion of two tubes which invaginate during stages 12-13. The posterior tube invaginates at the base of the fold between the maxillary and labial lobes. The anterior tube invaginates between the hypopharyngeal and mandibular lobes. The two grow towards one another and fuse to form a continuous canal at stage 13 (Tepass and Knust 1990; Younossi-Hartenstein et al. 1993). The anterior opening of this duct marks the dorsolateral region of the intercalary segment: most of the intercalary *engrailed* stripe is incorporated into the tentorium. The posterior opening of the tentorium marks the dorsolateral region of the maxillary segment.

16.3.1
Formation of the Atrium

The atrium forms from cells of the head ectoderm that move through the stomodeum into the interior. The ventral and lateral regions of the labral segment will form the roof of the atriopharyngeal cavity, whereas its floor is formed by derivatives of ventral and lateral regions of the intercalary and gnathal segments. In the atriopharyngeal cavity the original anterior-posterior order of segment primordia is reversed. Derivatives of the intercalary segment come to lie at the rear end, while those of the labial segment occupy the front end of the cavity.

The prospective proventriculus and oesophagus, as well as the anlagen of the stomatogastric nervous system, have already invaginated by early stage 12. Invagination of the ventral and lateral regions of the labral and intercalary segments follows during stages 12-13. The hypopharyngeal organ X, which differentiates within the hypopharyngeal lobe (a derivative of the intercalary segment) early in stage 13, comes to lie in the floor of the pharynx (hypopharynx) by the end of this stage. The original anterior opening of the tentorium, which is part of the lateral intercalary segment, reaches a position rostro-lateral to the organ X in the hypopharynx (Figs. 9.13-9.16, 16.1). The labral sense organ [epiphysis; see Jürgens et al (1986), for synonyms of sense organs and other larval head structures], located dorsally in the clypeolabral lobe at stage 12, shifts ventrally to occupy a position right in front of the stomodeum. By the end of stage 13, the former ventral region of the labral segment is inside the embryo where it forms most of the roof of the pharynx (epipharynx). The lateral region of the labral segment, along with the labral sense organ, moves in and forms the roof of the atrium in stage 15.

From stage 13 onward, the ventral regions of the gnathal segments also move into the stomodeum. The movement of the labium can easily be reconstructed by following the displacement of the salivary gland and its duct, and the two labial sensory complexes. The invagination of the salivary gland involves much of the ventral labial segment. Therefore, the labial lobe harbouring the presumptive labial sense organs comes to be located at a rather ventral position by stage 12. The formation of the salivary duct internalizes more of the ventral labium (and a small part of the ventral maxilla, see above), such that at stage 13 the former labial lobes approach and fuse with each other at the ventral midline. Invagination of the labrum and intercalary segment into the stomodeum (see above) pulls the gnathal segments forward so that the fusion of the labial lobes takes place right behind the stomodeum. The ventral labial tissue then moves into the stomodeum and forms the floor of the atrium. The labial sense organ comes to lie at the ventral midline immediately in front of the opening of the salivary duct.

Unfortunately, no good landmarks exist with which to follow the movement of the ventral regions of the maxillary and mandibulary segments into the stomodeum. Topographical arguments (these regions occupy a position between labium and intercalary segment/hypopharynx) suggest that the ventral maxillary and mandibulary segments form the lateral wall of the atrium (Fig. 16.1, 16.4). This suggestion is further supported by experimental data: the lateral atrium secretes the "Lateralgräten" of the cephalopharyngeal skeleton, which laser ablation studies have shown to be of mandibular origin, and the lateral bars of the H-piece, derived from the maxillary segment (Jürgens et al. 1986).

16.3.2
Formation of the Dorsal Pouch

Apparently the leading structure in head involution is the dorsal ridge. The dorsal ridge has been shown by HRP injections to originate mainly from the labial segment (Technau and Campos-Ortega 1985). During early stage 12 the dorsal ridge becomes visible above the labial bud (Fig. 16.1); progressively, during germ band shortening, the dorsal ridge moves dorsally to reach the midline and fuse with its complement on the other side in stage 13, forming the so-called dorsal fold (Schoeller 1964; Turner and Mahowald 1979). As a consequence of the formation of the dorsal fold, the tissue immediately anterior to it forms a crescent-shaped depression on either side, dorsal to the antennomaxillary sensory organ primordia, and is swept medioposteriorly. This depression in fact marks the beginning of the formation of the dorsal pouch. Thereafter, the dorsal fold can be observed to progressively slide over clypeolabrum and procephalic lobe. Following this process, the territory lateral and posterior to the depression is folded over the tissue that borders the depression medially and anteriorly, and the acron and the dorsal regions of the head segments become incorporated into the dorsal pouch.

During their progress over clypeolabrum and procephalic lobe, the dorsal folds advance simultaneously from caudal to rostral and from lateral to medial. Therefore, it is virtually impossible to see where specific parts of the dorsal head region come to lie once they have folded inward. However, some markers allow us to follow directly the topographic transformation of the dorsal head during involution. One such marker is the former posterior opening of the tentorium, which represents the dorsolateral region of the maxillary segment and can be stained with the anti-Crumbs antibody (Figs. 2.25, 2.29, 2.34, 2.40). After head involution, the tentorium opens into the dorsal pouch laterally, about halfway between the anterior opening of the pouch and its very posterior tip (Fig. 16.1). Thus, the former dorsal region of the maxillary segment becomes translocated into the lateral "wings" of the dorsal pouch. This will be important for reconstructing the development of the eye-antennal disc from this position (see below). From topological considerations, it becomes clear that the former dorsal regions of the remaining head segments become incorporated into the dorsal pouch as well. Thus, the *engrailed*-expressing cells which formerly covered the dorsal ridge, and which we interpret as the fused dorsal regions of the gnathal segments, come to occupy a position in the outer lining of the dorsal pouch. Another good landmark for tracing the incorporation of the acron into the dorsal pouch is Bolwig's organ (Fig. 9.17), which ultimately comes to lie beneath the outer layer of the dorsal pouch in a position near the vertical plates of the cephalopharyngeal skeleton (Steller et al. 1987).

In the course of its movement over the clypeolabrum, the dorsal fold recruits cells from the neighbouring territories and enlarges considerably in size. It seems as if most of these cells derive from the procephalic lobe; the clypeolabrum itself appears not to contribute very much to this process. Moreover, the clypeolabrum does not seem to move any great distance from the position that it occupied prior to head involution. This can be readily observed by following the position of the labral sensory organ during head involution, by means of the 22C10 antibody, which binds to cells of the organ from relatively early stages on. Experimental evidence (i.e. injections of HRP) indicates that the cells of the dorsal ridge, which are of labial origin, come to be located at the oral end of the dorsal pouch, anterior to the median tooth, at the dorsal rim of the atrium (Schmidt-Ott and Technau 1994; see Figs. 16.2C-E), and that the entire dorsal division of the dorsal pouch, including both the dorsal and the ventral cell sheets, is mostly formed from cells of the procephalic lobe.

16.4
Cell Death During Cephalogenesis

There is evidence to suggest that cell death may play a role in reshaping the dorsal head region. This can easily be shown in preparations probed for *engrailed* expression (Fig. 16.4). For example, as mentioned above, the *engrailed* stripe that marks the antennal segment before head involution covers a broad region between the clypeolabrum and the dorsal ridge. Subsequently, most of these *engrailed*-expressing cells will disappear from the surface. While a few of them are incorporated into the brain, the majority degenerate and are taken up by macrophages (Younossi-Hartenstein et al. 1993). Similarly, many cells in the depths of the folds surrounding the gnathal lobes degenerate. It is only in the head and, to a lesser extent, in the ventral tail region (see Chapter 6.1) that widespread cell death occurs at this late stage of embryogenesis. The spatial pattern of late cell death in the head, i.e. the occurrence of large clusters of degenerating cells at distinct positions, appears very different from the rather diffuse areas of cell death seen in the earlier embryo (Campos-Ortega and Hartenstein 1985). For this reason, we speculate that this late cell death may be involved in reshaping the head during head involution.

16.5
Retraction of the Labrum

The protuberances of the maxillary and mandibulary segments, that is to say, the lateral divisions of the gnathal segments, are not drawn into the stomodeum. Instead, they come to lie at the rostral tip of the head, to form a prominent part of the antennomaxillary complex. The antennal component of the complex derives from the lateral region of the antennal segment which also shifts to this position. During the early stages of head involution, the labrum forms the clypeolabral lobe at the rostral tip of the head. From stage 14 on, a fold appears on either side between the clypeolabrum and the adjacent posterior tissue, i.e. the antennal segment which is marked by the antennal sense organ. This "lateral fold" deepens while the labrum is retracting. At the end of this movement, the tip of the labrum (which will become the median tooth), lies behind the level of the antennomaxillary complex. As a result, the antennomaxillary complexes on either side come close to each other, pointing anteriorly.

As with most other morphogenetic movements, the cellular basis of head involution is not clear. However, dorsal closure of the metameric germ band might be causally related to head involution. Before closure, the epidermal sheath of each segment is about one-third narrower at dorsal than at ventral levels. In all thoracic and abdominal segments, dorsal epidermal cells enlarge throughout stages 14 to 17. One possibility, thus, is that the growth of dorsal epidermal cells in thoracic and abdominal segments causes an enlargement of the trunk, and that this enlargement contributes to the displacement of the dorsal fold over the procephalon and, consequently, to head involution.

16.6
The Embryonic Origin of Head Imaginal Discs

The imaginal head develops from three imaginal discs, labial, clypeolabral and eye-antennal (ref. to Fig. 8.8). In the embryo, these discs are difficult to discern morphologically because they are not physically separate from the surrounding larval primordia. However, an enhancer trap line is available in which groups of embryonic cells destined to form the imaginal discs of the larva are labelled from stage 13 on (Younossi-Hartenstein et al. 1993).

The labial discs originate on the former labial lobes, which by stage 14 have moved to the posterior rim of the stomodeal opening. The 20-25 cells of the presumptive labial disc are grouped around the sensory neurons of the labial sense organ. After head involution, these cells occupy a position in the lateral wall of the atrium, where they remain for the rest of embryogenesis (Poulson 1950). During the first larval instar, these cells invaginate to become the labial disc, connected to the lateral atrium only by a short peripodial stalk.

The primordium of the clypeolabral discs (Gehring and Seipel 1967) consists of two bilaterally symmetric "wings" connected by a median "cross-bridge". By stage 14, the cells of the presumptive clypeolabral buds form the rostral rim of the stomodeal opening, situated right in front of and lateral to the labral sense organ. During head involution, these cells are moved backward and form part of the atrial roof. Unlike the labial (and most other) imaginal discs, they remain in the atrial epithelium throughout larval life.

The eye-antennal disc has the most complex origin. In early stages of head involution, the cells of the presumptive eye-antennal disc cover an area that roughly coincides with the crescent-shaped depression of the incipient dorsal pouch. This field consists of approximately 60 cells on each side, extending dorso-posteriorly around the dorsal midline to meet its contralateral counterpart. Anteriorly, it is bounded by the maxillary invagination of the tentorium. During later stages of head involution, the cells of the presumptive eye-antennal discs are located in the lateral "wings" of the dorsal pouch, formed by the dorsal regions of the head segments (maxillary and antennal segments in particular) as well as the acron. The cells of the eye-antennal primordia can be readily distinguished from the other cells of the dorsal pouch, being columnar and thus different in shape. During stage 17, the elongated cell groups undergo complex movements and, as a consequence, are transformed into the primordial eye-antennal discs. This process starts when the posterior half of the dorsal pouch, i.e. the part behind the posterior tentorial arms in which the primordial disc cells are located, is compressed in the longitudinal axis. Consequently, the cells of the presumptive eye-antennal discs, originally arranged in longitudinal stripes, are crowded into small clusters which invaginate from the posterior-lateral end of the dorsal pouch to form the eye-antennal disc. These morphogenetic movements explain how groups of cells deriving from different primordia (acron, antennal, maxillary, and, possibly, mandibular, intercalary and labial) are incorporated into the eye-antennal disc.

Chapter 17
Some Aspects of Segmentation

17.1
Introduction

The body of an insect is subdivided into a number of units called segments, morphologically definable in terms of the pattern elements they contain. In *Drosophila*, segments are built up by derivatives of the mesoderm and ectoderm, e.g. muscles, tracheae, nervous system, epidermis, whilst endodermal derivatives are neither segmentally organized nor do they display any metamerical organization. Anatomically, boundaries between segments of insects are defined by intersegmental furrows and by apodemes for insertion of muscles and, indeed, in the trunk region of the animal one can readily define the pattern elements of segments, as well as the segmental boundaries, using anatomical criteria. This is not the case in the terminalia, i.e. the procephalon and the telson, where segments are rudimentary or have fused. Therefore, we will deal first with the segmented germ band, including aspects of the gnathal segments; the preoral segments and the tail will be discussed at the end of this chapter.

In the germ band of the *Drosophila* embryo, definitive segmental furrows are formed during germ band shortening, apodemes slightly later. However, a number of transient processes with clearly metamerical character occur prior to the establishment of the definitive segmental pattern, during earlier stages of embryogenesis. These include certain aspects of regional proliferation and cell movements within ectodermal and mesodermal territories, the invagination of tracheal pits and the formation of parasegmental furrows.

In the following account, we shall reconsider the pattern elements of larval segments and discuss differences and possible homologies between patterns in different segments. We shall attempt to establish relationships between the earlier, transient metamerical events and the intersegmental furrows and apodemes that appear later, when we discuss how definitive segmentation occurs.

17.2
Metamerical Pattern Elements of the Germ Band

Externally, the *Drosophila* larva is subdivided into three thoracic and eight abdominal segments. However, it has become customary to consider abdominal segment 8 (*a8*) and the rudimentary *a9* segment as forming part of the terminal region called the telson, caudal to the abdomen. Additionally, gnathal and preoral segments have become integrated in the atrium during head involution. During the phylotypic stage of embryogenesis (Sander 1983), presumptive gnathal and trunk segments share several characteristics. The participation of gnathal segments in atrium formation and the fusion of the caudalmost abdominal segments to form the terminal telson, however, imply important deviations from the pattern of the remaining segments of the trunk. For example, the borders of gnathal segments, i.e. furrows and apodemes, cannot be defined in the fully developed embryo because of the modifications undergone by the gnathal segments and their derivatives during embryonic development, as they come to be incorporated into the atrium and its vicinity; sensory organs of gnathal segments chiefly form large complexes of sensory neurons rather than isolated sensilla; muscles derived from gnathal or caudal abdominal mesoderm are organized in different patterns, etc. The gnathal segments and the telson will be considered at the end of this chapter. The pattern elements outlined below are shared by all thoracic and abdominal segments *a1-a7*. Hence, we consider these elements to constitute the basic repertoire of the trunk segment.

17.2.1.
Segmental borders

Segmental borders are defined in the epidermis by transverse apodemes serving as sites of insertion for longitudinal and oblique muscles, and by intersegmental furrows; muscle attachment sites and furrows are intimately related (ref. to Figs. 17.1, 17.2). Segmental furrows in the larva correspond to the furrows in the embryo; however, in the larva they are produced by muscular contraction whereas it is not clear what determines the formation of

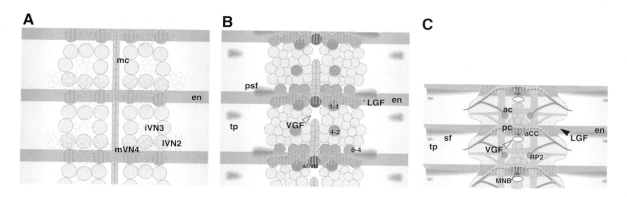

Fig. 17.1. Segmentation in the epidermal and neural primordium. **A - C** show drawings of surface views of flattened germ bands at stage 9 (**A**), 11 (**B**) and 13 (**C**). The ventral midline is marked by the double row of mesectoderm cells (*mc*). *Blue transverse stripes* indicate domains of *engrailed* expression; *blue circles* represent *engrailed*-positive neuroblasts, grey circles represent *engrailed*-negative neuroblasts. In **C**, the whole neural primordium is shaded *grey*; the neuropile (*ac* anterior commissure; *pc* posterior commissure) is shown in darker shade of *grey*. During stage 9, a metamerically organized pattern of proliferation in the ventral neurectoderm can be observed. Thus, the small mitotic domains whose cells (indicated by *open blue circles*) undergo mitosis 14 first (*mVN4*, *iVN1, 3, 4, lVN2*) are clustered at the anterior boundary of the *engrailed* stripe. The parasegmental furrow (*psf*), develops at this position during stage 11 (**B**). It delimits the parasegmental boundary across the germ band. At a more lateral and posterior level in each metamere, the tracheal primordia invaginate (tracheal pits, *tp*). During stage 11, the neuroblast layer becomes indented laterally, perhaps as a consequence of the adjacent parasegmental furrow pressing on it. Thus, the deepest point of this indentation (lateral ganglionic furrow, *LGF*) comes to lie at the level of the parasegmental furrow, corresponding to neuroblast row 6 (*en*-positive *NB 6-4*). The positions of three other neuroblasts, *1-1*, *4-2*, and *MNB* relative to the parasegmental furrow and *engrailed*-stripe are shown in **B**. During germ band retraction, the definitive segmental furrow (*sf*) appears in the developing epidermis. It is formed by the fusion of the parasegmental furrow with another, more posteriorly located indentation called the segmental slit. As a result of this complex movement, the bottom of the segmental furrow comes to lie 2-3 cell diameters posterior to the *psf*, at the posterior boundary of the *engrailed* stripe. By contrast, in the ventral cord, the lateral ganglionic furrow does not shift and remains coincident with the anterior boundary of the *engrailed* stripe. The posterior commissure develops at the level of the lateral ganglionic furrow; the anterior corner cell (*aCC*), progeny of NB 1-1, is located right behind the posterior commissure. A second set of metameric furrows, called ventral ganglionic furrows (*VGF*) have formed at this stage. These indentations are apparent in the ventral midline of the ventral cord and are formed by the vertical grooves which develop right behind the median neuroblast (*MNB*). Thus, these grooves and the *VGF* they form are offset by several cell diameters relative to the *LGF*. The two commissures are located in the middle between adjacent *VGFs*

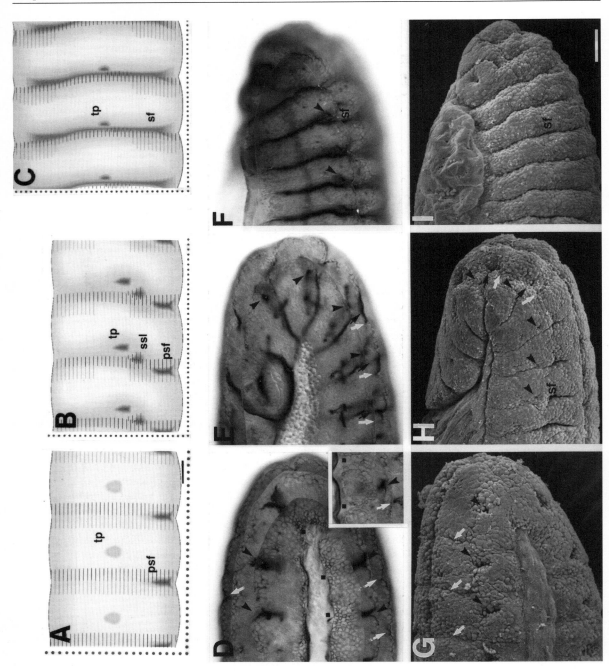

Fig. 17.2 Segment furrow formation in wild-type development is depicted by schematic drawings (**A, B, C**), the anti-Crumbs expression pattern (**D, E, F**) and scanning electron micrographs (**G, H, I**). All panels present lateral views with anterior to the left and dorsal to the top. In **A to C**, the expression patterns of the genes *wingless (blue)*, *engrailed (purple)*, and *sine oculis (orange)* are indicated. The distance between dots on the dotted lines flanking panels **A to C** corresponds roughly to an ectoderm cell diameter. At stage 11 (**A, D, G**), parasegmental furrows (*white* arrows, *psf*), appear ventrolaterally at the *engrailed / wingless* boundary. Tracheal pits (*arrows, tp*) form lateral and posterior to parasegment furrows. The dorsal ectoderm is indented by shallow grooves (squares) whose position relative to other landmarks is variable. In the main panel **D**, dorsal grooves are in line with tracheal pits; in the inset, they are in line with the parasegment furrows. During stage 12 (**B, E, H**), segmental slits (*ssl*) appear between parasegmental furrows and tracheal pits. These slits, which coincide with the *sine oculis* expression domains, become confluent with the parasegmental furrows during germ band retraction and form the segmental furrows (*sf*). Segmental furrow formation follows an anterior-posterior gradient. In the anterior germ band of the embryo, shown in **H**, segmental furrows are nearly complete; posteriorly, parasegmental furrows and segmental slits are still separated. At stage 13 (**C, F, I**), segmental furrows form deep indentations across the entire ectoderm. *Bar* 25 μm

segmental furrows in the embryo. Apodemes do not extend in a single longitudinal plane around the entire larval body, for dorsal and ventral subdivisions of the apodeme are actually staggered (Fig. 8.5). Strikingly, the dorsal part of the apodeme of any larval segment in thorax and abdomen is located posterior to the ventral part and, consequently, the lateral one-third of the segment comes to be oriented obliquely, from anteroventral to posterodorsal. In fact lateral extensions of the dorsal and ventral apodemes exist which together form a cuticular ridge of oval shape that connects the two subdivisions in the dorsolateral region of thoracic and abdominal segments. These ovals are considerably narrower in thoracic than in abdominal segments.

17.2.2
Ventral Cord

Whereas no ambiguities cloud the precise definition of metameric boundaries in the sheet of epidermal cells, this is a difficult task in the ventral cord due to its complicated three-dimensional structure. In the larva, the ventral cord is strongly condensed. Condensation of the ventral cord leads to loss of isotopic relationships between neuromeres and dermomeres. However, condensation starts in stage 14, well after distinct features of the neuropile, chiefly commissures, have been established (Fig. 17.1). Therefore, to determine topological relationships between commissures and metameric boundaries, and follow the fate of cells labelled with internal markers, such as the anti-Engrailed antibody, or in enhancer trap lines, is quite feasible before condensation of the ventral cord has started (Fig. 17.1). Consequently, any description of pattern elements relevant to segmentation of the ventral cord will as a rule rely on the situation found during stages 13 and 14, preceding ventral cord condensation.

Two conspicuous morphological landmarks can be seen to separate consecutive neuromeres in the stage 13 nerve cord, one is ventromedial and the other dorsolateral. Unfortunately, the assignment of pattern elements to particular neuromeres is made more difficult by the fact that the two landmarks are not in register. The ventromedial landmark, or ventral ganglionic furrow, is a sharp indentation that marks the position of the so-called interneuromeric channel, which is located between the prominent clusters of midline cells. Some of these midline cells, such as the median neuroblast and its progeny, and adjacent neurons expressing *engrailed,* are located right in front of this ventromedial indentation. The commissures are located in the middle of the two midline clusters. The dorsolateral landmark, or dorsal ganglionic furrow, is another, wider interneuromeric indentation that is shifted anteriorly with respect to the ventral one. The commissures lie at the level of the dorsolateral landmark. The interneuromeric channel becomes evident as germ band shortening is completed and ventral cord condensation starts. The position of the channel corresponds in lateral projection to the epidermal segmental furrow. Therefore, the ventromedial landmark is intersegmental. Judging from the position of *engrailed*-positive cells, the dorsolateral indentation is a parasegmental landmark.

In the stage 17 embryo, when condensation of the ventral cord is practically finished, the dorsolateral indentations can no longer be observed. However, intersegmental and segmental nerves, i.e. anterior and posterior fascicles, exit from the ventral cord together at approximately the level of the channels. That is to say, the nerves are located roughly at the position of the segmental boundaries.

17.2.3.
Non-innervated Cuticular Structures

The anterior portion of the cuticle in each segment in the trunk is covered ventrally by rows of short setae which form the so-called denticle belts (Fig. 8.1). Denticle belts comprise three (thoracic) or seven (abdominal) rows of relatively large denticles. In abdominal segments, the intersegmental boundary runs between the first and the second denticle row (Szabad et al. 1979; own observations); within thoracic levels, this boundary lies immediately in front of the first denticle row. Denticles of a given row all exhibit the same polarity; however, the different rows of denticles have different polarities. At ventral levels, the cuticle behind the denticle belts is totally devoid of hairs. Conversely, the dorsal and lateral parts of the cuticle are almost entirely covered by hairs of varying thickness, length, and polarity. With respect to hair morphology the dorsolateral cuticle can be subdivided into three regions, as explained in Chapter 8 (Figs. 8.1 - 8.4).

17.2.4.
Epidermal Sensory Organs

Three types of sensillum occur in all of the thoracic and abdominal segments (Fig. 9.3; Table 1): hair (trichoid) sensilla and papillae, both innervated by single neurons; and chordotonal organs which may have one, three or five scolopidia and as many innervating neurons. In the dorsal and ventral abdominal trichoid sensillum, two bristles occur close together, one large and the other very small. Each is innervated by a single neuron. Basiconical sensilla constitute a fourth type with club-shaped cuticular specialisations, which occur both in thoracic segments ('black dots') and in $a8$ and $a9$ (sensory cones); some, if not all, of these basiconical sensilla are multiply innervated. Finally, a fifth type are the trichoid sensilla, which are present only in thoracic segments and correspond to Keilin's organs, with five sensory neurons each. In addition to the sensilla, there are a number of multidendritic neurons in every segment.

Cell bodies of sensory neurons are arranged in clusters (Figs. 9.3 - 9.7). Dorsal, lateral and ventral cell clusters can be distinguished in each segment. The ventral cluster splits up into several distinct groups during late embryonic development, whilst somata of the lateral and dorsal clusters remain clustered. The special relationships of dorsal and lateral clusters are further emphasized by the fact that their axons course through a common nerve (see below). Cuticular specializations of sensory organs are aligned within a narrow transverse strip in the middle of each segment.

17.2.5
Peripheral Nerves

With one exception, the pattern of segmental innervation is invariant from $t1$ to $a7$ (Figs. 9.3, 11.7, 17.1). Each segment is innervated by a mixed nerve, with afferent and efferent fibres, formed by two fascicles. The anterior fascicle, or intersegmental nerve (Thomas et al. 1984) originates from two roots: the posterior root derives from the anterior commissure of the homotopic neuromere and the anterior from the posterior commissure of the neuromere in front. The anterior fascicle leads the axons of motoneurons to the dorsal muscles. However, Sink and Whitington (1991) have found that the axons of contralateral RP motoneurons travel through the anterior fascicle to innervate ventral muscles. In addition, the intersegmental nerve carries axons from the dorsal and lateral sensilla. The prothorax is an exception, in that there is a lateral triscolopidial chordotonal organ in the prothorax, the axons of which course in the anterior fascicle of the mesothoracic (*t2*) nerve. The posterior fascicle (segmental nerve) originates from a single root in the neuropile of the homotopic neuromere and contains fibres coming from the ventral sensilla and fibres that run to the pleural and ventral muscles (Johansen et al. 1989). Both fascicles leave the cortex of the ventral nerve cord very close to each other.

Nerves of *a8* and *a9* are different in that they form a single fascicle, which corresponds to the intersegmental nerve (anterior fascicle) of *t1-a7*.

17.2.6
Muscles and dorsal vessel

Although the entire mesodermal layer is segmentally organized during early stages of embryogenesis, the somatic musculature and the dorsal vessel are the only mesodermal derivatives that are overtly segmented in the larva. Muscles of trunk segments can be subdivided into dorsal, lateral and ventral groups (Figs. 7.1, 7.2). Five different patterns of muscles have been described, which correspond to those of *t1, t2, t3, a1-a7* and *a8-a9*. The reader is referred to Chapter 4 for details of these patterns.

The dorsal vessel forms a tube which is suspended from the dorsal epidermis by pairs of segmentally repeated alary muscles (see Chapter 5). Alary muscles insert at the intersegmental apodeme, from the *t3-a1* apodeme to the *a6-a7* apodeme. Each segmental set of cardioblasts comprises two rows of six cells each. The segmental organization of the cardioblasts is shown by a number of enhancer trap lines, which exhibit β-galactosidase activity in segmentally repeated subsets of cardioblasts. For example, one such line (E2-3-9; Bier et al. 1989; Hartenstein and Jan 1992) is expressed in each segment in two adjacent cardioblasts flanking the boundary between two adjacent segments. During their differentiation, the alary muscles establish contact with the cells labelled in this line.

17.3
Homologies and Differences Between Segments

We have tacitly assumed on several occasions in this book that structures of a given segment have homologues in other segments. Homology between groups of pattern elements has already been claimed to exist in abdominal and thoracic segments, e.g. dorsal, lateral and ventral groups of muscles or sensory organs, peripheral nerves etc., find counterparts in all segments of thorax and abdomen. However, difficulties arise when we try to establish homologies between individual pattern elements. And yet it might well be convenient for various purposes, for example, in comparing phenotypes of the various types of segmentation mutants or understanding the behaviour of appendages in a given experimental situation, to establish clearly whether certain pattern elements of a given segment have homologues in other segments.

With regard to muscles, homologies between groups of muscles are generally clear. However, we find no reasonable basis on which to assign homologies between individual muscles. To give an example, there are four dorsal internal oblique muscles in the metathorax, but only three in each of abdominal segments $a1$-$a7$. Since they occupy similar positions in the respective segments, one could postulate homology between the metathoracic dorsal internal oblique muscles 1-3 and the abdominal muscles of the same name; in this case one could argue that the metathoracic dorsal internal oblique muscle 4 lacks a homologue in abdominal segments. However, also on the basis of the location of the muscles under discussion, another possible homology could be postulated between metathoracic dorsal internal oblique 2-4 and abdominal dorsal internal oblique 1-3; in this case muscle number 1 would be missing in abdominal segments. It is not possible to conclude from topological considerations alone which alternative, if either, is correct. Unfortunately, other than their location, we have no criterion, functional or ontogenetic, which would help to establish on a sound basis the homology between individual muscles. The situation is slightly different in the case of ectodermal derivatives, particularly sensilla. As sensilla are suitable pattern elements for defining the spatial coordinates of segments - and have indeed been used for this purpose - possible homologies between sensilla of different segments will be considered below.

Morphologically, four criteria can be defined which allow one to postulate certain homologies between sensilla: location, structural organization, trajectory of the sensory axons towards the CNS, and developmental characteristics (lineage, whenever possible; time of formation). If two sensilla in different segments are identical with respect to all of these criteria we consider them to be unequivocally homologous; conversely, no homology exists if sensilla do not share any of these criteria. Difficulties in interpreting relationships between two sensilla appear when some of the criteria are shared by both, while others are not. The main problem, in our opinion, is to evaluate the significance of the criteria. On applying these criteria we find that unequivocal homology exists between corresponding sensilla in mesothorax and metathorax, and between those of the first seven abdominal segments. The prothoracic pattern of sensilla contains elements, the great majority of which have homologues in the other two thoracic segments. However, the relationships between most elements of the meso-metathoracic pattern and those of the abdominal pattern are ambiguous.

We will consider in some detail four types of sensillum which are actually among the most conspicuous elements of thoracic and abdominal patterns. These are the lateral pentascolopidial and monoscolopidial chordotonal organs of a_1-a_7, the dorsal and lateral triscolopidial chordotonal organs of t_1-t_3, the black dots and Keilin's organs of t_1-t_3. The abdominal lateral penta- and monoscolopidial chordotonal organs differentiate in two phases. In the first phase the three anterior scolopidia of the lateral pentascolopidial organ form; in the second phase the posterior two of the pentascolopidial, and the single element of the monoscolopidial chordotonal organ. The three anterior scolopidia appear at the same time and at approximately the same dorsoventral level as the meso-metathoracic dorsal triscolopidial chordotonal organs and the prothoracic lateral triscolopidial organ. During later stages, the triscolopidial organs in t_2 and t_3 shift to a more dorsal level. We speculate, thus, that the anterior group of three abdominal scolopidia and the thoracic triscolopidial chordotonal organs are homologues, for they share location, axonal trajectory, structural organization during development, and time of appearance. Keilin's organs of thoracic segments are composed of three hairs, two of which point anteriorly, the third posteriorly. A monoscolopidial chordotonal organ is present lateral to each Keilin's organ cells . In the abdominal segments, no trichoid sensilla occur at the location of Keilin's organs. Instead there are two papillae and two monoscolopidial chordotonal organs which are innervated by a nerve corresponding to that which, in the thoracic segments, innervates the Keilin's organ and the ventral thoracic monoscolopidial chordotonal organ. On the basis of similarity in position, axonal trajectory and number of cells involved, Keilin's organ and the thoracic monoscolopidial chordotonal organ together may be considered to be homologous to two ventral papillae and two monoscolopidial chordotonal organs at abdominal levels.

17.4
Embryology of Segmentation

17.4.1
Transient Metamery

A variety of processes and structures in the germ band exhibit a clearly metameric organization prior to the appearance of definitive segmental boundaries. Although the significance of most of these transient metamerical features for the process of definitive segmentation is unfortunately unknown, they are nonetheless very striking. Therefore, we would like to discuss them before dealing with various aspects of the definitive segmentation.

The first manifestation of metamery distinguishable in conventional histological preparations is found in the stage 8 embryo. Small clusters of cells, periodically arranged within the intermediate region of the neuroectoderm in the abdomen are seen to divide several minutes earlier than their neighbours (Fig. 14.5; Hartenstein and Campos-Ortega 1984). Foe (1989) describes the same mitotic clusters in abdominal metameres (domains $\partial 1416$ $\partial 1415$ and $\partial 1417$) and, in addition, other mitotic domains (d1421) within gnathal and thoracic metameres with similar characteristics. The second manifestation of metamery is found in the pattern of neuroblasts after the segregation of the first subpopulation. SI neuroblasts become arranged in a medial and a lateral row, along with two neuroblasts within the third, intermediate row. The intermediate row of neuroblasts will completed soon thereafter, after segregation of the second, SII subpopulation of neuroblasts. However, transient gaps, with a period equivalent to a prospective metamere, occur in the intermediate row of neuroblasts at abdominal levels (Fig. 11.3); these gaps correlate with the mitotic activity in the neuroectoderm, which is organized metamerically from the first postblastoderm mitosis on (Fig. 14.5, see Chapter 14).

A further transient sign of metamerical organization can be seen in the mesodermal layer (Fig. 2.12). At the end of their second mitosis, mesodermal cells rearrange into a regular epithelium. While this reorganization of the mesoderm is in progress, bulges appear transiently in the germ layer that exhibit a metamerical organization. This metamery of the mesoderm may in fact be related to the disposition of neuroblasts. One particular neuroblast, called MP2 (Doe 1992), lies at a different level from the remaining neuroblasts and the entire array takes on a sinusoidal appearance. This undulation of the neuroblast array could impinge on the overlying mesodermal layer and cause it to bulge with the same periodicity. In addition, the pattern of the third mesodermal mitosis is metamerically organized (see Chapter 14).

17.4.2
The Parasegmental Furrows

Definitive segmental boundaries, i.e. intersegmental furrows and transverse apodemes, are established by the end of germ band shortening (Fig. 17.2). However, prior to this stage, there are other furrows which subdivide the germ band into parasegments (Martinez-Arias and Lawrence 1985). The first metameric furrows are located ventrally and appear in the epidermal primordium at the end of the the second postblastodermal mitosis. They were called parasegmental because they define a metamerical register different from the segmental frame (Martinez-Arias and Lawrence 1985). These epidermal furrows, the first to appear, were called intersegmental furrows by Poulson (1950), Turner and Mahowald (1979) and Campos-Ortega and Hartenstein (1985). However, it was shown by using several different molecular markers (Martinez-Arias and Lawrence 1985; Ingham et al. 1985) that these furrows are in fact transitory structures that will be replaced during germ band shortening by the definitive intersegmental furrows. Thus, for example, the *engrailed* expression domain in each segment lies right behind the ventral, parasegmental furrows prior to germ band shortening; after germ band shortening, however, the *engrailed* domains are located in front of the intersegmental furrows. A thorough description of the evolution of epidermal furrows around the time of germ band retraction can be found in Martinez-Arias (1993). The concept of parasegments was thus introduced to designate metameric units which are the same size as, but out of register with, the segments as defined morphologically. Parasegmental units became evident when it was discovered that the realms of expression and function of many regulatory genes, such as the genes of the Bithorax complex, do not coincide with segments (Struhl 1984; Akam and Martinez-Arias 1985). The germ band is subdivided into 14 parasegments, comprising the primordia of the segments from the mandible to *a9*. Homeotic genes specifying the development of the tail end define two additional parasegments beyond PS14; due to the extreme modification of segmentation in the preoral head, it is unclear whether parasegments exist in this region of the embryo.

Parasegmental furrows form in ventral territories of the germ band, extending from the ventral midline to approximately 40% of the ventrodorsal perimeter. No clear furrows are present within dorsal territories at these earlier stages. In some embryos, periodical shallow indentations of the epidermis appear at its boundary with the amnioserosa (Fig. 17.2). These dorsal epidermal indentations are apparently out of register with the ventral furrows and in register with the tracheal pits (see below).

17.4.3
The epidermal segment boundary

During germ band retraction, the definitive intersegmental furrows are formed (Fig. 17.2). While the parasegmental furrows are still present in the ventral epidermal primordium, circumscribed grooves appear further lateral to them. These grooves, for which the term segmental slits has been proposed (Martinez-Arias 1993), represent the starting points for the segment furrows. Segmental slits lie approximately two cell diameters posterior to the ventrally adjacent parasegmental furrows. They lie slightly ventral, and one or two cell diameters anterior, to the openings of the tracheal pits. During germ band retraction, parasegmental furrows and segmental slits become confluent and a deep furrow is formed which traverses the entire epidermal primordium. When contacted by the clusters of fusing myoblasts, the row of cells at the base of this furrow forms the apodeme. Thus, the definitive segment boundaries are established.

Tracheal placodes and tracheal pits appear during stage 11 (Fig. 2.15, 2.17). With respect to the ventral subdivision of the segmental furrows, tracheal primordia are located within the anterior half of segments *t2* to *a8*, slightly dorsal to the furrows themselves (Figs. 17.1, 17.2).

17.4.4
Development of Neuromeres

The parasegmental furrows in the epidermis, as well as molecular markers expressed in the neuroblasts and their progeny throughout development (e.g. *engrailed*), permit study of the segmentation of the ventral cord (Fig. 17.1). The array of neuroblasts remains uniform during stages 9 and 10; the only sign of metamerical organization that can be recognized without using special markers are the MP2 cells, which, as mentioned above, lie at a deeper level than the other neuroblasts. During stage 11, when the ectoderm exhibits parasegmental furrows and other metamerically repeated structures, such as tracheal pits, the neuroblast layer can also be subdivided into morphologically distinct neuromeres by histological means. According to Doe (1992), neuroblasts are arranged in seven rows (see Chapter 11). The numbering system used in Doe's study to identify each neuroblast is largely based on the system introduced for the grasshopper neuroblast map (Bate 1978). Thus, for example, in this map the median neuroblast is located at the posterior neuromere boundary; the adjacent row of lateral neuroblasts is thereby defined as row 7. In *Drosophila*, the neuroblasts in rows 6 and 7 express *engrailed*. In stage 11, the epidermal parasegmental furrow is located at the level of neuroblast rows 5 and 6, and a clear indentation appears in the neuroblast array at neuroblast 5-2, defining a neuromeric boundary at this stage, which topologically corresponds to the parasegmental furrow (see 17.2.2). After germ band retraction, a pronounced furrow appears mid-ventrally behind the median neuroblast, approximately at the level of neuroblast row 2. These furrows apparently do not change their relative position with respect to markers expressed by the neuroblasts and their progeny, such as *engrailed*.

17.4.5
Metameric Development of the Mesodermal Derivatives

The cells of the mesodermal germ layer behave in a rather uniform manner during early stages of germ band elongation. During the first two mitoses, cells in the mesoderm divide parasynchronously. However, all of them divide within a relatively short period of time at each mitosis. We have mentioned above the transient appearance of metameric bulges in the mesodermal germ layer in stage 9, at the end of the second mesodermal mitosis, which may be related to the position of the MP2 cells in the array of neuroblasts. Derivatives of the mesoderm exhibit a clear metamerical organization from late stage 11 on. These derivatives include the founder cells of particular muscles, the visceral musculature, the fat body and the dorsal vessel.

17.5
Segmentation in the Head

In *Drosophila*, head segmentation is, as in many other insects, still an open question. The criteria used to define segments refer, chiefly, to repeated paired structures, such as the neuromeres, the coelomic sacs, the apodemes or the segmental protuberances (Matsuda 1964; Rempel 1975). Based on these criteria, and following embryological studies of several insect orders, Jürgens et al. (1986) proposed that the head of the *Drosophila* embryo consists of three pregnathal and three gnathal segments [however, see Schmidt-Ott and Technau (1992) for an alternative view]. In addition to these six cephalic segments, there is an anterior, non-segmented acron, which comprises most of the dorsal head capsule including the visual sense organ (eye), and most of the protocerebrum, particularly the visual centers. The first head segment is called the labral (or pre-antennal) segment and its neuromere gives rise to part of the protocerebrum. The clypeolabrum thus corresponds to the labral appendages, which are therefore fused medially; there is also a segmental apodeme. The appendages of the second head segment are the antennae and its neuromere is the deuterocerebrum. The third head segment, named intercalary (or pre-mandibular), lacks appendages; however, rudimentary limb buds seem to occur transiently in some insect species. The intercalary segment contributes the anterior arm to the tentorium, which is a specialized ectodermal structure that serves as the major muscle attachment site in the head. The neuromere of the third head segment is the tritocerebrum, which together with the protocerebrum and the deuterocerebrum constitutes the supraoesophageal ganglion (brain). The fourth to sixth head segments, or gnathal segments, meet all the criteria for segments. The three gnathal neuromeres are fused into the suboesophageal ganglion. The fourth cephalic (first gnathal) segment bears a pair of appendages, the mandibles, and well-formed mandibular apodemes. The fifth segment bears a pair of appendages, the maxillae, and contributes the posterior arm to the tentorium (Fig. 16.1). The sixth segment bears a pair of fused appendages called labia, while its ectodermal invaginations give rise to the salivary glands. It is not clear whether the salivary glands are serially homologous with the segmental apodemes of the other head segments. Within each segment, the neuromere derives from the ventral, sternal region, and the appendage from the lateral region. The dorsal regions of the head segments, although reduced in size, appear to contribute to the head capsule (Wada 1966). The mesoderm of the pregnathal territory derives from cells of the ventral furrow; the fate map (see Chapter 18, Fig. 18.1B) suggests that this mesoderm is topologically congruent with the ectodermal regions corresponding to the procephalon.

17.6
The Tail Region of the Larva

The rear end of the *Drosophila* larva is comparatively easy to describe in terms of segmental derivatives, as the tail primordia have remained contiguous during their morphogenetic movements (Jürgens 1987). It is formed from three morphologically distinct segment primordia (*a8*, *a9* and *a10-a11*) in front of the proctodeum, and a rudimentary non-segmental end-piece called the telson (Lohs-Schardin et al. 1979). The ventral denticle band of *a8* is separated by a thin strip of naked cuticle derived from *a8* and *a9*, from the posteriorly adjacent anal pads, mostly derived from *a11*, surrounding the anal opening. A single tuft of denticles representing the unsegmented telson of other insects (Jürgens and Hartenstein 1993) is flanked by the anal sense organs (caudal sense organs, derived from *a10*) along the posterior margin of the anal pads. Other sense organs of both the hair and papillar types are located at defined positions in the tail region: four anterior pairs of these sense organs are derived from *a8*, the two more posteriorly located ones are of *a9* provenance (Sato and Denell 1986; Jürgens 1987; Kuhn et al. 1992). The dorsal surface extends into two protuberances (mostly *a8*, a more posterior part is derived from *a9*) which bear the posterior spiracles (*a8*). Their openings are each surrounded by four groups of spiracular hairs. Internally, they are covered by a mesh of hairs named Filzkörper. The most posterior of the segmentally repeated *engrailed* stripes marks the posterior compartment of *a9*. Another patch of *engrailed*-expressing cells has been found in the dorsal (posterior) wall of the proctodeum (Ingham et al. 1985), which may thus represent a "cryptic segment" primordium. Abdominal segments *a8*, *a9* and *a10* can be clearly identified at the phylotypic stage. Posterior to *a10* lie the primordia of the anal pads which flank the anus. Genetic evidence suggests that the anal pads constitute a highly modified *a11* segment. On the basis of the pheotypic modifications in various segmentation mutants, the anal pads and several internal structures, such as the hindgut, can be considered part of the terminal region (Jürgens and Hartenstein 1993). The (true) telson is morphologically indistinct at the phylotypic stage (Sander 1983). The later development of the tail segments can be reconstructed on the basis of morphological landmarks, such as the sensory organs and the apodemes of segmental muscles, and molecular markers like *engrailed*. Most of the dorsal portion of *a8* gives rise to the posterior spiracle. The ventral regions of *a8* and *a9* fuse and are greatly reduced by cell death (see also Kuhn et al. 1992). The rudimentary *a10*, which has no *engrailed* stripe, occupies the narrow fringe between the anal pads and the posterior boundary of *a9*.

As a result of germ band retraction, the primordium of *a8* comes to lie anterior and dorsal to the other tail segment primordia (Turner and Mahowald 1979; Campos-Ortega and Hartenstein 1985). The amnioserosa covering the dorsally open embryo forms a wedge-shaped posterior end which abuts the dorsal edges of *a8* and *a9-a10*. The boundary between the latter two primordia has become indistinct. The amnioserosa is gradually displaced by the dorsal edges of the segment primordia during dorsal closure. Continued retraction of the dorsal tail region moves the posterior tip of the embryo to a ventro-posterior position as indicated by the location of the anal opening. This movement also compresses the anal pad primordia in the dorsoventral dimension, while, at the same time, they expand mediolaterally. The segment boundary between *a8* and *a9-a10* finally becomes indistinct.

Part of the anterior-posterior compression of the tail region is due to cell death. Up to stage 13, the ventral parts of *a8* and *a9* are each marked by a separate *engrailed* stripe. During stage 14-15, abundant cell death appears in this region. Following this stage, there is only a single narrow *engrailed* stripe for both *a8* and *a9*. Evidently, much of posterior *a8* and anterior *a9* has degenerated. There is no such morphogenetic cell death at more lateral and dorsal levels.

A major morphogenetic movement during later development of the tail region involves the formation of the posterior spiracles. Late in germ band retraction, a large depression appears dorsolaterally in the *a8* primordium, which signals the formation of the spiracle. The depression deepens to form the spiracular chamber which makes contact with the posterior end of the tracheal trunk. The surrounding tissue bulges out to form the protuberance that is to harbour the spiracular opening. Mid-dorsal tissue of the *a9* primordium is drawn in between the spiracles which themselves are derived from *a8*, as visualized by *engrailed* staining (DiNardo et al. 1985). Later, groups of cells surrounding the spiracular opening differentiate the spiracular hairs while the cells lining the spiracular chamber differentiate a network of hairs that form the Filzkörper (Lohs-Schardin et al. 1979; Jürgens 1987).

Chapter 18
A Fate Map of the Blastoderm

After a period of successive nuclear divisions, the blastoderm is generated from the cleavage nuclei and the egg cytoplasm. The cells of the blastoderm do not show any apparent variation in size and shape; yet there is evidence that these cells are already committed to developing into the various segments of the larval and imaginal body (Wieschaus and Gehring 1976; Steiner 1976; see review by Sander 1976). In this book, the regularity with which major events of *Drosophila* embryogenesis occur has been repeatedly emphasized, and this regularity, which was recognized a long time ago, has suggested to investigators the possibility of establishing a fate map of the blastoderm.

Fate mapping of the blastoderm of *Drosophila* has been repeatedly attempted with varying degrees of success. The maps obtained up to now have been concerned with only partial aspects of the entire fate map, and the reason for this incompleteness can be explained in terms of the method used in each instance. Three main methods were used in the past. The first consists of the analysis of gynandromorphs, and was invented by Sturtevant, who also obtained the first fate map (1929) by calculating the frequency with which two epidermal structures were of different genotype in gynandromorphs of *Drosophila simulans*. This approach was further developed by Garcia-Bellido and Merriam (1969) in *D. simulans* and by Hotta and Benzer (1972) in *D. melanogaster*. The resolution of fate-mapping and cell lineage studies with gynandromorphs has been considerably improved by the introduction of enzyme markers (Janning 1972, 1974, 1976; Kankel and Hall 1976; Szabad et al. 1979; Lawrence 1981), allowing some internal organs of both larva and imago to be scored.

The second method is the histological method, which relies on the analysis of staged embryos with techniques of normal histology. Essentially the histological method consists in tracing back to the blastoderm the location of the ancestors of larval organs, once these become recognizable in the developing embryo. We have already discussed (see Chapter 15) how the monolayer architecture of the blastoderm becomes transformed into the complicated structure of the young embryo by the morphogenetic movements of gastrulation and germ band elongation. Initially, only germ layers can be distinguished, but very soon distinct primordia appear (Sonnenblick 1950; Poulson 1950). Thus, the study of complete series of developing embryos with anatomical methods allows one to follow embryonic cells throughout development, both forwards and backwards, and to establish a fate map of larval organs. The histological method was used with great success by Poulson (1950) in *Drosophila melanogaster*, who constructed a fate map that was widely used for a long time.

The third approach relies on the apparent incapacity of blastoderm cells to regulate, and consists in destroying cells of the blastoderm by any of several different means (UV microbeam, Nöthiger and Strub 1972; Bownes and Kalthoff 1974; microcautery, Bownes and Sang 1974a; UV-laser microbeam, Lohs-Schardin et al. 1979; Jürgens et al. 1986; Jürgens 1987; mechanical, Bownes and Sang 1974b; Underwood et al. 1980b), and correlating the position of the initial lesions with the defects detected in the late embryo, the larva or the imago.

We have attempted fate-mapping of the *Drosophila* blastoderm by two different, complementary approaches. In the first one horseradish peroxidase (HRP) was injected into young embryonic cells and the distribution of the label was studied in the injected cells and/or their progeny in the late embryo (see Technau and Campos-Ortega 1985). With respect to the other approaches this method has the advantage of allowing one to establish the location of progenitor cells of virtually all organs of the larva in a rather direct way. In particular this method permits accurate assessment of the extent of morphogenetic cell movements. However, with this method it is difficult to make estimates of the number of cells that contribute to each of the larval primordia. The second approach used was the histological method, which provided all the quantitative estimates discussed below (see Hartenstein and Campos-Ortega 1985). In order to obtain the material for cell counts two techniques were used. The first involved staining embryos of increasing age with fuchsin (Zalokar and Erk 1977) in wholemounts, and noting the cell cycle phase of every embryonic cell. The second technique consisted in making planimetric reconstructions of complete series of sections of embryos of similar ages, by 'rolling-off' folded and invaginated tissues. Both techniques allowed us to obtain reliable quantitative data for the various embryonic stages studied, as well as to roughly assess the extent of displacement of larval anlagen that occurs during gastrulation and germ band elongation. This was possible because the early stages of *Drosophila* embryogenesis are characterised by a high degree of precision and numerical constancy, making comparisons between different specimens possible. The reader is referred to these accounts (Hartenstein and Campos-Ortega 1985; Technau and Campos-Ortega 1985; Hartenstein et al. 1985) for further details.   The aim of this chapter is to present the fate map that we obtained with these techniques, which contains information about the location and number of progenitors of larval organs.

18.1
The Fate Map

Fig. 18.1B shows the fate map. It is drawn on the lateral aspect of the blastoderm and indicates the locations of organ anlagen and the number of ancestor cells that give rise to several larval organs and to the imaginal optic lobes. The anlagen of other larval organs and imaginal discs could not be precisely located and were therefore omitted. The fate map is drawn on a computed planimetric reconstruction of the blastoderm. The reader is referred to Hartenstein et al. (1985) for the procedure followed to obtain this reconstruction. In order to enable direct comparison of the results of HRP injections with those of the histological analysis, the positions at which injections were performed in the early gastrula (see Technau and Campos-Ortega 1985) were translated onto the planimetric reconstruction of an embryo of similar age, and from there to the blastoderm, as indicated in Fig. 18.1A.

18.2
The Rationale of Fate Map Construction

The rationale followed to establish the fate map consisted of the following steps. Determining how many times embryonic cells have divided before the various larval organs become distinguishable, and then counting the number of cells of each larval primordium, permitted us to calculate the number of blastoderm cells in each anlage. Then by analysing the morphogenetic movements performed by cells of the blastoderm, and using

Fig. 18.1. A Fate map of the gastrula in a lateral projection (this is Fig. 10 of Technau and Campos-Ortega 1985). Circles mark sites of injection of HRP into cells of the early gastrula (given in % EL and % VD). Structures which were found labelled at fixation are indicated in the circles. Hatched circles indicate cells later found in the gut, injection at the sites of grey circles resulted in labelling of both epidermal and neural cells. *Arrowheads* indicate injection sites within the cephalic furrow. **B** Fate map of the *Drosophila* blastoderm (this is Fig. 1C of Hartenstein et al. 1985). A planimetric reconstruction of the left half of the blastoderm is shown. Hatched areas will invaginate at gastrulation. Numbers indicate the sizes of the different anlagen referred to one side of the blastoderm. The scale indicates EL% (0-10% and 90-100% values are distorted due to the reconstruction procedure). The inset at the bottom right corner shows the size of a *t1-a8* hemisegment anlage; each consists of 32 dorsal epidermal progenitor cells (*dEpi*), 60 ventral epidermal (neurogenic, *vNR*) progenitor cells, of which 13-16, depending on the segment considered, will segregate as neuroblasts (*Nbl*), and 4 mesectodermal cells (*me*). The anlagen are indicated as follows. *am* anterior midgut; *as* amnioserosa; *c.* clypeolabrum; *dEpi* dorsal epidermis; *dr* dorsal ridge; *es* oesophagus; *mt* Malpighian tubules; *MS* mesoderm; *ol* optic lobes; *ph* pharynx; *vc* procephalic lobe; *pm* posterior midgut; *pNR* procephalic neurogenic region; *pr* proctodeum; *sg* salivary gland; *tr* trachea; *vNR* ventral neurogenic region; *C1-C3* gnathal segments; *T1-T3* thoracic segments; *A1-A10* abdominal segments. See text for further details

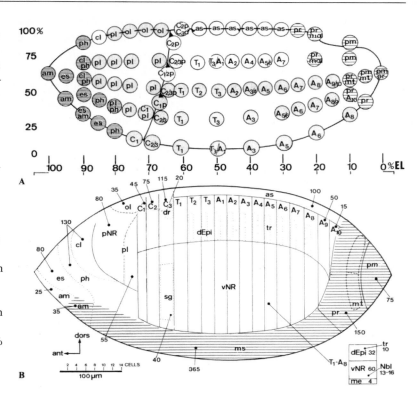

some additional assumptions, to be discussed for each particular case, enabled us to translate these data to the surface of the blastoderm.

The number of cells in each larval organ can be determined once boundaries between developing organs become distinguishable. Since most boundaries are well defined, generally no ambiguities were encountered in counting cells, nor in counting the number of divisions. Thus we are confident that the values obtained for the sizes of blastoderm anlagen are rather good approximations. For example, we found that at the cellular blastoderm stage cells are arranged in a monolayer of fairly regular hexagons with an average diameter of 7.5 mm (in formaldehyde-fixed material). Cell counts of the entire complement of four different fuchsin-stained embryos could be readily made; they yielded approximately 5000 blastoderm cells (5010 ± 88). If the sizes of the various blastoderm anlagen, calculated by counting the mitotic divisions and the number of cells of the larval organs, are added up, the resulting total number of 4980 cells closely corresponds to the average figure of 5010±88 cells counted directly in four different blastoderms. However, translating the course of larval organ boundaries to the blastoderm is a different matter, for this procedure relies on indirect assumptions. In several cases boundaries become distinguishable at relatively late stages; for example, the tracheal pits, the salivary glands or the Malpighian tubules. In other cases the prospective location of boundaries becomes evident at early stages because of regional differentiation in the proliferation pattern, e.g. intersegmental boundaries, or because of morphogenetic movements, e.g. boundaries between germ layers. In any case, however, when boundaries become visible the original topological relationships of blastoderm cells have undergone considerable distortion because of intervening morphogenetic movements. Although morphogenetic movements can be analyzed fairly well, both by reconstructing embryos of slightly different ages with methods of normal anatomy (see Hartenstein and Campos-Ortega 1985), and by following the fate of embryonic cells injected with HRP (see Technau and Campos-Ortega, 1985), the *Drosophila* embryo is fairly large and these movements rather complex. Thus, there always remains a certain degree of uncertainty in tracing the course of larval organ boundaries back to the blastoderm, because this operation is in each case somewhat arbitrary. Thus the boundaries between larval organ anlage indicated in Fig. 18.1 are to be understood as approximations.

18.3
The Size of the Blastoderm Anlagen

18.3.1
The Mesoderm and the Endoderm

The first boundary distinguishable in development is the boundary between germ layers, i.e. between the primordia invaginated at gastrulation (mesoderm and endoderm) and the ectoderm; boundaries between the primordia of dorsal and ventral epidermis, of epidermis and hindgut, of epidermis and foregut, of hindgut and posterior midgut, of central nervous system and epidermis, and of consecutive segments, appear later on. Finally, particular organs, e.g. salivary glands, tracheal tree, Malpighian tubules etc., become evident.

The lateral extent of the blastoderm region that gives rise to mesoderm and endoderm - which invaginates at gastrulation -

was determined from counts on transverse sections and fuchsin-stained wholemounts. This region was found to comprise 15% of the ventrodorsal egg perimeter on either side. About 1250 (1256±19) cells invaginate during gastrulation, of which approximately 450 (449±28) form the amnioproctodeal invagination whereas the remaining 800 form the ventral furrow. A transient mesodermal segmentation occurs in stage 9 which permits one to determine the size of each of the prospective myomeres t1-a8; this was found to be approximately 180 mesodermal cells per segment. Since each mesodermal cell has performed two mitoses by this stage, the mesoderm anlage of each segment should contain about 45 blastoderm cells.

The boundaries between mesodermal and endodermal anlagen (posterior midgut-mesoderm-anterior midgut), as well as the boundary between proctodeum and posterior midgut, were determined chiefly from quantitative considerations and the results of HRP injections. First of all, we assumed that the pole cells retain their original position relative to the underlying cells during germ band elongation. On this assumption, the bottom of the amnioproctodeal invagination, which carries the pole cells and will form the posterior midgut, must derive from the posterior pole of the blastoderm. The walls of the amnioproctodeal invagination can then be related to the posterior pole, and the number of cells can be counted. The sizes of the anlagen of proctodeum (see below) and posterior midgut were found to be 300 and 150 cells, respectively. On the basis of the shape of the amnioproctodeal invagination we assume that the anlage of the posterior midgut has a regular shape around the posterior egg pole and, thus, locate the border with the proctodeum anlage at a position corresponding to an area containing 150 cells. All these assumptions are strongly supported by the results of HRP injections.

The anterior midgut-mesodermal border and the proctodeal-mesodermal border were determined in a similar way. The endodermal anterior midgut becomes clearly distinguishable from the mesoderm relatively late in development, after the third mitosis. At this time the anterior midgut comprises 560 cells, thus its blastoderm anlage must have 70 cells.

18.3.2
The Hindgut and the Terminalia

The proctodeum forms part of a sac, the bottom of which is the primordium of the posterior midgut. This means that the anlage of the proctodeum (its size has been considered in the previous paragraph) must be organized like a ring around the anlage of the posterior midgut. The results of HRP injections fully support this hypothesis.

Immediately after the second postblastoderm mitosis of proctodeal cells, four symmetrically arranged pouches appear near the proctodeal (hindgut)-midgut border. These are the anlagen of the Malpighian tubules which will gradually grow into long slender tubes. It has not been determined how many proctodeal cells contribute to the formation of the Malpighian tubes. However, a considerable percentage, if not all, of these cells undergo a third postblastoderm mitosis at about 220 min.

After the emergence of the Malpighian tubules the proctodeum decreases greatly in diameter. Thus the border between the anlage of the anal plate and proctodeum, on the one hand, and of *a8*, on the other hand, becomes distinguishable. The anlage of the terminalia consists of two wings that flank the funnel-shaped proctodeal opening. A transverse groove subdivides the anlage on either side into a large rostral part (*a9*) and a small caudal one (the so-called *a10*). Both together contain 260 (261±8) cells.

18.3.3
The Amnioserosa

The amnioserosa derives from a narrow strip at the dorsal midline of the blastoderm in register with the anlage of the metameric germ band, an arrangement that is evident from both normal anatomy (Poulson 1950) and HRP injections. At about 110 min the amnioserosa can be clearly distinguished, due to the characteristic shape of its cells, and comprises 200 (201±10) cells. Amnioserosa cells have not been observed to divide at all during embryonic development and the length of the anlage should correspond to that of the metameric germ band, for the amnioserosa is in register with all segments. Therefore the width of the anlage can be extrapolated from the final number of amnioserosa cells.

18.3.4
The Thoracic and Abdominal Epidermis

Amnioserosa and ectoderm together comprise about 3800 cells (3784±36); since the amnioserosa comprises 200, about 3600 cells correspond to the ectoderm, of which approximately 1000 (1005±36) are located in front of the cephalic furrow, including its anterior lip. The primordia of the dorsal and ventral epidermis become conspicuously different during the course of the first postblastodermal mitosis. The anlage of the dorsal trunk epidermis contains 920 cells and that of the ventral (neurogenic) region 1770 cells, including the posterior lip of the cephalic furrow. Assuming a vertical course of intersegmental boundaries, 180-190 blastoderm cells can be assigned to each of the segments *t1-a8*. About 26-32 of these cells, depending on the segment, will segregate as neuroblasts, the remainder will develop into epidermis, sensory organs and tracheae. Since intersegmental boundaries in the larva are oriented perpendicular to the midline, and since there is no indication of intervening morphogenetic movements which could derange this orientation during development, we assumed that the boundaries between the segmental anlagen are oriented orthogonal to the ventral midline of the blastoderm as well. This assumption is strongly supported by the results of HRP injections. The entire anlage of the metameric germ band was then subdivided in the various segmental anlagen according to the number of cells in each segment. The anteroposterior extent of each segmental anlage corresponds to less than 3 cells, and that of the entire metameric germ band to roughly 40 cells.

However, in Chapter 17 we stated that segmental boundaries exhibit a sharp bend in their dorsoventral course; further, intersegmental furrows originate from dorsal and ventral subdivisions of the germ band which are apparently out of register. Thus, it might well be that the blastoderm anlagen of segmental boundaries show the same sharp bend and are not as straight as assumed in Fig. 18.1B. It must be pointed out that the small size of the segmental anlagen precludes reliable delineation of the actual course of these boundaries on the blastoderm by experimental means.

18.3.5
The Gnathal Segments

The dorsal ridge, gnathal buds, and salivary glands (all arising from various parts of the gnathal segments) become evident late in development. The sizes of their blastoderm anlagen have been defined based on quantitative considerations, and their locations derive chiefly from the results of HRP injections. Cell counts performed in embryos after germ band shortening yielded the following results. The presumptive dorsal ridge and neighbouring epidermis together comprise 180 cells, whereby the majority contribute to the dorsal ridge; the gnathal buds and the epidermis ventral to them (sternum) amount to 290 (286±42) cells (mandible), 490 (494±25) cells (maxilla), and 690 (692±20) cells (labium, including salivary glands); these figures are the sums from both sides of the embryo. Since these cells have divided twice, the size of the blastoderm anlagen of the gnathal segments can be calculated, adding the complement of gnathal neuroblasts - 10, 32 and 34, for the mandibular, maxillar and labial neuroblasts, respectively (Hartenstein and Campos-Ortega 1984) - to the result of dividing the figures above by 4; approximately 30 more cells belong to the dorsal ridge, which have to be added to labium. This results in about 90, 150 and 230 cells for the anlage of the mandibular, maxillary and labial segments, respectively (both sides together).

18.3.6
The Procephalon

Four regions can be morphologically distinguished in the embryonic procephalon from early stages on. These are the procephalic neurogenic region (containing precursors of supraoesophageal neuroblasts and of epidermal cells of the procephalic lobe), the anlage of the foregut (oesophagus, pharynx and hypopharynx) and clypeolabrum, a narrow dorsomedian strip of 'parietal cells' (some of which become integrated in the optic lobe primordium), and a narrow vertical strip of cells located between the procephalic neurogenic region and the anterior border of the germ band; chiefly on the basis on quantitative estimations it was assumed that the latter territory will be integrated into the epidermis of the procephalic lobe. Since boundaries between the above regions become evident before distortions due to foregut invagination and anteroventral shift of the gnathal epidermis occur, these boundaries can be directly projected onto the blastoderm after compensating for minor alterations caused by gastrulation. The boundaries drawn in our map between oesophagus, pharynx and clypeolabrum are inferred chiefly from the results of HRP injections, the size of these anlagen is based on cell counts.

The entire procephalon was found to comprise about 810 blastoderm cells. Of these, 470 will form the foregut. The anlage of the oesophagus was calculated to comprise 80 cells, those of pharynx, hypopharynx and clypeolabrum 130 cells. Some 50 cells that invaginate with the stomodeum, and therefore belong to the ectoderm, have been found by HRP injections (Technau and Campos-Ortega 1985) to become incorporated into the proventriculus and anterior midgut.

In the procephalon the neurogenic ectoderm has been found to comprise approximately 80 cells. Some irregularities (discussed in Hartenstein and Campos-Ortega 1985) prevent us from conclusively establishing the actual proportion of neurogenic versus epidermogenic cells in the procephalon, and make it impossible to exclude a derivation of some neuroblasts from neighbouring procephalic territories (see Technau and Campos-Ortega 1985). Nevertheless, most neuroblasts of the supraoesophageal ganglion certainly derive from the area histologically defined as the procephalic neurogenic region. The dorsomedian cells mentioned above, together with the cells between the procephalic neurogenic region and the gnathal segments, will contribute to the walls of the frontal sac.

The interested reader will find a more extensive discussion of the issue of fatemapping in *Drosophila* in Hartenstein and Campos-Ortega (1985), Technau and Campos-Ortega (1985) and Hartenstein et al. (1985).

References

Abrams JM, White K, Fessler LJ, Steller H (1993) Programmed cell death during *Drosophila* embryogenesis. Development 117: 29-43

Akam M, Martinez-Arias A (1985) The distribution of *Ultrabithorax* transcripts in *Drosophila* embryos. EMBO J 4: 1689-1700

Allen GE (1975) The introduction of *Drosophila* into the study of heredity, 1900-1910. Isis 66: 322-333

Allen GE (1978) Thomas Hunt Morgan. The man and his science. Princeton Univ Press, Princeton, NJ

Allis CD, Underwood EM, Caulton JH, Mahowald AP (1979) Pole cells of *Drosophila melanogaster* in culture. Normal metabolism, ultrastructure and functional capabilities. Dev Biol 69: 451-465

Anderson DT (1962) The embryology of *Dacus tryoni* (Frogg) (Diptera, Trypetidae (=Trephitidae)), the Queensland fruit-fly. J Embryol Exp Morph 10: 148-292

Anderson DT (1972) The development of holometabolous insects. In: Counce SJ, Waddington CH (eds) Developmental Systems: Insects, Vol. 1. Academic Press, New York, pp 165-241

Bate CM (1978) Development of sensory systems in arthropods. In: Jacobson M (ed) Handbook of Sensory Physiology, Vol. IX, Springer Verlag, Berlin-Heidelberg, pp 1-53

Bate CM (1990) The embryonic development of larval muscles in *Drosophila*. Development 110: 791-804

Bate CM (1993) Mesoderm. In: Bate CM, Martinez-Arias A (eds) The Development of *Drosophila melanogaster*. Cold Spring Harbor Laboratory Press, Cold Spring Harbor, New York, pp 1013-1090

Bate CM, Martinez-Arias A (1991) The embryonic origin of imaginal discs in *Drosophila*. Development 112: 755-761

Bate CM, Rushton E, Currie DA (1991) Cells with persistent *twist* expression are the embryonic precursors of adult muscles in *Drosophila*. Development 113: 79-89

Bauer V (1904) Zur inneren Metamorphose des Centralnervensystems der Insecten. Zool Jhb Abt Anat Ontog Tiere 20: 123-150

Becker T, Technau GM (1990) Single cell transplantation reveals interspecific cell communication in *Drosophila* chimeras. Development 109: 821-832

Beer J, Technau GM, Campos-Ortega JA (1987) Lineage analysis of transplanted individual cells in embryos of *Drosophila melanogaster*. IV. Commitment and proliferative capabilities of mesodermal cells. Roux's Arch Dev Biol 196: 220-230

Berendes HD, Ashburner M (1978) The salivary glands. In: Ashburner M, Wright TRF (eds) The Genetics and Biology of *Drosophila*, Vol. 2b. Academic Press, New York, pp 453-498

Bier E, Vaessin H, Shepherd S, Lee K, McCall K, Barbel S, Ackerman L, Carretto R, Uemura T, Grell E, Jan LY, Jan YN (1989) Searching for pattern and mutation in the *Drosophila* genome with a P-*lacZ* vector. Genes Dev 3: 1273-1287

Bodenstein, D. (1950) The postembryonic development of *Drosophila*. In: Demerec M (ed) Biology of *Drosophila*. Wiley, New York, pp 275-367

Bodmer R, Carretto R, Jan YN (1989) Neurogenesis of the peripheral nervous system in *Drosophila* embryos. Neuron 3: 21-32

Bodmer R, Jan YN (1987) Morphological differentiation of the embryonic peripheral neurons in *Drosophila*. Roux's Arch Dev Biol 196: 69-77

Bossing T, Technau GM (1994) The fate of the CNS midline progenitors of *Drosophila* as revealed by a new method for single cell labelling. Development 120: 1895-1906

Bossing T, Technau GM, Doe CQ (1996a) *huckebein* is required for glial development and axon finding in the NB 1-1 and NB 2-2 lineages in the *Drosophila* CNS. Mech Dev 55: 53-64

Bossing T, Udolph G, Doe CQ, Technau GM (1996b) The embryonic CNS lineages of *Drosophila melanogaster*. I. The lineages derived from the ventral half of the truncal neuroectoderm. Dev Biol 179: 41-64

Bownes M, Kalthoff K (1974) Embryonic defects in *Drosophila* eggs after partial UV irradiation at different wavelength. J Embryol Exp Morph 31: 329-345

Bownes M, Sang JH (1974a) Experimental manipulation of early *Drosophila* embryos. I. Adult and embryonic defects resulting from microcautery at nuclear multiplication and blastoderm stages. J Embryol Exp Morph 32: 253-272

Bownes M, Sang JH (1974b) Experimental manipulation of early *Drosophila* embryos. II. Adult and embryonic defects resulting from the removal of blastoderm cells by pricking. J Embryol Exp Morph 32: 273-285

Bownes M (1975) A photographic study of development in the living embryo of *Drosophila melanogaster*. J Embryol exp Morph. 33: 789-801

Bownes M (1982) Embryogenesis. In: Ransom R (ed) Handbook of *Drosophila* Development. Elsevier, Amsterdam, New York, Oxford, pp 66-94

Brewster R, Bodmer R (1995) Origin and specification of type II sensory neurons in *Drosophila*. Development 121: 364-375

Broadie K, Skaer H leB, Bate M (1992) Whole-embryo culture in *Drosophila*: development of embryonic tissues in vitro. Roux's Arch Dev Biol 201: 364-375

Broadus J, Skeath JB, Spana EP, Bossing T, Technau GM, Doe CQ (1995) New neuroblast markers and the origin of the aCC/pCC neurons in the *Drosophila* central nervous system. Mech Devel 53: 393-402

Brookman JJ, Toosy AT, Shashidhara LS, White RAH (1992) The 412 retrotransposon and the development of gonadal mesoderm in *Drosophila*. Development 116: 1185-1192

Bueznow DE, Holmgren R (1995) Expression of the *Drosophila gooseberry* locus defines a subset of neuroblast lineages in the central nervous system. Dev Biol 170: 338-349

Callaini G, Riparbelli MG, Dallai R (1995) Pole cell migration through the gut wall of the *Drosophila* embryo: analysis of cell interactions. Dev Biol 170: 365-375

Campos-Ortega JA (1982) Development of the nervous system. In: Ransom R (ed) Handbook of *Drosophila* development. Elsevier, Amsterdam, New York, Oxford, pp 153-181

Campos-Ortega JA (1993) Early neurogenesis in *Drosophila melanogaster*. In: Bate CM, Martinez-Arias A (eds) The Development of *Drosophila melanogaster*. Cold Spring Harbor Laboratory Press, Cold Spring Harbor, New York, pp 1091-1129

Campos-Ortega JA, Haenlin M (1992) Regulatory signals and signal molecules in early neurogenesis of *Drosophila melanogaster*. Roux´s Arch Dev Biol 201: 1-11

Campos-Ortega JA, Hartenstein V (1985) Development of the nervous system. In: Kerkut GA, Gilbert LI (eds) Comprehensive Insect Physiology, Biochemistry and Pharmacology, Vol 5. Pergamon Press, Oxford, pp 49-84

Canal I, Ferrus A (1986) The pattern of early neuronal differentiation in *Drosophila melanogaster*. J Neurogenet 3: 293-319

Carr JN, Taghert PH (1988) Formation of the transverse nerve in moth embryos. I. A scaffold of nonneuronal cells prefigures the nerve. Dev Biol 130: 487-499

Chen TY (1929) On the development of imaginal buds in normal and mutant *Drosophila melanogaster*. J Morphol 47: 135-199

Chu-LaGraff Q, Doe CQ (1993) Neuroblast specification and formation regulated by *wingless* in the *Drosophila* CNS. Science 261: 1594-1597

Chu-LaGraff Q, Schmid A, Leidel J, Brönner G, Jäckle H, Doe CQ (1995) *huckebein* specifies aspects of CNS precursor identity required for motoneuron axon pathfinding. Neuron 15: 1041-1051

Chu-Wang I-W, Axtell RC (1972) Fine structure of the terminal organ of the house fly *Musca domestica* L. Z Zellforsch 127: 287-305

Cohen SM (1993) Imaginal disc development. In: Bate CM, Martinez-Arias A (eds) The Development of *Drosophila melanogaster*. Cold Spring Harbor Laboratory Press, Cold Spring Harbor, New York, pp 609-685

Cohen B, Wimmer EA, Cohen SM (1991) Early development of leg and wing primordia in the *Drosophila* embryo. Mech Dev 33: 229-240

Cohen B, Simcox AA, Cohen SM (1993) Allocation of the imaginal primordia in the *Drosophila* embryo. Development 117: 597-608

Costa M, Sweeton D, Wieschaus E (1993) Gastrulation in *Drosophila*: cellular mechanisms of morphogenetic movements. In: Bate CM, Martinez-Arias A (eds) The Development of *Drosophila melanogaster*. Cold Spring Harbor Laboratory Press, Cold Spring Harbor, New York, pp 425-465

Counce SJ (1963) Developmental morphology of polar granules in *Drosophila*, including observations of pole cell behaviour and distribution during embryogenesis. J Morph 112: 129-146

Crews ST, Thomas JB, Goodman CS (1988) The Drosophila *single-minded* gene encodes a nuclear protein with similarity to the per gene product. Cell 52: 143-151

Crossley AC (1978) The morphology and development of the *Drosophila* muscular system. In: Asburner M, Wright TRF (eds) The genetics and biology of *Drosophila*, Vol. 2b. Academic Press, New York, pp 499-560

Dambly-Chaudière C, Ghysen A (1986) The sense organs in the *Drosophila* larva and their relationship to the embryonic pattern of sensory neurons. Roux Arch Dev Biol 195: 222-228

Dambly-Chaudiere C, Jamet E, Burri M, Bopp D, Basle, K, Hafen E, Dumont N, Spielmann P, Ghysen A, and Noll M (1992) The paired-box gene *pox neuro*: a determinant of poly-innervated sense organs in *Drosophila*. Cell 69: 159-172

Dambly-Chaudiere C, Ghysen A (1987) Independent subpatterns of sense organs require independent genes of the achaete-scute complex in *Drosophila* larvae. Genes Dev 1: 297-306

Diederich RJ, Pattatucci AM, Kaufman TC (1991) Developmental and evolutionary implications of *labial*, *Deformed* and *engrailed* expression in the *Drosophila* head. Development 113: 273-281

DiNardo S, Kuner JM, Theis J, O´Farrell PH (1985) Development of embryonic pattern in *Drosophila melanogaster* as revealed by accumulation of the nuclear engrailed protein. Cell 43: 59-69

Doe CQ (1992) Molecular markers for identified neuroblasts and ganglion mother cells in the *Drosophila* nervous system. Development 116: 855-863

Doe CQ, Hiromi Y, Gehring WJ, Goodman CS (1988). Expression and function of the segmentation gene *fushi-tarazu* during *Drosophila* neurogenesis. Science 239: 170-175

Ede D, Counce SJ (1956) A cinematographic study of the embryology of *Drosophila melanogaster*. Wilhelm Roux Arch Ent Mech Org 148: 402-415

Edgar BA, O'Farrell PH (1989) Genetic control of cell division patterns in the *Drosophila* embryo. Cell 57: 177-187

Foe VA (1989) Mitotic domains reveal early commitment of cells in *Drosophila* embryos. Development 107: 1-22

Foe VA, Alberts BM (1983) Studies of nuclear and cytoplasmic behaviour during the five mitotic cycles that precede gastrulation in *Drosophila* embryogenesis. J Cell Sci 61: 31-70

Foe VA, Odell GM, Edgar BA (1993) Mitosis and morphogenesis in the *Drosophila* embryo: point and counterpoint. In: Bate CM, Martinez-Arias A (eds) The Development of *Drosophila melanogaster*. Cold Spring Harbor Laboratory Press, Cold Spring Harbor, New York, pp 149-300

Fredieu JR, Mahowald AP (1989) Glial interactions with neurons during *Drosophila* embryogenesis. Development 106: 739-748

Fujita SC, Zipursky SL, Benzer S, Shotwell SL (1982) Monoclonal antibodies against the *Drosophila* nervous system. Proc Natl Acad Sci USA 79: 7929-7933

Fullilove SL, Jacobson AG (1971) Nuclear elongation and cytokinesis in *Drosophila montana*. Dev Biol 26: 560-577

Fullilove SL, Jacobson AG (1978) Embryonic development: descriptive. In: Ashburner M, Wright TRF (eds) The Genetics and Biology of *Drosophila*, Vol. 2c. Academic Press, New York, pp 105-227

Garcia-Bellido A, Merriam JR (1969) Cell lineage of the imaginal discs in *Drosophila* gynandromorphs. J Exp Zool 170: 61-76

Gehring WJ, Seippel S (1967) Die Imaginalzellen des Clypeo-Labrums und die Bildung des Rüssels von *Drosophila melanogaster*. Rev Suisse Zool 74: 589-596

Gergen JP, Coulter D, Wieschaus E (1986) Segmental pattern and blastoderm cell identities. Symp Soc Dev Biol 43: 195-220

Ghysen A, O`Kane C (1989) Detection of enhancer-like elements in the genome of *Drosophila*. Development 105: 35-52

Ghysen A, Dambly-Chaudiere C, Aceves E, Jan LY, Jan YN (1986) Sensory neurons and peripheral pathways in *Drosophila* embryos. Roux's Arch Dev Biol 195: 281-289

Goldschmidt R (1927) Physiologische Theorie der Vererbung. J Springer, Berlin

Goldschmidt R (1938) Physiological genetics. McGraw-Hill, New York and London

González-Gaitán M, Jäckle H (1995) Invagination centers within the *Drosophila* stomatogastric nervous system anlage are positioned by Notch-mediated signaling that is spatially controlled through *wingless*. Development 121: 2313-2325

González-Gaitán M, Rothe M, Wimmer EA, Taubert H, Jäckle H (1994) Redundant functions of the genes *knirps* and *knirps-related* for the establishment of anterior *Drosophila* head structures. Proc Natl Acad Sci USA 91: 8567-8571

Goodman CS, Spitzer NC (1979) Embryonic development of identified neurons: differentiation from neuroblast to neurone. Nature 280: 208-214

Goodman CS, Bate CM (1981) Neuronal development in the grasshopper. Trends Neurosci 4: 163-169

Goodman CS, Doe CQ (1993) Embryonic development of the *Drosophila* nervous system. In: Bate CM, Martinez-Arias A (eds) The Development of *Drosophila melanogaster*. Cold Spring Harbor Laboratory Press, Cold Spring Harbor, New York, pp 1131-1206

Goodman CS, Bate CM, Spitzer NC (1981) Embryonic development of identified neurons: Origin and transformation of the H-cell. J Neurosci 1: 94-102

Goodman CS, Bastiani MJ, Doe CQ, du Lac S, Helfand SL, Kuwada JY, Thomas JB (1984) Cell recognition during neuronal development. Science 225: 1271-1279

Gorczyca MG, Phillips RW, Budnik V (1994) The role of *tinman*, a mesodermal cell fate gene, in axon pathfinding during the development of the transverse nerve in *Drosophila*. Development 120: 2143-2152

Goriely A, Dumont N, Dambly-Chaudiere C, Ghysen A (1991) The determination of sense organs in *Drosophila*: effects of the neurogenic mutations in the embryo. Development 113: 1395-1404

Green P, Hartenstein-Younossi A, Hartenstein V (1993) The embryonic development of the *Drosophila* visual system. Cell Tissue Res 273: 583-598

Grenningloh G, Rehm EJ, Goodman CS (1991) Genetic analysis of growth cone guidance in *Drosophila*: Fasciclin II functions as a neuronal recognition molecule. Cell 67: 45-57

Guillemin K, Groppe J, Dücker K, Treisman R, Hafen E, Affolter M, Krasnow MA (1996) The prune gene encodes the *Drosophila* serum response factor and regulates cytoplasmic outgrowth during terminal branching of the tracheal system. Development 122: 1353-1362

Haget A (1977) L'embryologie des insectes. In: Grasse P-P (ed) Traite de Zoologie, Vol VIII, Fs V-B. Masson, Paris, pp 1-387

Halter DA, Urban J, Rickert C, Ner SS, Ito K, Travers AA, Technau GM (1995) The homeobox gene *repo* is required for the differentiation and maintenance of glia function in the embryonic nervous system of *Drosophila melanogaster*. Development 121: 317-332

Hama C, Ali Z, Kornberg TB (1990) Region-specific recombination and expression are directed by portions of the *Drosophila engrailed* promoter. Genes Dev 4: 1079-1093

Harbecke R, Janning W (1989) The segmentation gene *Krüppel* of *Drosophila melanogaster* has homeotic properties. Genes Dev 3: 114-122

Hartenstein V (1988) Development of *Drosophila* larval sensory organs: spatio-temporal pattern of sensory neurones, peripheral axonal pathway and sensilla differentiation. Development 102: 869-886

Hartenstein V (1993) Atlas of *Drosophila* Development. Cold Spring Harbor Laboratory Press, Cold Spring Harbor, New York

Hartenstein V, Campos-Ortega JA (1984) Early neurogenesis in wild-type *Drosophila melanogaster*. Roux's Arch Dev Biol 193: 308-325

Hartenstein V, Campos-Ortega JA (1985) Fate-mapping in wild-type *Drosophila melanogaster*. I. The spatio-temporal pattern of embryonic cell divisions. Roux's Arch Dev Biol 194: 181-195

Hartenstein V, Campos-Ortega JA (1986) The peripheral nervous system of mutants of early neurogenesis in *Drosophila melanogaster*. Roux`s Arch Dev Biol 195: 210-221

Hartenstein V, Jan YN (1992) Studying *Drosophila* embryogenesis with P-*lacZ* enhancer trap lines. Roux`s Arch Dev Biol 201: 194-220

Hartenstein V, Posakony JW (1990) Sensillum development in the absence of cell division: the sensillum phenotype of the *Drosophila* mutant *string*. Dev Biol 138: 147-158

Hartenstein V, Technau GM, Campos-Ortega JA (1985) Fate-mapping in wild-type *Drosophila melanogaster*. III. A fate map of the blastoderm. Roux's Arch Dev Biol 194: 213-216

Hartenstein V, Rudloff E, Campos-Ortega JA (1987) The pattern of proliferation of the neuroblasts in the wild-type embryo of *Drosophila melanogaster*. Roux's Arch Dev Biol 196: 473-485

Hartenstein AY, Rugendorff A, Tepass U, Hartenstein V (1992) The function of the neurogenic genes during epithelial development in the *Drosophila* embryo. Development 116: 1203-1220

Hartenstein V, Tepass U, Gruszynski-deFeo E (1994) Embryonic development of the stomatogastric nervous system in *Drosophila*. J Comp Neurol 350: 367-381

Hartenstein V, Yanoussi-Hartenstein A, Lekven A (1994) Delamination and division in the *Drosophila* neurectoderm: spatiotemporal pattern, cytoskeletal dynamics, and common control by neurogenic and segment polarity genes. Dev Biol 165: 480-499

Hay B, Ackerman L, Barbel S, Jan LY, Jan YN (1988a) Identification of a component of *Drosophila* polar granules. Development 103: 625-640

Hay B, Jan LY, Jan YN (1988b) A component of Drosophila polar granules is encoded by *vasa* and has extensive sequence similarity to ATP-dependent helicases. Cell 55: 577-587

Hertweck H (1931) Anatomie und Variabilität des Nervensystems und der Sinnesorgane von *Drosophila melanogaster* (Meigen). Z wiss Zool 139: 559-663

Hildreth PE, Lucchesi JC (1963) Fertilization in *Drosophila*. Dev Biol 6: 262-278

Hoch M, Broadie K, Jäckle H, Skaer H (1994) Sequential fates in a sigle cell are established in the neurogenic cascade in the Malpighian tubules of *Drosophila*. Development 120: 3439-3450

Hofbauer A, Campos-Ortega JA (1990) Proliferation pattern and early differentiation of the optic lobes in *Drosophila melanogaster*. Roux`s Arch Dev Biol 198: 264-274

Hotta Y, Benzer S (1972) Mapping of behaviour in *Drosophila* mosaics. Nature 240: 527-535

Huettner AF (1924) Maturation and fertilization in *Drosophila melanogaster*. J Morphol 39: 249-265

Huettner AF (1927) Irregularities in the early development of the *Drosophila melanogaster* egg. Z Zellforsch mikr Anat 4: 599-610

Huettner AF (1933) Continuity in the centrioles of *Drosophila melanogaster*. Z Zellforsch mikr Anat 19: 119-134

Illmensee K, Mahowald AF, Loomis MR (1976) The ontogeny of germ plasm during oogenesis of *Drosophila*. Dev Biol 49: 40-65

Imaizumi T (1958) Recherches sur l'expression des facteurs letaux hereditaires chez l'embryon de la *Drosophile*. Sur l'embryogenese et le mode des letalités au cours du developpement embryonaire. Cytologia 23: 270-285

Ingham P, Martinez-Arias A, Lawrence PA, Howard K (1985) Expression of *engrailed* in the parasegments of *Drosophila*. Nature 317: 634-636

Irvine KD, Wieschaus E (1994) Cell intercalation during *Drosophila* germ band extension and its regulation by pair-rule segmentation genes. Devlopment 120: 827-841

Ito K, Urban J, Technau GM (1994) Distribution, classification and development of *Drosophila* glial cells during late embryogenesis. Roux's Arch Dev Biol 204: 284-307

Jacobs JR, Goodman CS (1989) Embryonic development of axon pathways in the *Drosophila* CNS. I. A glial scaffold appears before the first growth cones. J Neurosci 9: 2402-2411

Jacobs JR, Hiromi Y, Patel NH, Goodman CS (1989) Lineage, migration, and morphogenesis of longitudinal glia in the Drosophila CNS as revealed by a molecular lineage marker. Neuron 2: 1625-1631

Jaglarz MK, Howard KR (1994) Primordial germ cell migration in *Drosophila melanogaster* is controlled by somatic tissue. Development 120: 83-89

Jan LY, Jan YN (1982) Antibodies to horseradish peroxidase as specific neuronal markers in *Drosophila* and in grasshopper embryos. Proc Natl Acad Sci USA 72: 2700-2704.

Jan YN, Jan LY (1993) The peripheral nervous system. In: Bate CM, Martinez-Arias A (eds) The Development of *Drosophila melanogaster*. Cold Spring Harbor Laboratory Press, Cold Spring Harbor, New York, pp 1207-1244

Jan YN, Ghysen A, Barbel CS, Jan LY (1985) Formation of neuronal pathways in the imaginal discs of *Drosophila melanogaster*. J Neurosci 5: 2453-2464

Janning W (1972) Aldehyde oxidase as a cell marker for internal organs in *Drosophila melanogaster*. Naturwiss 59: 516-517

Janning W (1974) Entwicklungsgenetische Untersuchungen an Gynandern von *Drosophila melanogaster*. I. Die inneren Organe der Imago. Roux's Arch Dev Biol 174: 313-332

Janning W (1976) Entwicklungsgenetische Untersuchungen an Gynandern von *Drosophila melanogaster*. IV. Vergleich der morphologischen Anlagepläne larvaler und imaginaler Strukturen. Roux's Arch Dev Biol 179: 349-372

Janning W, Lutz A, Wissen D (1986) Clonal analysis of the blastoderm anlage of the Malpighian tubules in *Drosophila melanogaster*. Roux's Arch Dev Biol 195: 22-32

Johansen J, Halpern ME, Keshishian H (1989) Axonal guidance and the development of muscle fibre-specific innervation in *Drosophila* embryos. J Neurosci 9: 4318-4332

Jürgens G (1987) Segmental organization of the tail region in the embryo of *Drosophila melanogaster*. A blastoderm fate map of the cuticle structures of the larval tail region. Roux's Arch Dev Biol 196: 141-157

Jürgens G, Hartenstein V (1993) The terminal regions of the body pattern. In: Bate CM, Martinez-Arias A (eds) The Development of *Drosophila melanogaster*. Cold Spring Harbor Laboratory Press, Cold Spring Harbor, New York, pp 687-746

Jürgens G, Lehmann R, Schardin M, Nüsslein-Volhard C (1986) Segmental organization of the head in the embryo of *Drosophila melanogaster*. A blastoderm fate map of the cuticle structures of the larval head. Roux's Arch Dev Biol 195: 359-377

Kam Z, Minden JS, Agard DA, Sedat JW, Leptin M (1991) *Drosophila* gastrulation: analysis of cell shape changes in living embryos by three-dimensional fluorescence microscopy. Development 112: 365-370

Kankel DR, Ferrus A, Garen SH, Harte PJ, Lewis PE (1980) The structure and development of the nervous system. In: Ashburner M, Wright TRF (eds) The Genetics and Biology of *Drosophila*, Vol. 2d. Academic Press, New York, pp 295-368

Kankel DR, Hall JC (1976) Fate mapping of nervous system and other internal tissues in genetic mosaics of *Drosophila melanogaster*. Dev Biol 48: 1-24

Keilin D (1944) Respiratory systems and respiratory adaptations in larvae and pupae of Diptera. Parasitology 36: 1-66.

King RC, Aggarwal SK, Bodenstein D (1966) The comparative submicroscopic morphology of the ring gland of *Drosophila melanogaster* during the second and third larval instars. Z Zellforsch 73: 272-285

Klämbt C, Goodman CS (1991) The diversity and pattern of glia during axon pathway formation in the *Drosophila* embryo. Glia 4: 205-213

Klämbt C, Jacobs JR, Goodman CS (1991) The midline of the *Drosophila* central nervous system. A model for the genetic analysis of cell fate, cell migration, and growth cone guidance. Cell 64: 801-815

Knust E, Schrons H, Grawe F, Campos-Ortega JA (1992) Seven genes of the *Enhancer of split* complex of *Drosophila melanogaster* encode helix-loop-helix proteins. Genetics 132: 505-518

Kuhn DT, Sawyer M, Packert G, Turenchalk G, Mack JA, Sprey TE, Gustavson E, Kornberg TB (1992) Development of the *Drosophila melanogaster* caudal segments involves supression of the ventral regions of A8, A9 and A10. Development 116: 1-20

Lasko P, Ashburner M (1988) The product of the *Drosophila* gene *vasa* is very similar to eukaryotic initiation factor A4. Nature 335: 611-617

Lasko P, Ashburner M (1990) Posterior localization of vasa protein correlates with, but is not sufficient for, pole cell development. Genes Dev 4: 905-921

Lawrence PA (1981) The cellular basis of segmentation. Cell 26: 3-10

Lawrence PA (1966) Development and determination of hairs and bristles in the milkweed bug (*Oncopeltus fasciatus*) (Cygaeidae, Hemptera). J Cell Sci: 475-498

Lawrence PA, Johnston P (1986) Observations on cell lineage of internal organs of *Drosophila*. J Embryol exp Morph 72: 251-266

Lees AD, Waddington CH (1942) The development of the bristles in normal and some mutants of *Drosophila melanogaster*. Proc Roy Soc Lond Ser B 131: 87-110

Leiss D, Hinz U, Gasch A, Mertz R, Renkawitz-Pohl R (1988) β3 tubulin expression characterizes the differentiating mesodermal germ layer during *Drosophila* embryogenesis. Development 104: 525-531.

Leptin M, Grunewald B (1990) Cell shape changes during gastrulation in *Drosophila*. Development 110: 73-84

Leptin M (1994) Control of epithelial cell changes. Curr Biol 4: 709-712

Lohs-Schardin M, Cremer C, Nüsslein-Volhard C (1979) A fate map of the larval epidermis of *Drosophila melanogaster*: localized cuticle defects following irradiation of the blastoderm with an ultraviolet laser microbeam. Dev Biol 73: 239-255

Lundell MJ, Hirsh J (1994) Temporal and spatial development of serotonin and dopamine neurons in the *Drosophila* CNS. Dev Biol 165: 385-396

Mahowald AP (1962) Fine structure of pole cells and polar granules in *Drosophila melanogaster*. J Exp Zool 151: 201-215

Mahowald AP (1963a) Electron microscopy of the formation of the cellular blastoderm in *Drosophila melanogaster*. Exp Cell Res 32: 457-468

Mahowald AP (1963b) Ultrastructural differentiation during formation of the blastoderm in the *Drosophila melanogaster* embryo. Dev Biol 8: 186-204

Mahowald AP (1971) Polar granules of *Drosophila*. III. The continuity of polar granules during the life cycle of *Drosophila*. J Exp Zool 176: 329-344

Mahowald AP, Illmensee K, Turner FR (1976) Interspecific transfer of polar plasm between *Drosophila* embryos. J Cell Biol 70: 358-373

Manning G, Krasnow MA (1993) Development of the *Drosophila* tracheal system. In: Bate CM, Martinez-Arias A (eds) The Development of *Drosophila melanogaster*. Cold Spring Harbor Laboratory Press, Cold Spring Harbor, New York, pp 609-685

Martinez-Arias A (1993) The larval epidermis. In: Bate CM, Martinez-Arias A (eds) The Development of *Drosophila melanogaster*. Cold Spring Harbor Laboratory Press, Cold Spring Harbor, New York, pp 517-608

Martinez-Arias A, Lawrence PA (1985) Parasegments and compartments in the *Drosophila* embryo. Nature 313: 639-642

Matsuda R (1964) Morphology and evolution of the insect head. Mem Amer Ent Inst 4: 1-334

Menne TV, Klämbt C (1994) The formation of commissures in the *Drosophila* CNS depends on the midline cells and on the *Notch* gene. Development 120: 123-133

Merritt DJ, Whitington PM (1995) Central projections of sensory neurons in the Drosophila embryo correlate with sensory modality, soma position and proneural gene function. J Neurosci 15: 1755-1767

Myers PZ, Bastiani MJ (1993a) Cell-cell interactions during the migration of an identified commissural growth cone in the embryonic grasshopper. J Neurosci 13: 115-126

Myers PZ, Bastiani MJ (1993b) Growth cone dynamics during the migration of an identified commissural growth cone. J Neurosci 13: 127-143

Nelson RE, Fessler LI, Takagi Y, Blumberg B, Keene DR, Olson PF, Parker CG, Fessler JH (1994) Peroxidasin: a novel enzyme-matrix protein of *Drosophila* development. EMBO J 13: 3438-3447

Nonidez JF (1920) The internal phenomena of reproduction in *Drosophila*. Biol Bull 39: 207-230

Nordlander RH, Edwards JS (1969) Postembryonic brain development in the monarch butterfly, *Danaus plexippus plexippus* L. I. Cellular events during brain morphogenesis. Roux´s Arch Dev Biol 162: 197-217

Nöthiger R, Strub S (1972) Imaginal defects after UV-microbeam irradiation of early cleavage stages of *Drosophila melanogaster*. Rev Suisse Zool 79: 267-279

Nüsslein-Volhard C, Wieschaus E, Jürgens G (1982) Segmentierung bei *Drosophila*: Eine genetische Analyse. Verh dtsch zool Ges 1982. Gustav Fischer, Stuttgart, pp 91-104

O´Kane CJ, Gehring WJ (1987) Detection in situ of genomic regulatory elements in *Drosophila*. Proc Natl Acad Sci USA 84: 9123-9127

Panzer S, Weigel D, Beckendorf SK (1992) Organogenesis in *Drosophila melanogaster*: embryonic salivary gland determination is controlled by homeotic and dorsoventral patterning genes. Development 114: 49-57

Parks S, Wieschaus E (1991) The *Drosophila* gastrulation gene *concertina* encodes a Ga-like protein. Cell 64: 447-458

Patel NH, Snow PM, Goodman CS (1987) Characterization and cloning of fasciclin III: a glycoprotein expressed on a subset of neurons and axon pathways in *Drosophila*. Cell 48: 975-988

Patel NH, Scafer B, Goodman CS, Holmgren R (1989) The role of segment polarity genes during *Drosophila* neurogenesis. Genes Dev 3: 890-904

Penzlin H (1985) Stomatogastric nervous system. In: Kerkut GA, Gilbert LI (eds) Comprehensive Insect Physiology, Biochemistry and Pharmacology, Vol 5. Pergamon Press, Oxford, pp 371-406

Pesacreta TC, Byers T, Dubreuil R, Kiehart DP, Branton D (1989) *Drosophila* spectrin: the membrane skeleton during embryogenesis. J Cell Biol 108: 1697-1709

Poodry CA (1980) Epidermis: Morphology and development. In: Ashburner M, Wright TRF (eds) The Genetics and Biology of *Drosophila*, Vol. 2d. Academic Press, New York, pp 443-498

Poulson DF (1937a) Chromosomal deficiencies and the embryonic development of *Drosophila melanogaster*. Proc Natl Acad Sci USA 23: 133-137

Poulson DF (1937b) The embryonic development of *Drosophila melanogaster*. Actual Scient 498: 1-51

Poulson DF (1940) The effect of certain X-chromosome defficiencies on the embryonic development of *Drosophila melanogaster*. J Exp Zool 83: 271-235

Poulson DF (1943) Induced chromosome deficiencies. Their effect on normal individuals. Yale Sci Magazine 17: 3-5

Poulson DF (1945) Chromosomal control of embryogenesis in *Drosophila*. Amer Nat 79: 340-363

Poulson DF (1950) Histogenesis, organogenesis and differentiation in the embryo of *Drosophila melanogaster* Meigen. In: Demerec M (ed) Biology of *Drosophila*. Wiley, New York, pp 168-274

Prokop A, Technau GM (1991) The origin of postembryonic neuroblasts in the ventral nerve cord of *Drosophila melanogaster*. Development 111: 79-88.

Rabinowitz M (1941a) Studies on the cytology and early embryology of the egg of *Drosophila melanogaster*. J Morphol 69: 1-49

Rabinowitz M (1941b) Yolk nuclei in the egg of *Drosophila melanogaster*. Anat Rec 81: 80-81

Rempel JG (1975) The evolution of the insect head: the endless dispute. Quaest Entomol 11: 7-25

Reuter R (1994) The gene *serpent* has homeotic properties and specifies endoderm versus ectoderm within the *Drosophila* tail. Development 120: 1123-1135

Reuter R, Leptin M (1994) Interacting functions of *snail*, *twist* and *huckebein* during the early development of germ layers in *Drosophila*. Development 120: 1137-1150

Reuter R, Scott MP (1990) Expression and function of the homeotic genes *Antennapedia* and *Sex combs reduced* in the embryonic midgut of *Drosophila*. Development 109: 289-303

Rickoll WL (1976) Cytoplasmic continuity between embryonic cells and the primitive yolk sac during early gastrulation in *Drosophila melanogaster*. Dev Biol 49: 304-310

Rickoll WL, Counce SJ (1980) Morphogenesis in the embryo of *Drosophila melanogaster* - germ band extension. Roux's Arch Dev Biol 188: 163-177

Rizki TM (1978a) The circulatory system and associated cells and tissues. In: Ashburner M, Wright TRF (eds) The Genetics and Biology of *Drosophila*, Vol. 2b. Academic Press, New York, pp pp 397-452

Rizki TM (1978b) Fat body. In: Ashburner M, Wright TRF (eds) The Genetics and Biology of *Drosophila*, Vol. 2b. Academic Press, New York, pp 561-601

Roonwal ML (1937) Studies on the embryology of the African migratory locust, *Locusta migratoria migratorioides* (Reiche and Frm.) (Orthoptera, Acrididae). II. Organogeny. Phil Trans Roc Soc B 227: 157-244

Rubin GM, Spradling AC (1982) Genetic transformation of *Drosophila* with transposable element vectors. Science 218: 348-353.

Rugendorff A, Younossi-Hartenstein A, Hartenstein V (1994) Embryonic origin and differentiation of the *Drosophila* heart. Roux's Arch Dev Biol 203: 266-280

Ruiz-Gomez M, Ghysen A (1993) The expression and role of a pro-neural gene, *achaete*, in the development of the larval nervous system of *Drosophila*. EMBO J 12: 1121-1130

Samakovlis C, Hacohen N, Manning G, Sutherland D, Guillemin K, Krasnow MA (1996) Development of the *Drosophila*: tracheal system occurs by a series of morphologically distinct but genetically coupled branching events. Development 122: 1395-1407

Sander K (1976) Specification of the basic body pattern in insect embryogenesis. Adv Insect Physiol 12: 125-238

Sander K (1983) The evolution of patterning mechanisms: gleanings from insect embryogenesis and spermatogenesis. In: Goodwin BC, Holder N, Wylie CG (eds) Development and Evolution. Cambridge Univ Press, pp 137-159

Sander K (1985a) August Weissmanns Untersuchungen zur Insektenentwicklung 1862-1892. In: Sander K (ed) August Weismann und die theoretische Biologie des 19. Jahrhunderts. Freiburger Universitätsblätter 87/88: 43-52

Sander K (1985b) Fertilization and egg activation in insects. In: Metz C, Monroy A (eds) Biology of Fertilization, Vol VII. Academic Press, New York, pp 409-430

Sato T, Denell RE (1986) Segmental identity of caudal cuticular features of *Drosophila melanogaster* larvae and its control by the bithorax complex. Dev Biol 116: 78-91

Schmidt-Ott U, Technau GM (1992) Expression of *en* and *wg* in the embryonic head and brain of *Drosophila* indicates a refolded band of seven segment remnants. Development 116: 111-125

Schmidt-Ott U, Technau GM (1994) Fate-mapping in the procephalic region of the embryonic *Drosophila* head. Roux's Arch Devl Biol 203: 367-373

Schmidt-Ott U, González-Gaitán M, Jäckle H, Technau GM (1994) Number, identity, and sequence of the *Drosophila* head segments as revealed by neural elements and their deletion patterns in mutants. Proc Natl Acad Sci USA 91: 8363-8367

Schoeller J (1964) Recherches descriptives et experimentales sur la cephalogenese de *Calliphora erythrocephala* (Meigen) au cours des developpements embryonnaire et postembryonnaire. Arch Zool Exp Gen 103: 1-216

Schoeller-Raccaud J (1977) La cephalogenese larvaire des dipteres. In: Grasse P-P (ed) Traité de Zoologie, Vol VIII, Fs V-B. Masson, Paris, pp 262-279

Singh RC, Singh K (1984) Fine structure of the sensory organs of the *Drosophila melanogaster* Meigen larva. Int J Insect Morph Embryol 13: 255-273

Sink H, Whitington PM (1991) Location and connectivity of abdominal motoneurons in the embryo and larva of *Drosophila melanogaster* . J Neurobiol 12: 298-311

Skear H leB (1989) Cell division in Malpighian tubule development in *Drosophila melanogaster* is regulated by a single tip cell. Nature 342: 566-569.

Skear H leB (1993) The alimentary canal. In: Bate CM, Martinez-Arias A (eds) The Development of *Drosophila melanogaster*. Cold Spring Harbor Laboratory Press, Cold Spring Harbor, New York, pp 941-1012

Smith AV, Orr-Weaver TL (1991) The regulation of the cell cycle during *Drosophila* embryogenesis: the transition to polyteny. Development 112: 997-1008

Snodgrass RE (1935) Principles of insect morphology. McGraw-Hill, New York

Sonnenblick BP (1941) Germ cell movements and sex differentiation of the gonads in the *Drosophila* embryo. Proc Nat Acad Sci USA 27: 484-489

Sonnenblick BP (1950) The early embryology of *Drosophila melanogaster*. In: Demerec M (ed) Biology of *Drosophila*. Wiley, New York, pp 62-167

Sonnenfeld MJ, Jacobs JR (1994) Mesectodermal cell fate analysis in *Drosophila* midline mutants. Mech Dev 46: 3-13

Spradling AC, Rubin GM (1982) Transposition of cloned P-elements into *Drosophila* germline chromosomes. Science 218: 341-347.

Springer CA, Rutschky CW (1969) Comparative study of the embryological development of the median cord in hemiptera. J Morphol 129: 375-400

Steiner E (1976) Establishment of compartments in the developing leg imaginal discs of *Drosophila melanogaster*. Wilhelm Roux's Arch Entw Mech 180: 9-30

Steller H, Fischbach KF, Rubin GM (1987) *Disconnected*, a locus required for neuronal pathway formation in the visual system of *Drosophila*. Cell 50: 1139-1153

Stern C (1968) Genetic mosaics in animal and man. In: Genetic mosaics and other essays. Harvard Univ Press, Cambridge, Mass., pp 27-129

Stortkuhl KF, Hofbauer A, Keller V, Gendre N and Stocker RF (1994) Analysis of immunocytochemical staining patterns in the antennal system of *Drosophila melanogaster*. Cell Tiss Res 275: 27-38

Strasburger M (1932) Bau, Funktion und Variabilität des Darmtraktus von *Drosophila melanogaster* Meigen. Z wiss Zool 140: 539-649

Struhl G (1984) Splitting the bithorax complex of *Drosophila*. Nature 308: 454-457

Sturtevant AH (1929) The *claret* mutant type of *Drosophila simulans*: a study of chromosome elimination and cell-lineage. Z wiss Zool 135: 323-356

Sweeton D, Parks S, Costa M, Wieschaus E (1991) Gastrulation in *Drosophila*: the formation of the ventral furrow and posterior midgut invaginations. Development 112: 775-789

Szabad J, Schüpbach T, Wieschaus E (1979) Cell lineage and development in the larval epidermis of *Drosophila melanogaster*. Dev Biol 73: 256-271

Tautz D, Pfeifle C (1989) A non-radioactive in situ hybridization method for the localization of specific RNAs in *Drosophila* embryos reveals translational control of the segmentation gene *hunchback*. Chromosoma 98: 81-85

Technau GM, Campos-Ortega JA (1985) Fate-mapping in wild-type *Drosophila melanogaster*. II. Injections of horseradish peroxidase in cells of the early gastrula stage. Roux's Arch Dev Biol 194: 196-212

Technau GM, Campos-Ortega JA (1986a) Lineage analysis of transplanted individual cells in embryos of *Drosophila melanogaster*. II. Commitment and proliferative capabilities of neural and epidermal cell progenitors. Roux's Arch Dev Biol 195: 445-454

Technau GM, Campos-Ortega JA (1986b) Lineage analysis of transplanted individual cells in embryos of *Drosophila melanogaster*. III. Commitment and proliferative capabilities of pole cells and midgut progenitors. Roux's Arch Dev Biol 195: 489-498

Tepass U, Hartenstein V (1994a) Epithelium formation in the *Drosophila* midgut depends on the interaction of endoderm and mesoderm. Development 120: 579-590

Tepass U, Hartenstein V (1994b) The development of cellular junctions in the *Drosophila* embryo. Dev Biol 161: 563-596

Tepass U, Hartenstein V (1995) Neurogenic and proneural genes control cell fate specification in the *Drosophila* endoderm. Development 121: 393-405

Tepass U, Knust E (1990) Phenotypic and developmental analysis of mutations at the *crumbs* locus, a gene required for the development of epithelia in *Drosophila melanogaster*. Roux´s Arch Dev Biol 199: 189-206

Tepass U, Fessler LI, Aziz A, Hartenstein V (1994) Embryonic origin of hemocytes and their relationship to cell death in *Drosophila*. Development 120: 1829-1837

Thomas JB, Bastiani MJ, Bate M, Goodman CS (1984) From grasshopper to *Drosophila*: a common plan for neuronal development. Nature 310: 203-207

Thomas JB, Crews ST, Goodman CS (1988) Molecular genetics of the *single-minded* locus: a gene involved in the development of the *Drosophila* nervous system. Cell 52: 133-141

Tix S, Bate M, Technau GM (1989a) Pre-existing neuronal pathways in the leg imaginal discs of *Drosophila*. Development 107: 855-862

Tix S, Minden J, Technau GM (1989b) Pre-existing neuronal pathways in the developing optic lobes of *Drosophila*. Development 105: 739-746

Truman JW, Bate CM (1988) Spatial and temporal patterns of neurogenesis in the central nervous system of *Drosophila melanogaster*. Dev Biol 125: 145-157.

Turner FR, Mahowald AP (1976) Scanning electron microscopy of *Drosophila* embryogenesis. I. The structure of the egg envelopes and the formation of the cellular blastoderm. Dev Biol 50: 95-108

Turner FR, Mahowald AP (1977) Scanning electron microscopy of *Drosophila melanogaster* embryogenesis. II. Gastrulation and segmentation. Dev Biol 57: 403-416

Turner FR, Mahowald AP (1979) Scanning electron microscopy of *Drosophila melanogaster* embryogenesis. III. Formation of the head and caudal segments. Dev Biol 68: 96-109

Udolph G, Prokop A, Bossing T, Technau GM (1993) A common precursor for glia and neurons in the embryonic CNS of *Drosophila* gives rise to segment specific lineage variants. Development 118: 765-775

Udolph G, Lüer K, Bossing T, Technau GM (1995) Commitment of CNS progenitors along the dorsoventral axis of *Drosophila* neuroectoderm. Science 269: 1278-1281

Underwood EM, Caulton JH, Allis CD, Mahowald AP (1980a)
Developmental fate of pole cells in *Drosophila melanogaster*.
Dev Biol 77: 303-314

Underwood EM, Turner FR, Mahowald AP (1980b) Analysis of cell
movements and fate mapping during early embryogenesis in
Drosophila melanogaster. Dev Biol 74: 286-301

Wada S (1966) Analyse der Kopf-Hals-Region von *Tachycines* (Sal-
tatoria) in morphogenetische Einheiten. II. Mitteilung: Experi-
mentell-teratologische Befunde am Kopfskelett mit Berück-
sichtigung des zentralen Nervensystems. Zool Jb Anat 83: 235-
326

Wahl B (1914) Über die Kopfbildung cyclorapher Dipterenlarven
und die postembryonale Entwicklung des Miscidenkopfes. Arb
Zool Inst Univ Wien 20: 159-272

Warn RM, Magrath R (1982) Observations by a novel method of
surface changes during the syncytial blastoderm stage of the
Drosophila embryo. Dev Biol 89: 540-548

Weigel D, Jürgens G, Küttner F, Seifert E, Jäckle H (1989) The
homeotic gene *fork head* encodes a nuclear protein and is
expressed in the terminal regions of the *Drosophila* embryo.
Cell 57: 645-658

Weismann A (1863) Die Entwicklung der Dipteren im Ei. Z wiss
Zool 13: 107-220

Wheeler WM (1891) Neuroblasts in the arthropod's embryo. J
Morphol 4: 337-343

Wheeler MW (1893) A contribution to insect embryology. J Morphol 8: 1-160

White K, Kankel DR (1978) Patterns of cell division and cell movement in the formation of the imaginal nervous system in *Drosophila melanogaster*. Dev Biol 65: 296-321

Whitten J (1980) The tracheal system. In: Ashburner M, Wright
TRF (eds) The Genetics and Biology of *Drosophila*, Vol. 2d.
Academic Press, New York, pp 499-541

Wieschaus E, Gehring W (1976) Clonal analysis of primordial disc
cells in the early embryo of *Drosophila melanogaster*. Dev Biol
50: 249-263

Wieschaus E, Nüsslein-Volhard C (1986) Looking at embryos. In:
Ransom R (ed) *Drosophila:* a practical approach. IRL Press,
Oxford, pp 199-227

Wieschaus E, Sweeton D, Costa M (1991) Convergence and extension during germ band elongation in *Drosophila* embryos. In:
Keller R (ed) Gastrulations: Movements, patterns and molecules. Plenum Press, New York, pp 213-223

Wright TRF (1970) The genetics of embryogenesis in *Drosophila*. Adv Genetics 15: 262-395

Young PE, Pesacreta TC, Kiehart DP (1991) Dynamic changes in the distribution of cytoplasmic myosin durin *Drosophila* embryogenesis. Development 111: 1-14

Younossi-Hartenstein A, Hartenstein V (1993) The role of the trachea and musculature during pathfinding of *Drosophila* embryonic sensory axons. Dev Biol 158: 430-447

Younossi-Hartenstein A, Tepass U, Hartenstein V (1993) Embryonic origin of the imaginal discs of the head of *Drosophila melanogaster*. Roux's Arch Dev Biol 203: 74-82

Younossi-Hartenstein A, Hartenstein V (1996) Pattern, time of birth and morphogenesis of sensillum precursors in *Drosophila*. J comp Neurol, in press

Younossi-Hartenstein A, Nassif C, Hartenstein V (1996) Early neurogenesis of the *Drosophila* brain. J comp Neurol 370: 313-329

Zalokar M (1970) Fixation of *Drosophila* eggs without pricking. Drosophila Inf Serv 47: 128

Zalokar M (1976) Autoradiographic study of protein and RNA formation during early development of *Drosophila* eggs. Dev Biol 49: 425-437

Zalokar M, Erk I (1976) Division and migration of nuclei during early embryogenesis of *Drosophila melanogaster*. J Micr Biol Cell 25: 97-106

Zalokar M, Erk I (1977) Phase-partition fixation and staining of *Drosophila* eggs. Stain Technology 52: 89-95

Zawarzin A (1912) Histologische Studien über Insekten. III. Über das sensible Nervensystem der Larven von *Melolontha vulgaris*. Z wiss Zool 100: 447-458

Index

DATE DUE

JAN 1 0 2004